Front endpaper map.

British commercial mapmaker James Wyld published an intricately detailed map of North America in 1823; this is the portion of the map that covers what is today Washington and Oregon but also includes that part eastward into what is today Montana. Wyld depicted the topography with hachures, short lines aligned up and down the slope, which give an excellent idea of the shape of the land because of the shading effect. Certainly Wyld's map shows, correctly, a complex Rocky Mountain range, and *Lewis and Clarke's [sic] Route across the Mountains* is noted. Before Lewis and Clark the Rocky Mountains had been thought to be perhaps a single range of mountains that would be relatively easy to cross. The course of the *R. Columbia* is reasonably accurate, but the course of the Willamette, here shown by its earlier name the *Multnomah River,* is not, and still derives from Native reports noted by Lewis and Clark. The course of the Snake River—*Lewis's R. South Fork*—is more accurately shown but will soon be improved based on information gathered by the Hudson's Bay Company, which began regular forays up this river to try to trap out all the beaver and make it unremunerative for the competing American fur traders—the famous mountain men such as Jedediah Smith. The boundary line shown at bottom is the 42°N line agreed to in the Transcontinental Treaty with Spain in 1819; today this forms the boundary between Oregon and California.

HISTORICAL ATLAS
OF
WASHINGTON & OREGON

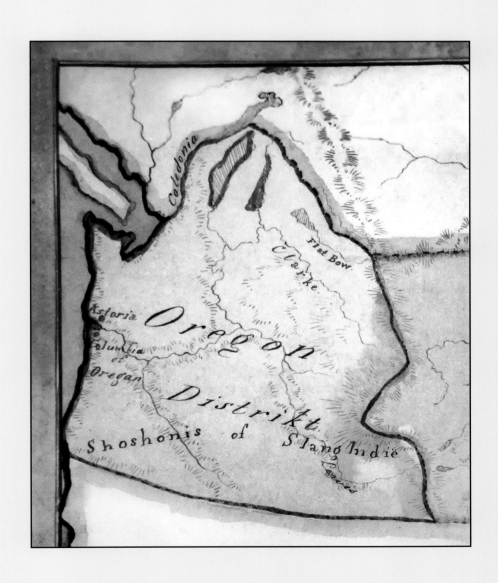

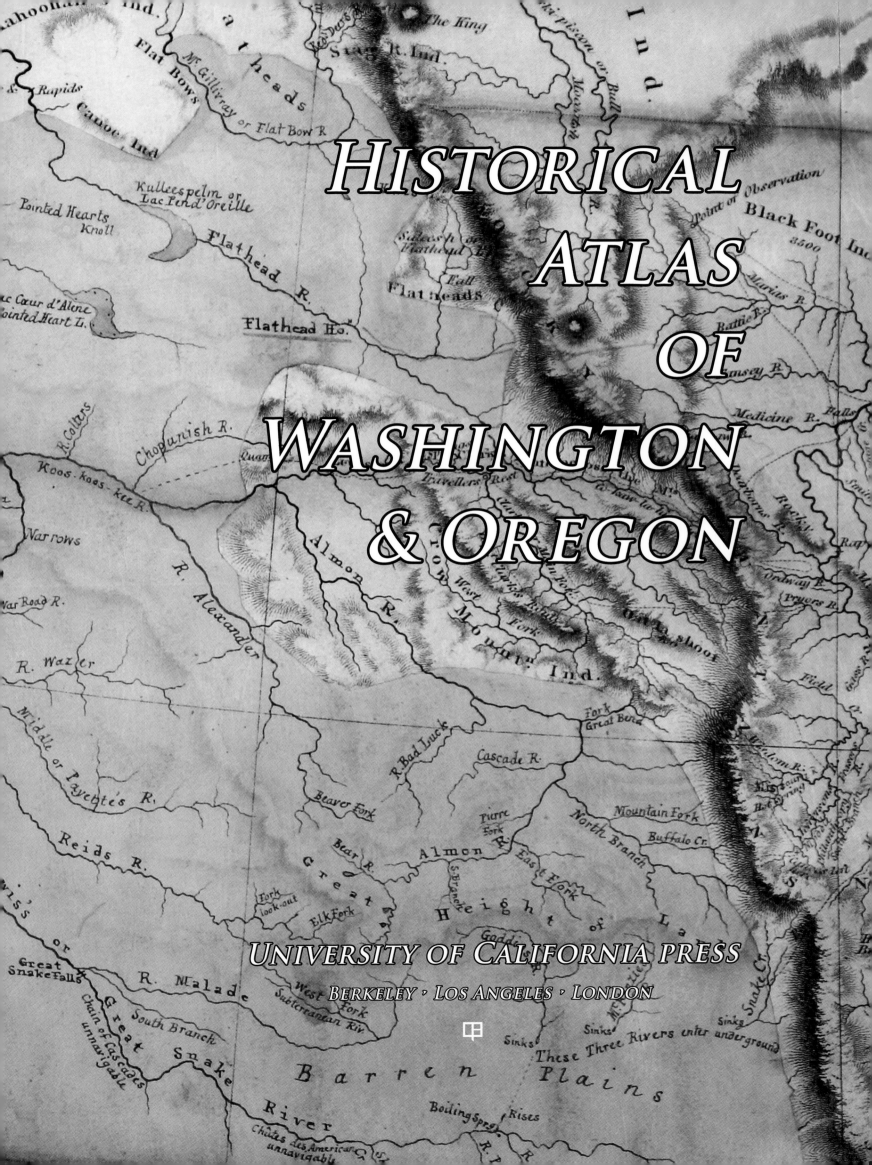

HISTORICAL
ATLAS
OF
WASHINGTON
& OREGON

UNIVERSITY OF CALIFORNIA PRESS

BERKELEY · LOS ANGELES · LONDON

University of California Press, one of the most distinguished university presses in the United States, enriches lives around the world by advancing scholarship in the humanities, social sciences, and natural sciences. Its activities are supported by the UC Press Foundation and by philanthropic contributions from individuals and institutions. For more information, visit www.ucpress.edu.

University of California Press
Berkeley and Los Angeles, California

University of California Press, Ltd.
London, England

Originated by Douglas & McIntyre, an imprint of D&M Publishers Inc., 2323 Quebec Street, Suite 201, Vancouver BC Canada V5T 4S7
www.douglas-mcintyre.com

Library of Congress Control Number 2011922318

ISBN 978-0-520-26615-5 (cloth : alk. paper)

Manufactured in China by C&C Offset Printing Co., Ltd.

19 18 17 16 15 14 13 12 11
10 9 8 7 6 5 4 3 2 1

Printed on acid-free paper

Design and layout: Derek Hayes
Editing and copyediting: Iva Cheung
Index: Judith Anderson
Jacket design: Lia Tjandra

To contact the author:
www.derekhayes.ca / derek@derekhayes.ca

ACKNOWLEDGMENTS

Many people contributed to this book, and I want to thank them all. I should specially mention Greg Walter; James Walker; Robin Inglis; Greg Lange, Washington State Archives, Puget Sound Branch; John Cloud, National Oceanic and Atmospheric Administration (NOAA) Central Library; Patricia Solomon, Oregon Department of Transportation; Carla Rickerson and Nicolette Bromberg, University of Washington Special Collections; Trevor Bond, Washington State University Manuscripts, Archives, and Special Collections; Dennis Freeman, College of the Siskiyous Library; Dan Linscheid, Yamhill County Surveyor; Jon Jablonski, University of Oregon; Jody Gripp and Bob Schuler, Tacoma Public Library; Dave Hastings, Lupita Lopez and Mary Hammer, Washington State Archives, Olympia; Ruth Steele, Center for Pacific Northwest Studies, Western Washington University; Mary Hansen, City of Portland Archives; Carolyn Marr, Museum of History and Industry, Seattle; Chris Warner; Barry Lawrence Ruderman; and Henry Ewert.

Thanks also to my editor, Iva Cheung, who has done a sterling job in helping improve the text; to Sheila Levine at the University of California Press for her continued support; and to Scott McIntyre of D&M Publishers for his ongoing enthusiasm for my books.

I have done my best to locate image copyright holders, but if I have inadvertently omitted credit for your image, please contact me, and a correction will be made in any future editions of this book.

MAP 1 (*half-title page*).
A Dutch map of North America drawn about 1838 carried this version of *Oregon Distrikt*, with the northern boundary of Oregon following the *Caledonia* River, known by that time to be fictitious (see page 42).

MAP 2 (*title page, background*).
The Hudson's Bay Company had a habit of updating published maps by sticking on patches with their latest geographical knowledge. This 1824 map has a very large patch stuck on it displaying a newly plotted course for the Snake River, the result of expeditions into the Snake Country that year as part of a strategy to drive out American fur traders (see page 32).

MAP 3 (*below*).
Maps come from many sources. This four-page letter and crude map were written and drawn by James O. Raynor, a Methodist circuit riding minister. He had been in Oregon for several years when he wrote this letter to his brother William in Granville, Ohio. He wrote it from Jacksonville on 9 March 1859, so it is likely that he wasn't yet aware of the official statehood of Oregon (14 February 1859). The letter contains some information on his circuit and the difficulties of travel across the Cascades. At one point he describes some of the people to whom he preaches as the ". . . descendants of those who killed and eat captain Cook."

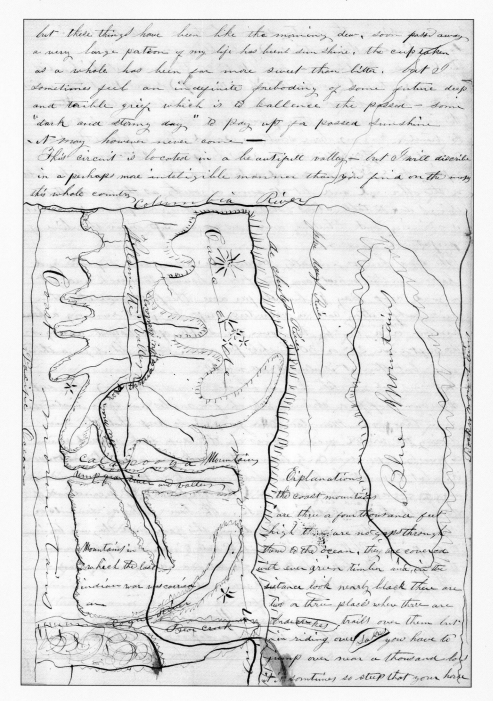

Contents

Mapping Washington and Oregon

Many books have been written about the history of Washington and Oregon, but most overlook the maps produced that depict the events at the time they occurred. This is a shame, because historical maps can tell us a great deal about the way the makers of history—and the public for whom the maps might be made—thought and acted. This book is an attempt to redress that imbalance.

As is to be expected in a book that tries to cover the whole extent of history in a single volume, even if for a single region, the maps and events covered are at best a personal selection, though I have tried to ensure that all the major events and developments have been included. Topics have been selected based on their historical importance, geographical significance—and, critically, the availability of surviving maps. In compiling a book such as the present volume, one soon realizes that there are far more maps that did not survive compared with those that did. This is a problem common to most historical documents.

MAP 4 (*below*).
This map of Oregon, published in 1904, is full of land survey detail. The survey grid, begun in Oregon in 1851, now covers all but the Cascade Mountains and parts of the arid southeast, and land grants of both railroad companies and military wagon road builders are shown. The latter are discussed in more detail on pages 78, 86, and others. At *right* is a transit used by surveyors; this one dates from about 1900, but the basic design had been around for a hundred years before that. It was a type of theodolite that could be flipped over to enable back-sighting.

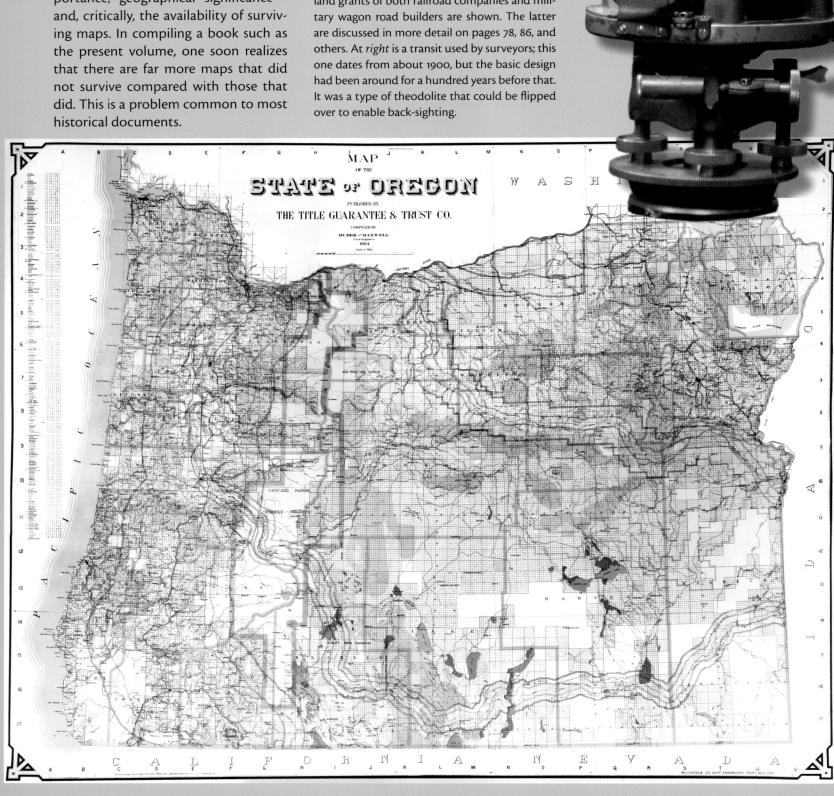

In addition to selecting maps for their historical significance, I have also included some just for their interest—they might be pictorial, sketch, whimsical—and especially if artistic in nature, for maps, in many cases, are art (MAP 5, *right*).

The text, which includes often extensive captions that are integral to the book, has been kept to a minimum in order to allow more space for what this book is all about—historical maps. For more details a bibliography of sources consulted or recommended is included.

So here—illustrated with contemporary maps—is the span of Washington and Oregon's history: the explorers, the fur traders, the promoters and the missionaries, the gold seekers and the engineers; the settling of the Willamette and the founding of Seattle, the campaigns against the Indians, the coming of the railroad, and the building of a road system. Here are the tales of the leveling of Seattle and of the fight against saloons in Portland; the schemes and dreams of bridge and canal builders and city planners; the making of atomic bombs and the plans for the evacuation of our cities; the defining of a boundary and the building of an empire.

This is history with maps, from traveling the Oregon Trail to opening the Oregon Coast, from wars to wagon roads, from irrigation to aviation, and from power to parks. The maps include all types—exploration maps, surveys, engineering drawings, proposals for additions and changes, advertising maps, and maps for public information, tourism, and promotion. There are military maps, maps to sell, and maps to persuade—550 in all. I hope that you will have as much fun studying the maps as I have had collecting them and assembling them for this book.

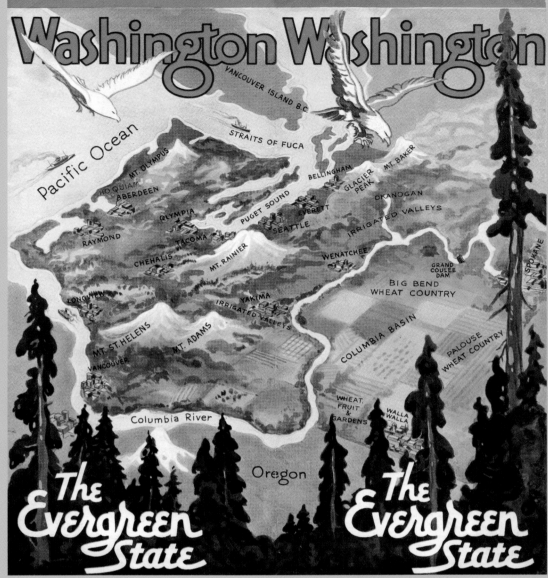

MAP 5 (*above*).
A different type of map altogether, this pictorial perspective map of Washington State appeared on the cover of a tourist brochure in 1931. Here visual appeal, rather than accuracy, is the aim, yet the map nevertheless contains a good deal of information.

MAP 6 (*left*).
Flanked by Indians and wagon trains, a map of the original extent of the American part of the Oregon Country (which covered the region east to the Rockies) appeared on a postage stamp in 1936 to celebrate the centennial of EuroAmerican migration to Oregon. In 1836 a missionary group led by Marcus Whitman traveled the *Oregon Trail*, although the first non-missionary settlers did not arrive until 1840, and the first wagon train two years after that (see page 52). The entire Oregon Country was under joint occupancy with Britain until 1846, when the forty-ninth parallel boundary was extended to the Strait of Georgia. *Walla Walla*, shown on this map, was the fur-trading post established by the North West Company in 1806 and subsequently taken over by the Hudson's Bay Company. *Astoria* was the post established in 1810 by John Jacob Astor's Pacific Fur Company but was sold, under duress, to the North West Company in 1812, when the American fur traders anticipated the arrival of a British warship during the War of 1812 (see page 30).

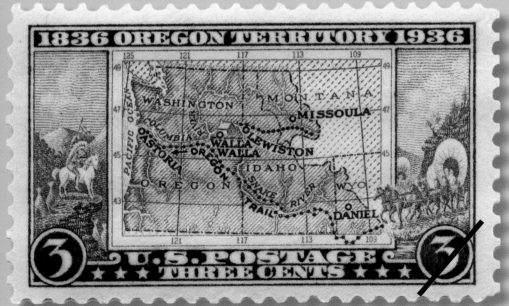

THE FIRST INHABITANTS

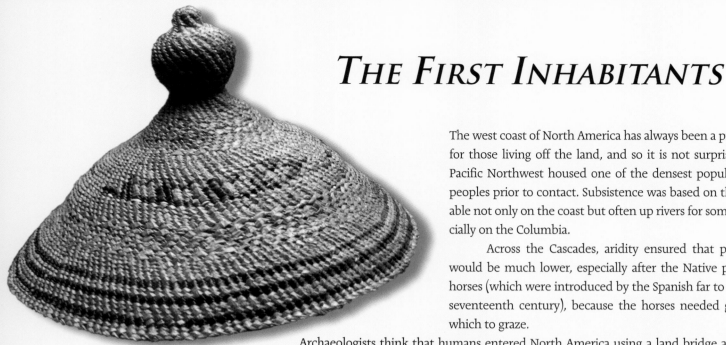

The west coast of North America has always been a providential place for those living off the land, and so it is not surprising that coastal Pacific Northwest housed one of the densest populations of Native peoples prior to contact. Subsistence was based on the salmon, available not only on the coast but often up rivers for some distance, especially on the Columbia.

Across the Cascades, aridity ensured that population levels would be much lower, especially after the Native peoples acquired horses (which were introduced by the Spanish far to the south in the seventeenth century), because the horses needed greater areas on which to graze.

Archaeologists think that humans entered North America using a land bridge across the Bering Strait during a lowering of sea levels about 13,000 years ago. Recent evidence now suggests this migration could have been even earlier. Humans could also have entered North America as early as 30,000 years ago by traveling along the coast in small boats. Yet, footprints that may be 40,000 years old were recently found near Mexico City. So although we know humans migrated here, we still do not know exactly when.

The earliest known habitation in Washington and Oregon is at Sequim, on the Strait of Juan de Fuca, where remains have been dated at 12,000 years old. West of the Cascades the earliest sites are the 11,000-year-old site at Lind Coulee, near Warden, and the 10,000-year-old Marmes Rockshelter, at the confluence of the Snake with the Palouse River and now under water behind the Lower Monumental Dam.

Above.
Typical coastal Native headwear, this one from the Clatsop around the mouth of the Columbia River. The hat is on display at Fort Clatsop, Oregon.

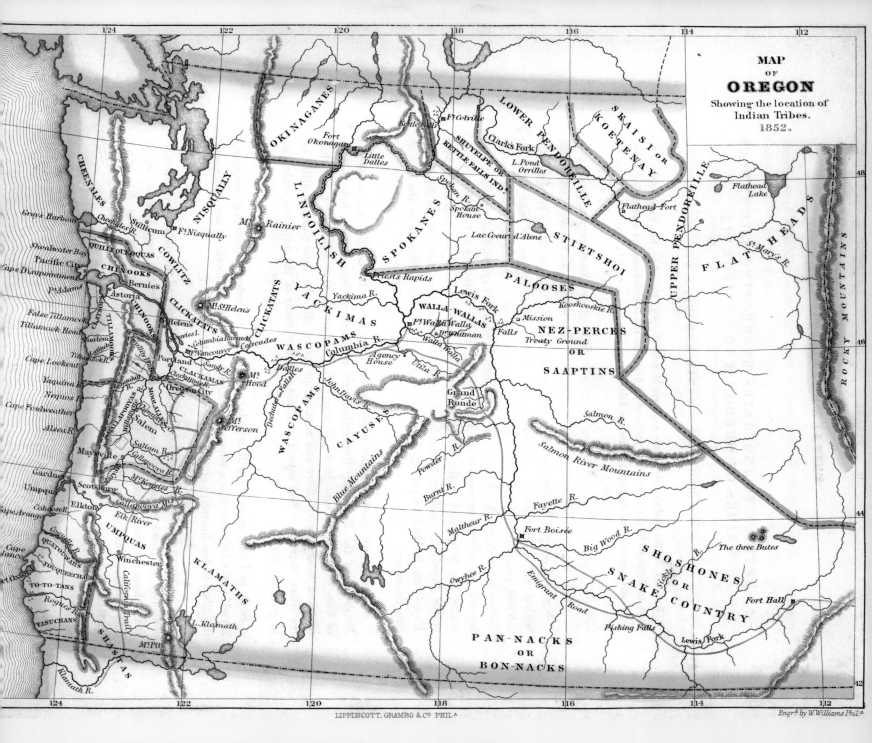

Confusion as to the origins of humans in Washington and Oregon increased in 1996 when human remains, of an individual soon
dubbed Kennewick Man, were found on the banks of the Columbia
River near Kennewick, Washington. The bones have been radiocarbon-
dated to about 9,300 years ago; the date is not the surprise, but the
apparent origin is—Kennewick Man is most closely related to the
Ainu of northeast Asia. It is possible that Kennewick Man arrived in
the Pacific Northwest after being shipwrecked and drifting across the
Pacific, a well-documented possibility because of the configuration of
currents in that ocean. For example, the Hudson's Bay Company returned three Japanese fishermen to Japan, via London, after rescuing
them near Cape Flattery in 1833.

The Makah village site at Ozette, just south of Cape Flattery,
yielded thirty-seven steel blades. This village, excavated in the 1970s,
had been buried by a landslide about 1700, predating European contact. The blades could have been acquired through trading networks
or from shipwrecks of Oriental origin.

Map 7 (*above*).

An 1852 map showing the distribution of Indian tribes in Oregon, then
reaching to the Rocky Mountains and the area of both today's Washington
and Oregon. A map drawn a year earlier, showing the distribution of Indian
tribes in the Willamette Valley, is Map 129, *page 58*. Governments at all levels were becoming interested in Indian distribution at this time, but largely
with the idea of forcing them onto confined areas—reservations—so as to
make the remaining land available for EuroAmerican settlement.

Left. Much of the Native sustenance and culture of the Pacific Northwest
was based on the salmon, and no place epitomized the effect of the salmon
like Celilo Falls. Native peoples caught salmon here for centuries, but in 1957
the falls were submerged in the waters behind The Dalles Dam. This photo
was taken in 1926. The easy availability of food allowed the Native community at Celilo to grow into a major settlement, and in addition it became
the center of a large trading network. Celilo was the oldest continuously in-
habited settlement in North America—perhaps for 15,000 years—until the
works of EuroAmericans destroyed it. The bridge faintly visible in the background is the Spokane, Portland & Seattle Railway bridge, completed in 1910
to carry the railroad's line across the river and south to Bend (see page 168).

However, regardless of the exact origins of the aboriginal peoples of Washington and Oregon, they occupied the land long before EuroAmericans arrived. The latter brought with them diseases that the Native peoples had no immunity against, and their populations plummeted. Early fur traders brought smallpox to the coastal Indians; Hudson's Bay Company traders brought the so-called intermittent fever—which was probably malaria—to a swath of the Lower Columbia and Willamette valleys in 1830–33. Measles arrived via the Oregon Trail in 1847 and likely motivated the Cayuse attack on pioneer missionary Marcus Whitman (see page 40). A Native population in 1780 of perhaps 55,000 in the area that is now Washington and Oregon was reduced to something like 13,000 by 1855, and was a factor in the Indian uprisings of that time (see page 64).

Map 8 (*below*).

Isaac I. Stevens was not only the first governor of Washington Territory and surveyor of part of the northern route of the Pacific Railroad Surveys (see page 80) but also federal superintendent of Indian Affairs. As such he gained extensive knowledge of the distribution of the Native groups in the territory, information gathered at this time principally for military purposes. The Indian attack on the new settlement of Seattle had taken place in 1856, only a year before this map was drawn, and other conflicts would occur in the interior a year later (see page 64). Note the tables of population and other statistics. The inset table and Stevens's signature have been moved from the right-hand side of the map.

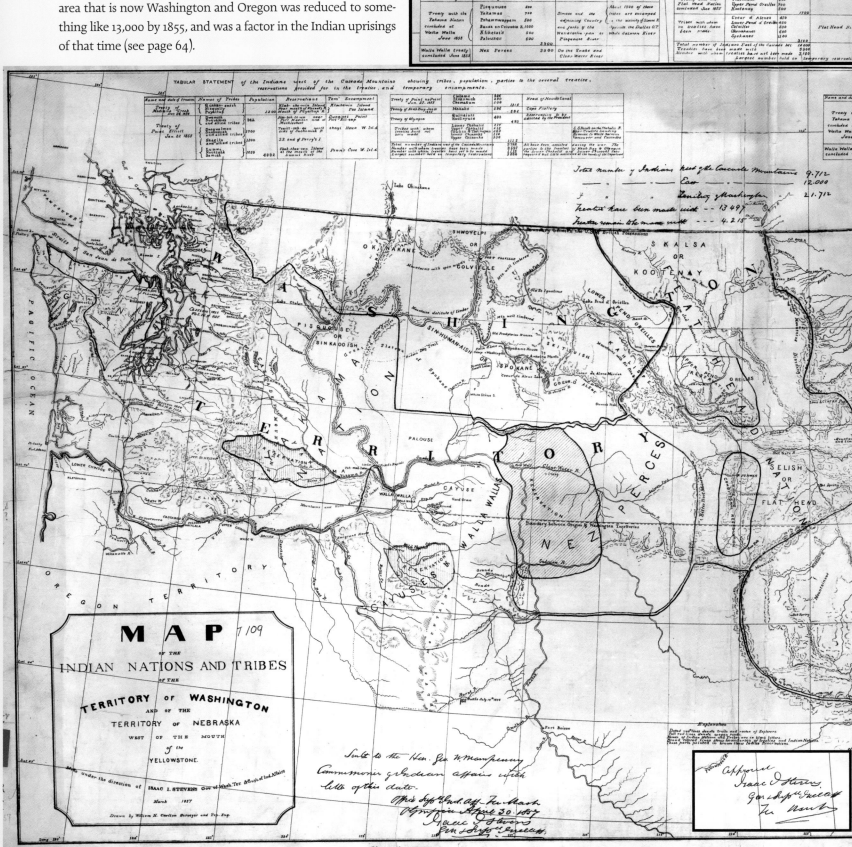

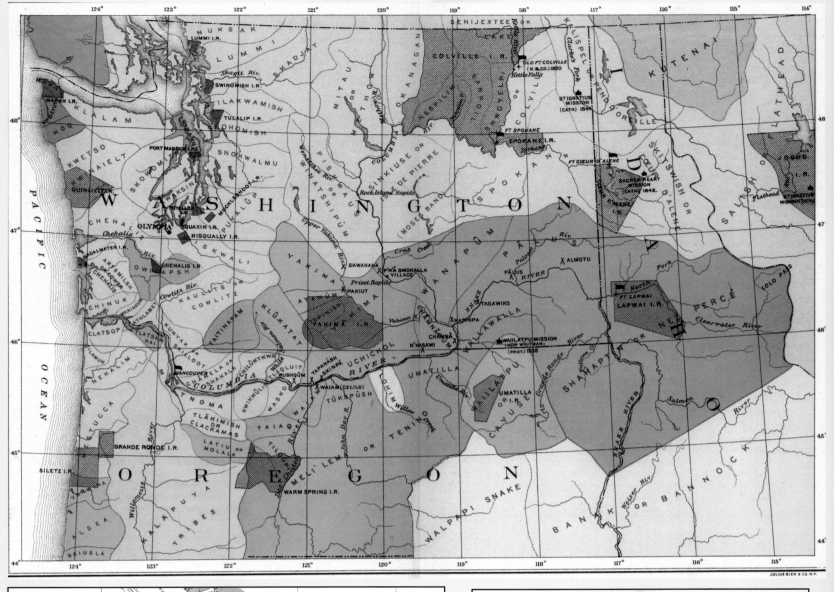

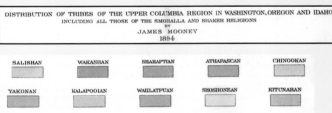

DISTRIBUTION OF TRIBES OF THE UPPER COLUMBIA REGION IN WASHINGTON, OREGON AND IDAHO
INCLUDING ALL THOSE OF THE SMOHALLA AND SHAKER RELIGIONS
BY
JAMES MOONEY
1894

| SALISHAN | WAKASHAN | SHAHAPTIAN | ATHAPASCAN | CHINOOKAN |
| YAKONAN | KALAPOOIAN | WAIILATPUAN | SHOSHONEAN | KITUNAHAN |

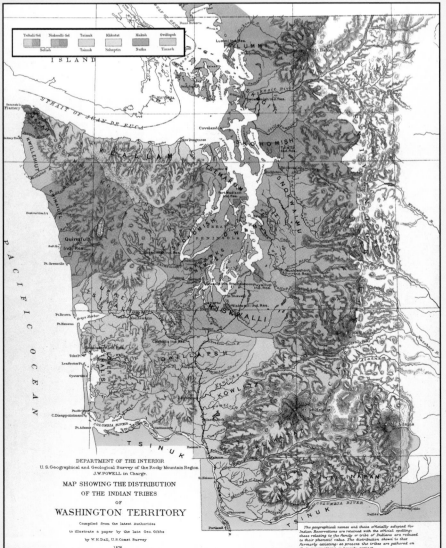

DEPARTMENT OF THE INTERIOR
U.S. Geographical and Geological Survey of the Rocky Mountain Region
J.W. POWELL in Charge.

MAP SHOWING THE DISTRIBUTION
OF THE INDIAN TRIBES
OF
WASHINGTON TERRITORY

Compiled from the latest Authorities
to illustrate a paper by the late Gen. Gibbs
by W.H. Dall, U.S. Coast Survey.
1876.

MAP 9 (*above*).

The U.S. Bureau of Ethnology produced this map of the distribution of the Indian tribes of Washington and northern Oregon in 1894. The bureau was led by John Wesley Powell, eminent geologist and ethnographer also famous for his exploration of the Colorado River. Note the increased numbers and smaller areas of the coastal tribes compared with the interior and even the Willamette Valley, which did not have good salmon runs above the Willamette Falls at Oregon City. The title and key have been moved and are shown separately below the map.

MAP 10 (*left*).

A map showing the distribution of Indians by tribe in Western Washington and part of Oregon, published in 1876 by the Department of the Interior's U.S. Geographical and Geological Survey of the Rocky Mountain Region under John Wesley Powell. Most of the coastal tribes are in Salish language groups, shown here in red and orange. As with all these old maps of Indian distribution, the nomenclature reflects the terminology used at the time.

11

THE NORTHWEST IMAGINED

The northwest coast of North America was one of the last major areas of the world to be explored by Europeans. The region was about as far away as it was possible to get from Europe, requiring either a voyage around the dangerous tip of South America or round that of Africa and across both the Indian and Pacific oceans. And there was no promise of gold or riches. The only incentive was the possibility of finding the western end of a Northwest Passage; the latter, if found, would shorten the distance from Europe to the riches and spices of the East, but the search for the Northwest Passage was mainly conducted, naturally enough, from the east side of North America.

Indeed, although there are a number of possible or apocryphal voyages recorded, the first documented European voyage that certainly reached the Northwest north of Cape Mendocino did not take place until 1774 (see page 19). In the meantime mapmakers used their imaginations—sometimes applied to possible but not documented voyages—to fill up the empty spaces. And the Northwest was the most prominent empty space on the map of the world for two centuries. Such maps at one point reached bizarre levels, with the Northwest being shown in virtually unrecognizable shapes (see pages 16–17).

For centuries some maps showed land extending from North America right across the Pacific, the result of erroneous sightings of land by various early mariners. Most notably these include the Dutch East India Company's Maarten de Vries, who in 1543 reported a continent extending east from the Kuril Islands when in fact it was just

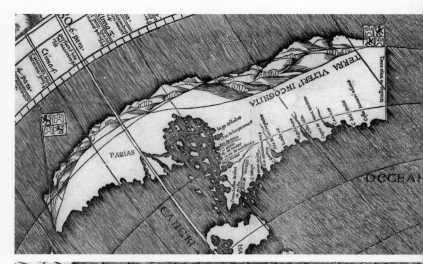

MAP 11 (*below*).
This map from the Gerard de Jode atlas of 1578 reflects Spanish knowledge of Baja California and otherwise fits into the Mercator concept of the world (MAP 14, *far right*), with little detail on the Northwest Coast.

MAP 12 (*above, top*, and *above*).
Martin Waldseemüller's groundbreaking world map of 1507 was the first to show North America as a separate continent—and the first to use the name *America*. As such, it was the first map to depict a west coast, though this seems to be entirely imaginary in shape, as the inset map makes clear. At *top* is part of the main map, while *above* is the inset.

MAP 13 (*below, right*).
The Northwest shown as a plate in the first printed atlas of America, published by Cornelius Wytfliet in 1597. The configuration of the land follows that of Cornelis de Jode's map drawn four years earlier (MAP 14, *far right*) but with less embellishment.

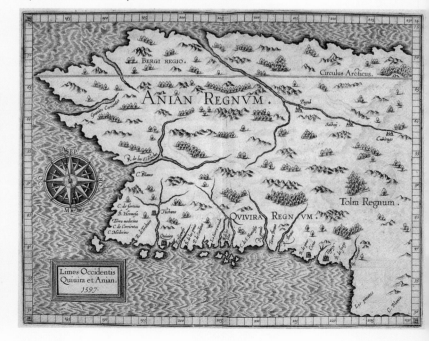

another small island; this was shown on maps as Company Land or Jesso. João da Gama reported mid-ocean land in 1589 or 1590 on a voyage from Malacca to Acapulco, leading Vitus Bering to waste valuable time looking for it in 1741 when searching for the west coast of North America.

Early mapmakers seemed always to abhor a vacuum, filling up the empty spaces on their maps with contrivances and artistic drawings—"elephants for want of towns" in Jonathan Swift's famous quatrain—but some mapmakers, especially French ones, went one better, filling their informational gaps in the American Northwest with interpretations of accounts of fictitious or undocumented voyages—or even outright hoaxes.

Names appearing on early maps were often transposed from other places, especially Spanish names from the Southwest, which was known much more intimately far earlier than the Northwest, since the latter lacked the incentive for exploration—a report of gold.

Until the late eighteenth century the Northwest remained a supposed Spanish territory, with only one incursion by another country—the English in 1579 in the shape of Sir Francis Drake. Much has been made of the possible sites where Drake careened his ship for repairs, and the claims include places on the Oregon Coast, but the issue remains inconclusive. There is also a bit of a mystery surrounding the maps showing Drake's progress up the West Coast, with a supposed plot to hide the truth so that Spain might think he had found the elusive Northwest Passage. A few interesting maps were created as a result, and none were more tantalizing than one that appeared in

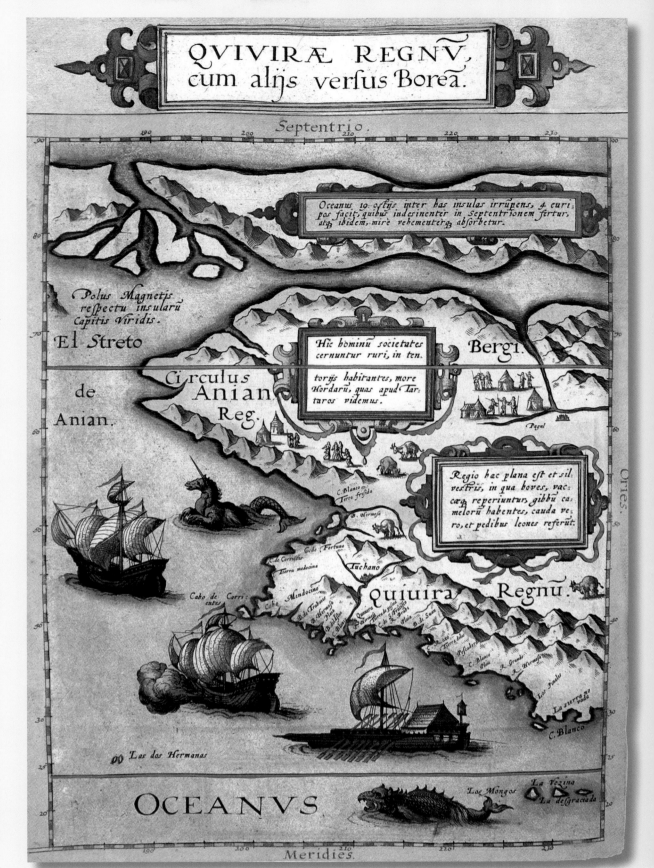

MAP 14 (*above*).
This beautiful example of the mapmaker's art was created in 1593 by Dutch cartographer Cornelis de Jode as a plate in an atlas. It bears almost no relation to reality. It reflects what was then considered to exist in the Northwest: the western end of a Northwest Passage running across the top of the continent, a concept initiated by Gerard Mercator on his 1569 world map and itself based on medieval ideas. The western entrance to this passage was something explorers searched for until George Vancouver's survey of 1792 finally demonstrated there was no entrance in temperate latitudes. *El Streto de Anian* (the Straits of Anian) leading to the passage were a staple of maps of the region for centuries. Anian was a name transposed from the works of Marco Polo. Farther south is *Quivira*, another transposition, this one of a mythical land of gold that Spanish explorers searched for in the Southwest for two centuries. The land in fact had cities of pueblo buildings, but in the Spanish mind all cities were the result of wealth. The sea—another potential blank on the map—was easier to fill, with classical ships and mythical sea beasts.

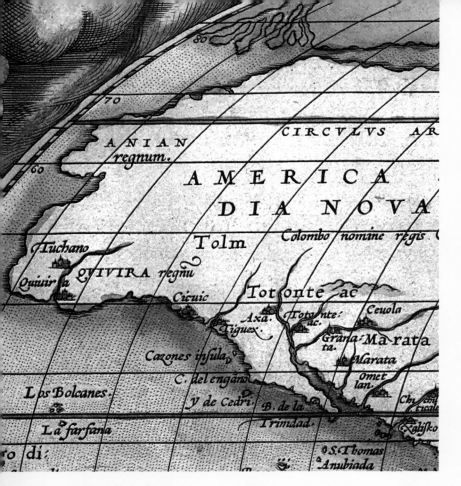

Map 15 (*left*).
Part of a world map published in 1570 by famous Dutch mapmaker Abraham Ortelius. Washington and Oregon are roughly in the position of Quivira on this map, itself a transposition of the name applied by the Spanish to part of the Southwest where they expected to find gold. Maps of the sixteenth century were not generally ones you would want to navigate by!

an atlas published in 1647 by Robert Dudley, who, because his father was one of the financial backers of Drake's voyage, was quite possibly in a position to have "inside" information. The map shows an undeniable similarity to the actual configuration of the coast of Washington between the Columbia River and Cape Flattery (Map 19, *far right*).

Two so-called apocryphal voyages found themselves over-represented on late-eighteenth-century maps of the Pacific Northwest. The first, that of Juan de Fuca, possibly took place, though no evidence has yet surfaced, and the second, that of the supposed Spanish admiral Bartholomew de Fonte, is known to have been a hoax.

Juan de Fuca was a Greek pilot in the service of Spain. Also known as Apostolos Valerianos, he is reputed to have told his story in 1596 to Michael Lok, an English promoter of exploration of the Northwest Passage. Fuca told him that he had been a pilot on a Spanish voyage to find and fortify the Strait of Anian (the supposed western entrance to the Northwest Passage). On the second of two voyages Fuca said he found a "broad inlet" between 47° and 48°N that he entered and sailed around in for twenty days. At the entrance—but on the north side—he recorded seeing a rock pillar. Receiving no reward from the Spanish, he approached Lok, perhaps in the hopes of financial reward for the information. Details such as the pillar suggest an

Map 16 (*below*).
Another part of a world map, this one by Dutch mapmaker Joducus Hondius printed in 1595. The track of Drake on the West Coast peters out, as though altered. It is not known exactly how far north Drake sailed, though it appears that he was looking for the western entrance to the Northwest Passage as a shortcut back to England to escape from the Spanish, who he thought were pursuing him. His ship, after all, was stuffed with Spanish gold plundered from a galleon returning to New Spain from Manila. The West Coast is *Nova Albion*—New England, and Drake's voyage staked a British claim to Oregon. *Inset* is Drake's elusive harbor, *Portus Novæ Albion*, which some have claimed to be on the Washington or Oregon coasts (Whale Cove, just south of Lincoln Beach, is one theory) but which was more likely at Point Reyes, California.

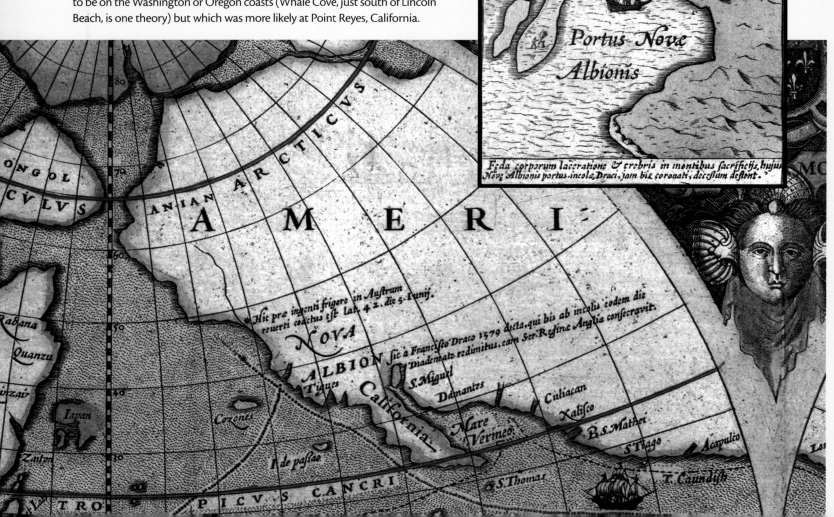

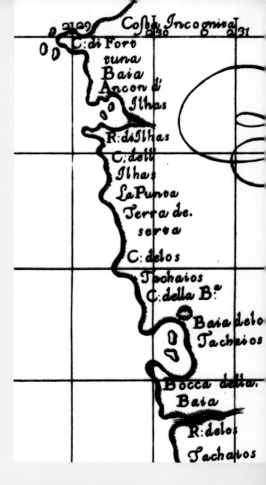

MAP 17 (*above*).
Fuca Pillar at the entrance to the Strait of Juan de Fuca is shown on a 1932 map. *Inset*: photo of the pillar.

MAP 19 (*right*).
The Northwest Coast depicted on a map by Robert Dudley published as part of a sea atlas *Dell' Arcano del Mare* (Concerning the Secrets of the Sea) in 1647. Dudley's father was one of the financial backers of Drake's voyage and thus, it is assumed, had inside knowledge of Drake's findings. *C: di Fort* has been interpreted as Cape Flattery, *Ancon d'Ilhas* as Grays Harbor, and *Baia de los Tachaios* as the mouth of the Columbia. But geographical similarity does not in itself prove Drake mapped the coasts of Washington and Oregon, though the possibility remains. Dudley's atlas was the first sea atlas to cover the entire world and the first to use Mercator's projection, on which straight lines are lines of constant bearing, an invaluable aid for mariners.

element of truth to the story, but to date no evidence of his voyage has been found in Spanish archives. However, we do know that the Spanish habitually shrouded their explorations in secrecy to prevent other nations from taking advantage of them, and it is entirely possible that records were lost. In any event, the cartographic interpretations of the Fuca voyage have left us with a legacy of sometimes amusing maps of bizarre geography (see overleaf).

Details of a supposed 1640 voyage by Admiral de Fonte to the Northwest and then through channels and straits leading to Hudson Bay appeared in a London magazine, *Monthly Miscellany, or Memoirs for the Curious,* in 1708. Although now known as a hoax, it was taken seriously at the time and once again led to interpretations of the voyage being incorporated into wonderfully absurd maps (shown overleaf).

Another addition to early maps resulted from the voyage of Martín de Aguilar (or d'Aguilar), captain of a ship with Sebastián Vizcaíno, who sailed north in 1602–03 looking for harbors for the Manila galleon, which often arrived distressed after crossing the Pacific. Separated, Aguilar's ship was pushed farther north than Vizcaíno's and reported an opening. It was probably the Rogue River in southern Oregon, but it could have been the Columbia, and in any case mapmakers soon connected it to the River of the West thought to flow across half the continent (see MAP 27, *page 18*).

MAP 18 (*below*).
The notion that America was first discovered by the Chinese is a widely held theory, though one with no factual evidence whatsoever. A Buddhist monk, Hui Shen, is said to have made a long voyage about 499 AD to "where the sun rises." The idea appeared as early as the eighteenth century, noting *Fou-Sang,* or some variant spelling, on the Northwest Coast. This is a simple one printed in Britain in 1781 to illustrate information on Spanish voyages in 1775 (see page 20). There are numerous records of Japanese or Chinese wrecks on the Northwest Coast, showing at least that the voyage was possible.

MAP 20 (*below*).
The distance overland to the Northwest was grossly underestimated for centuries, and no map depicts this more dramatically that this one, drawn in 1744 by fur trader Joseph La France, supposedly at first in the dirt on the floor of a London tavern. At least the map is honest in its notation *Unknown Coast*; this map reflects the idea that the Northwest Coast connected with the west side of Hudson Bay, at right. But the rivers and lakes of the interior, well-mapped by fur traders, were for many years shown far too close to the West Coast, an error caused by their inability to correctly measure longitude.

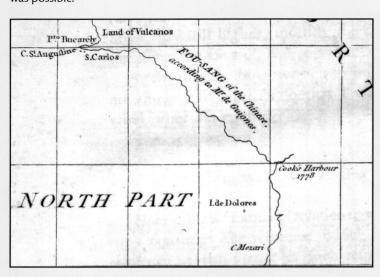

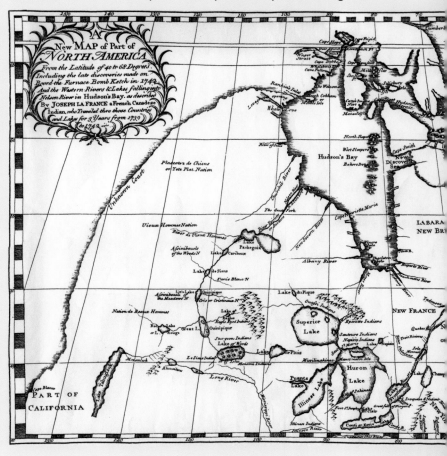

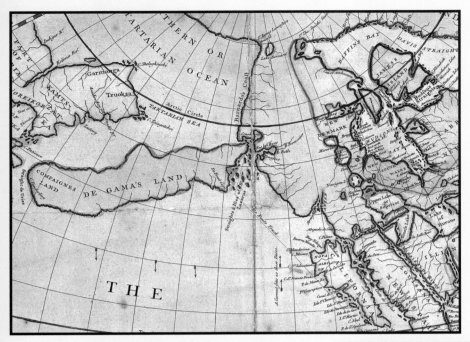

in the position of the Strait of Juan de Fuca. Note *De Fonte Track* offshore to the *Straights & Isles of St. Lazarus*, an archipelago that also features in the de Fonte hoax. *Hudson's Bay* and the Great Lakes are depicted far too far west; *Upper Lake or Superior* is almost in Oregon.

MAP 22 (*below, left*), MAP 23 (*below*), MAP 24 (*right, top*), MAP 25 (*right, bottom left*), and MAP 26 (*right, bottom right*).
Interpretations of the Juan de Fuca, Martín de Aguilar, and Bartholomew de Fonte accounts by French mapmakers of the mid-eighteenth century, respectively: Gilles and Didier Robert de Vaugondy in 1762; Jean Janvier in 1762; the Robert de Vaugondys again, in 1755; Jean Covens and Corneille Mortier about 1780 (by which time they should have known better!); and Jean Nolin in 1783 (likewise). All have the characteristic *Mer de l'Ouest* (Sea of the West) covering what should be Washington and more, an interpretation of Juan de Fuca's sea that he reported sailing in

Seas and straits scattered around the Northwest characterize this selection of bizarre maps, produced in response to accounts of the supposed voyages of Juan de Fuca, Bartholomew de Fonte, and others. Many of these maps leave one scratching one's head—where *are* Washington and Oregon? (See previous pages for the stories behind the names.)

MAP 21 (*above*).
Perhaps the finest and at the same time most ridiculous depiction of *De Gama's Land* was on this 1748 map produced for the Hudson's Bay Company and then—all but six copies—destroyed as being inaccurate. *Straights* and *Anian* are approximately

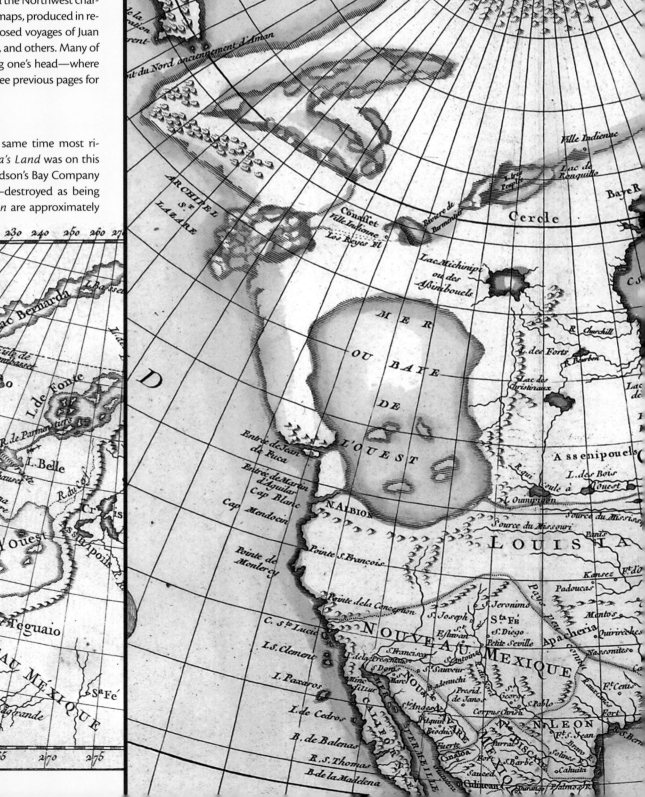

for twenty days. It *could* have been the Strait of Georgia and Puget Sound, of course; they are bodies of water in the right places. Four of the five maps also show a second entrance to the Sea of the West, the *Entreé de Martin d'Aguilar en 1603* (Map 24), Aguilar's opening translated into a strait by armchair geographers who had never been anywhere near the Northwest Coast. All the maps also show straits leading to Hudson Bay, the result of the Bartholomew de Fonte hoax. Other fantasies such as the Strait of Anian and Quivira are also evident on these maps.

Origan

Many theories exist as to the origins of the word "Oregon," but none has been proven. What is documented with some exactitude, however, is its first application—and this was on a map.

Ever since French explorers got excited about Native stories of a river flowing west to a great sea—actually into Lake Winnipeg—depictions of a River of the West flowing to the Pacific began to appear on maps, the product of a hope for an easy route west rather than any basis in fact. It variously entered the ocean at the Strait of Juan de Fuca or Martín de Aguilar's entrance, a bay some thought to be the Columbia River, discovered by one of the ships of the Spanish exploration of 1602, led by Sebastián Vizcaíno to search for refuges for the Manila galleon.

Jonathan Carver, a British explorer, published a book in 1778 that became a widely read bestseller, and it contained not only maps that showed the River of the West (MAP 29, below), but another that showed it rising just west of Lake Superior with the notation *Heads of Origan* (MAP 30, right). This appears to be the first published use of the name, which Carver seems to have obtained from Major Robert Rogers, commandant of the British Fort Mackinac, at the northern end of Lake Michigan. Variously spelled as *Ourigan*, *Origan*, or *Oregon*, the name appeared first as a name for a River of the West, then as the Columbia, and showed up as late as 1798 on a map of the Columbia by British mapmaker Aaron Arrowsmith, a published form of William Broughton's survey of 1792 (see page 25). Not long after, the name was applied to the country surrounding the river; the Oregon Country was the territory jointly occupied by Britain and the United States after 1818.

MAP 27 (*above, top*).
An *Opening discovered by Martin D'Aguilar in 1603* [*sic*; 1602] appears at the mouth of a *River of the West* flowing west from Lake Winnipeg on this 1768 map by British mapmaker Thomas Jefferys.

MAP 28 (*above*).
On this 1749 French map a *Fl[euve] de l'Ouest* (River of the West) is shown flowing to the Pacific through Lake Winnipeg from *Lac des Bois* (Lake of the Woods) and the Great Lakes system, an extension of the portage route used by French fur traders to Lake Superior at the time.

MAP 29 (*below*) and MAP 30 (*right*).
From Jonathan Carver's book come these two maps. The main map shows a River of the West flowing to both of the openings discovered by Juan de Fuca and Aguilar; MAP 30 shows the *Heads of Origan* flowing west from a small lake near *White Bear Lake,* which is in today's Minnesota. The small lake is in the position that Bald Eagle Lake is on modern maps. Carver was a believer in the then widely held concept of symmetry—that rivers flowed from the center of the continent symmetrically: the St. Lawrence to the east, the Mississippi to the south; the Bourbon River to Hudson Bay; and the River Oregon, or River of the West, to the Pacific. His Oregon River fitted entirely with the theory.

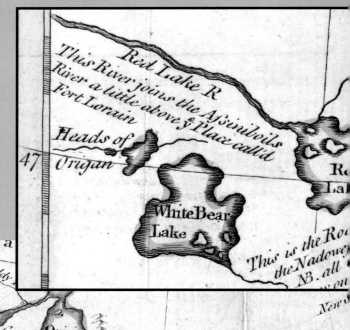

THE FIRST EXPLORATIONS

The first European explorations of the coasts of Washington and Oregon were prompted by the Russians. In 1741 Vitus Bering and Aleksei Chirikov had sailed east from Kamchatka and independently reached the coast of the Alaska Panhandle. When news of these voyages finally reached the Spanish court, they caused much consternation and were quickly construed as a threat to perceived Spanish domination of the Pacific out of a base in New Spain, today's Mexico. The situation was not helped by the publication in 1759 of a book written by a Spanish monk in Italy with the title *I Moscoviti nella California*—The Muscovites in California.

Although the wheels of government moved slowly, the result was that Spain decided to move north, colonizing Alta California as far north as San Francisco Bay beginning in 1769. In addition, several expeditions were organized to sail north to investigate any possible Russian incursions, and it was these expeditions that became the first documented European voyages to the Pacific Northwest.

In late 1773 Juan Pérez, one of the most competent of the Spanish naval officers at the time, received orders to sail a new ship, the *Santiago,* north from New Spain to 60° in search of "foreign settlements." In January 1774 Pérez sailed, reaching today's Dixon Entrance, at the southern extremity of the Alaska Panhandle, a latitude of 54°40´N, a point that would later become the northern limit of Spanish claims to the coast. The map from this important voyage (MAP 32, *right*) naturally enough shows little detail of the Washington and Oregon coasts but nevertheless was the first to depict them based on observation.

The following year another expedition was ordered north, this time with Pérez, having failed to reach his ordered 60°N in 1774, as second officer to Bruno de Hezeta y Dudagoitia, once more on the *Santiago.* The

MAP 31 (*below, left*).
Juan Francisco de la Bodega y Quadra's map from his voyage to Alaska, stopping off at the Washington coast, in 1775. There the *Entrada descubierta por D^n Bruno Ezeta* (the Columbia River) is shown, as is *Rada de Bucareli* (Grenville Bay), but not much other detail; indeed the coast north of that point reflects the offshore course necessary to sail north. Much more information is then given for the Alaska Panhandle.

MAP 32 (*below*).
Important as the first map of the Northwest Coast drawn as the result of exploration is this map by Spanish pilot José de Cañizares (whose name might be written, as on the map, in the older style *Josef de Cañizarez*), following the return of Juan Pérez to San Blas in New Spain. The map was found only in 1989 in the U.S. National Archives; no one knows how it might have found its way to that repository. The map shows the West Coast from San Francisco Bay (at bottom) to the northern tip of Haida Gwaii and the southern tip of the Alaska Panhandle (at top). The only Washington or Oregon detail shown is *Pta de Sta Rosalia* and *Cerro de Sta Rosalia*—Mount Olympus.

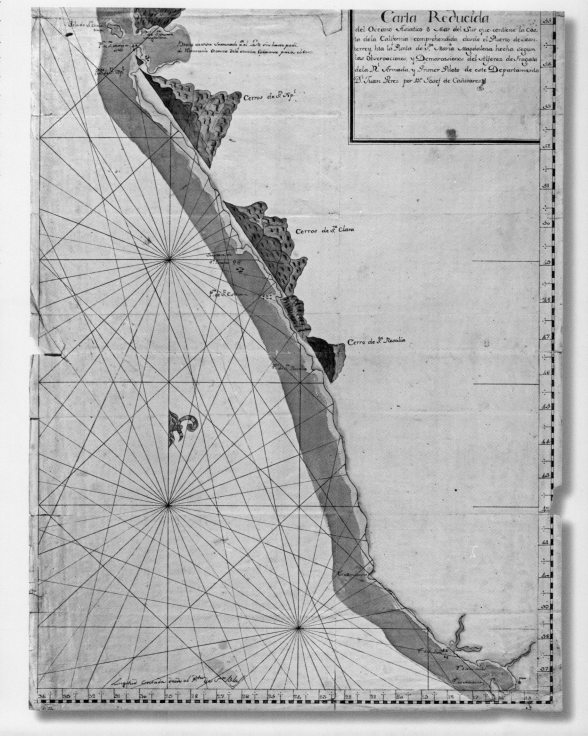

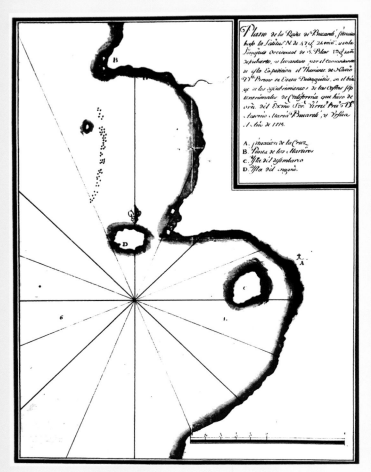

MAP 33 (*above*).

The first map of Washington, Bruno de Hezeta's map of the coast, with Cape Elizabeth, at *B*, named *Punta de los Martires* (Martyrs' Point). The name commemorates his seven crew members who were killed here. *A* is where a cross was erected to claim possession on the shores of Grenville Bay, which Hezeta called the *Rada de Bucareli* (Bucareli's Roadstead), after the viceroy of New Spain, Antonio María de Bucareli.

expedition, this time ordered to reach 65°N, also included a smaller, only 30-foot-long ship, the *Sonora,* under Juan Francisco de la Bodega y Quadra.

It was this expedition that made the first regional maps of any part of Washington and Oregon. Trying to obtain water and carry out a ceremony of possession on 14 July 1775 at Grenville Bay, just south of Cape Elizabeth on the Washington coast, Bodega lost seven men to an attack by three hundred Quinault (MAP 33, *above*).

Soon the two ships became separated, likely because Hezeta wanted to return south while Bodega y Quadra was determined to continue north. Bodega y Quadra made an epic voyage to Alaska, while Hezeta, whose men were suffering from scurvy, decided to return to New Spain, which he achieved by sailing south along the coast. In so doing, Hezeta found the mouth of the Columbia River but seems not to have recognized it as such, writing in his *diario* that he had "discovered a large bay" (MAP 35, *right*) that he named Bahía de la Asunción; this was shown on more general maps as an *entrada,* or entrance (MAP 31, *previous page,* and MAP 34, *above, right*).

Although Spain would send more expeditions north, the next country on the Northwest Coast was

MAP 34 (*above*).

Drawn in 1787, this Spanish map shows *Punta de los Martires* (Martyrs' Point, today Cape Elizabeth) and the *Entrada* (seen by) *Dn. Bruno Hezeta,* the mouth of the Columbia River. *C[ab]o de S[a]n Roque* is Cape Disappointment. *Ysla de los Dolores*—the Isle of Sorrows (at the mouth of the Hoh River)—was named by Hezeta in memory of his dead sailors.

MAP 35 (*below*).

Hezeta's map of his *Bahía de la Asunción,* the first map of the mouth of the Columbia. Cape Disappointment, at *B,* indexed as *Cabo de San Roque,* is shown as an island. North is to the left.

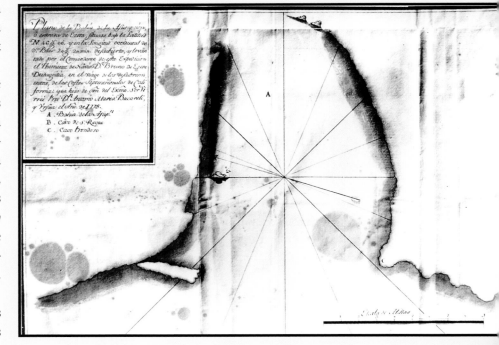

Britain, in the shape of the celebrated Captain James Cook, on his third voyage. Cook was looking not for Russians but for the elusive Northwest Passage and was charged with finding its western end or proving it did not exist. Although Cook had little effect on the map of Washington and Oregon, he did name Cape Foulweather in Oregon and Cape Flattery at the entrance to the Strait of Juan de Fuca, which he missed in bad weather. Not noted on his initial chart (MAP 37, *right*), the names were added to the map of the Northwest Coast published in 1784 (MAP 38, *right, inset*).

In the wake of Cook came a number of fur traders looking for sea otter pelts, and again, although most concentrated their efforts on the more crenulated coastline to the north, two are of particular note here. Charles Barkley, with his young wife, Frances, sailed to the Northwest Coast on a fur-trading mission in 1787. In July, sailing south along the coast of Vancouver Island, he found the entrance to a wide strait, which he believed must be the same one that Juan de Fuca discovered in 1592 (see page 15) and which he promptly named—or perhaps renamed—the Strait of Juan de Fuca. Knowing that James Cook's maps showed no such strait, he was very surprised. MAP 36 (*below*) is a contemporary chart showing his discovery. Barkley later lost six of his men to Indian attack at the mouth of the Hoh River, a tragedy very similar to the one Bodega y Quadra and Hezeta suffered twelve years before.

Barkley's charts were acquired by another fur trader, John Meares, whose relevance to Washington history is that in 1788 he renamed Juan Pérez's Cerro de San Rosalia, calling it Mount Olympus (MAP 39, *below, right*). Continuing south, Meares was unable to find the entrance to the Columbia, despite knowing Hezeta's report of it, and gave the area the name Deception Bay instead. The fact that Meares, an experienced Northwest sailor, could miss the mouth of the river, even knowing of its position, shows how well concealed the river's mouth was, the effect of shoals and sandbars.

In 1788, the Spanish, alarmed once more about incursions into their territory and determining also to find the western entrance to the Northwest Passage before any other power did so, launched a series of voyages to the Northwest, focusing more on Alaska and British Columbia, as this was where any passage was thought to be located. They occupied Nootka, on the west coast of Vancouver Island, and in 1789 they arrested two British fur traders considered to be trespassing. This action provoked an international incident, which Britain seized upon to force Spain to

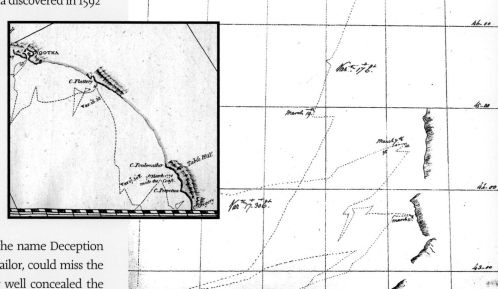

MAP 37 (*above*) and MAP 38 (*above, inset*).
A map of James Cook's brief contact with the Oregon and Washington coasts in March 1778 from the journal of one of his officers, James Burney, and the map published posthumously in 1784. Names added on the published version are *C[ape] Gregory* (Cape Arago); *C. Perpetua*, near Yachats; *C. Foulweather*, just south of Lincoln Beach; and *C. Flattery*, which he recorded in his journal, "flatered us with hopes of finding a harbour." The latter is now thought to have been Point of Arches, eight miles south of Cape Flattery.

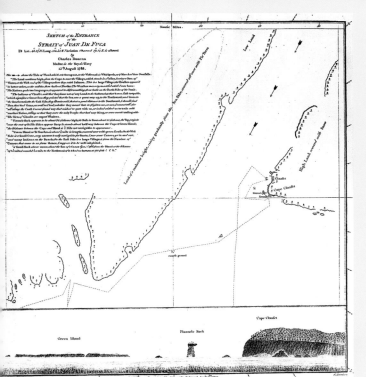

MAP 36 (*left*).
This *Sketch of the Entrance of the Strait of Juan de Fuca* was drawn in 1788 by another fur trader, Charles Duncan, from Charles Barkley's 1787 original chart of his find. The map was published in 1790 by Alexander Dalrymple, a British promoter of the concept of the Northwest Passage. The drawing below the map shows *Pinnacle* Rock (now Pillar Rock). *Cape Claaset* is Cape Flattery. Since Juan de Fuca was reputed to have described a rock pillar at the entrance to his strait (albeit on the north side), it is easy to see how discovery of the waterway might be attributed to him.

MAP 39 (*right*).
This detail from one of John Meares's maps, published in 1789, shows *Mt. Olympius*.

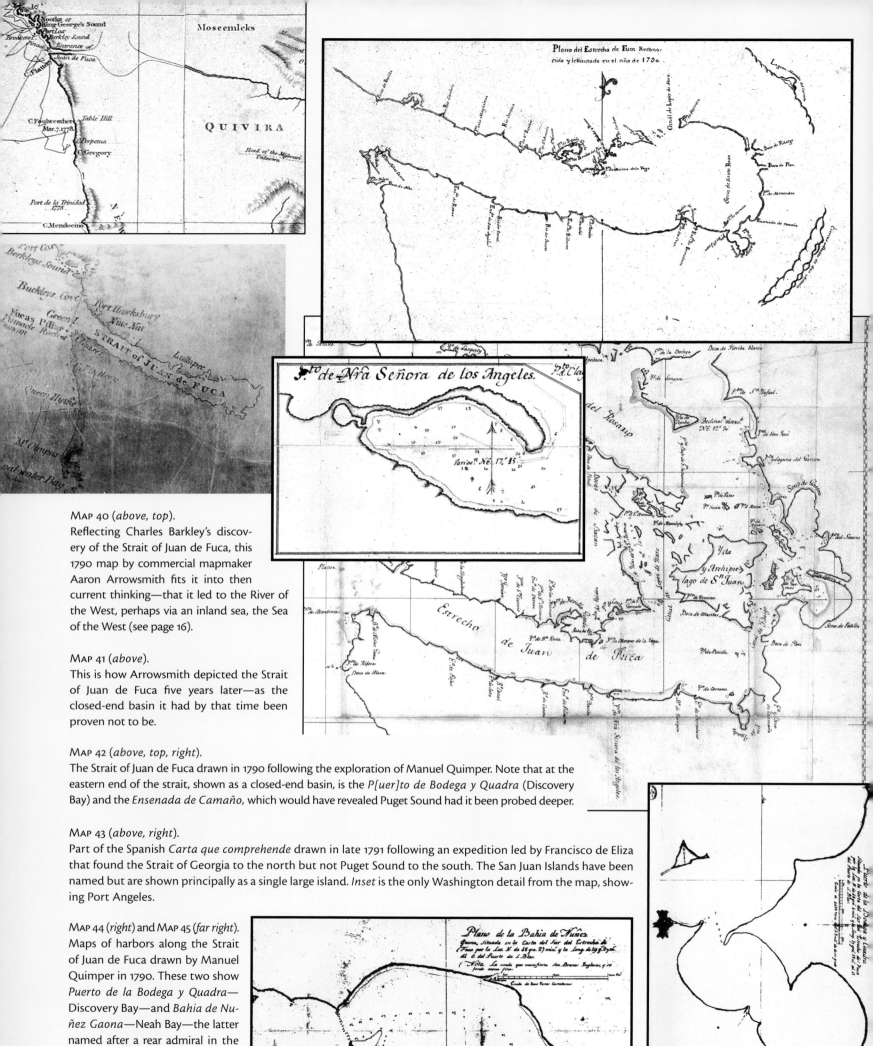

MAP 40 (*above, top*).
Reflecting Charles Barkley's discovery of the Strait of Juan de Fuca, this 1790 map by commercial mapmaker Aaron Arrowsmith fits it into then current thinking—that it led to the River of the West, perhaps via an inland sea, the Sea of the West (see page 16).

MAP 41 (*above*).
This is how Arrowsmith depicted the Strait of Juan de Fuca five years later—as the closed-end basin it had by that time been proven not to be.

MAP 42 (*above, top, right*).
The Strait of Juan de Fuca drawn in 1790 following the exploration of Manuel Quimper. Note that at the eastern end of the strait, shown as a closed-end basin, is the *P[uer]to de Bodega y Quadra* (Discovery Bay) and the *Ensenada de Camaño*, which would have revealed Puget Sound had it been probed deeper.

MAP 43 (*above, right*).
Part of the Spanish *Carta que comprehende* drawn in late 1791 following an expedition led by Francisco de Eliza that found the Strait of Georgia to the north but not Puget Sound to the south. The San Juan Islands have been named but are shown principally as a single large island. *Inset* is the only Washington detail from the map, showing Port Angeles.

MAP 44 (*right*) and **MAP 45** (*far right*).
Maps of harbors along the Strait of Juan de Fuca drawn by Manuel Quimper in 1790. These two show *Puerto de la Bodega y Quadra*—Discovery Bay—and *Bahia de Nuñez Gaona*—Neah Bay—the latter named after a rear admiral in the Spanish navy.

22

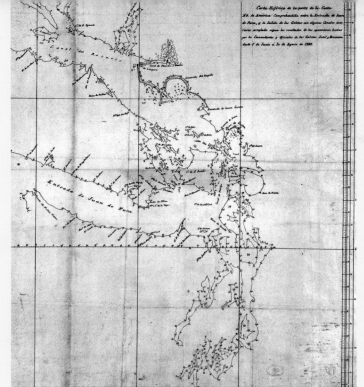

MAP 46 (*right*).
The Spanish explorers Dionisio Alcalá Galiano and Cayetano Valdés met and shared information with George Vancouver in 1792. This is one of their initial 1792 survey maps, on which Vancouver's survey of Puget Sound, which the Spanish did not visit, has been added. More detail of Vancouver Island was mapped by the Spanish, and this and other information was given to Vancouver in return.

accept in 1790, by the Nootka Convention, the principle of sovereignty through occupation rather than mere discovery. The agreement also involved transfer of lands at Nootka, and it was this that George Vancouver was dispatched to deal with in 1791.

In the meantime, the Spanish probed the Strait of Juan de Fuca, in 1790 mapping the entire strait (MAP 42, *left, top*) and in 1791 reaching the Strait of Georgia and mapping the San Juan Islands (MAP 43, *left*). They established a base at Neah Bay, near the entrance to the strait, naming it Nuñez Gaona (MAP 44, *far left, bottom*). Thus, the region was quite well known—to some, at any rate—by the time George Vancouver arrived on the scene in 1792.

Vancouver was also searching for the western entrance to the Northwest Passage and had been instructed to closely trace the "continental shore" to ensure no inlets were that entrance. In so doing he probed south at the eastern end of the Strait of Juan de Fuca, discovering Admiralty Inlet and Puget Sound, names, along with many others, bestowed by Vancouver and still in use today (MAP 49, *overleaf*). In May 1792 Vancouver's ship *Discovery* sailed as far south as Restoration Point, opposite the future site of Seattle; the rest of the inlet was surveyed in smaller boats. One was headed by Joseph Whidbey, ship's master, and Lieutenant Peter Puget, whose name Vancouver bestowed on the southern extremities of the sound that he surveyed and whose name would one day be applied to the whole.

MAP 47 (*right*).
The pen-and-ink preliminary chart of the Northwest Coast from George Vancouver's 1792 survey, drawn by Lieutenant Joseph Baker in July of that year. It contains the first mapping of Puget Sound (MAP 49, *overleaf*, is an enlargement) but does not show the mouth of the Columbia River, which Vancouver missed but was told about by Robert Gray (see overleaf).

MAP 48 (*right*).
The published chart of the Oregon Coast, annotated as *Part of the Coast of New Albion,* from George Vancouver's survey in April 1792 after making a landfall just over a hundred miles north of San Francisco and then sailing north. Part of the October 1792 survey of the Columbia River is visible at top.

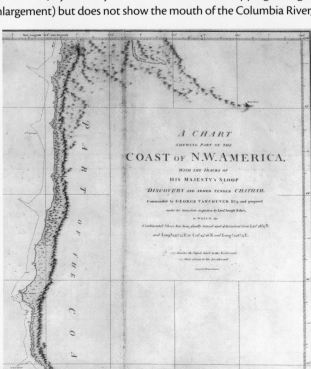

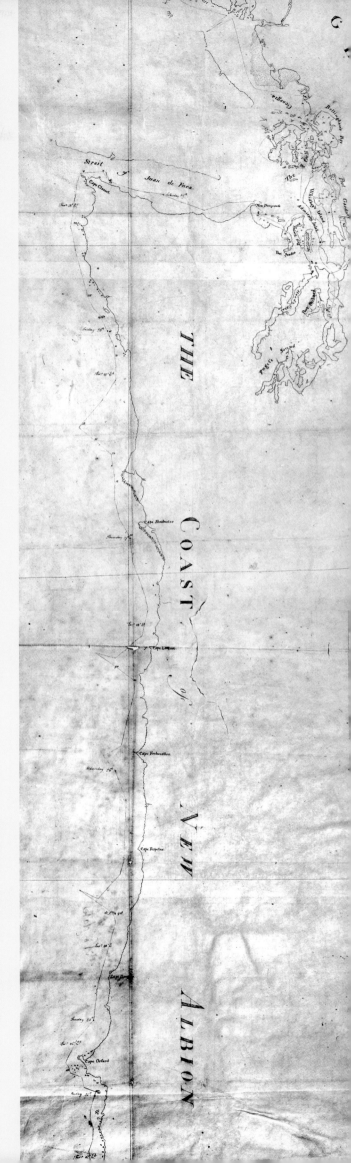

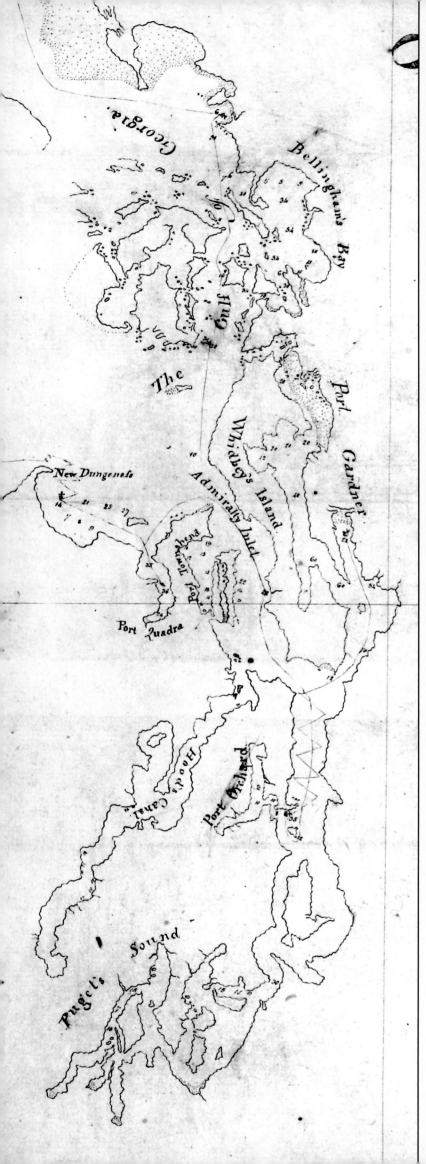

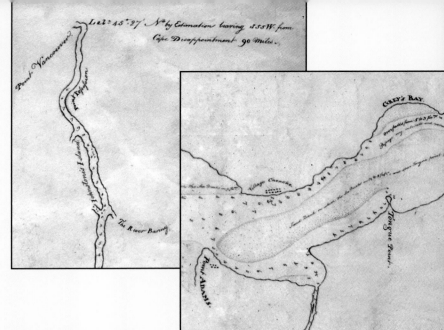

MAP 49 (*left*).

Detail of Puget Sound from the July 1792 preliminary chart of Vancouver's mapping of the Northwest Coast, drawn by Joseph Baker. This represents the first time the inlet had been mapped. Note that *Puget's Sound* was originally only intended to apply to the southern portion of the sound, with the northern part named, as the extreme northern part is today, *Admiralty Inlet*. Many features were named by Vancouver: *Whidbey's Island*, after Joseph Whidbey, master of *Discovery*, who proved it was an island by exploring through Deception Pass; *Bellingham's Bay*, after William Bellingham, controller of the British Navy storekeeper's accounts; *Hood's Canal*, after Admiral Lord Hood, Lord Commissioner of the Admiralty; *Port Orchard*, after ship's clerk Henry Masterman Orchard, who first sighted the bay; and *Port Gardner*, after Rear Admiral Sir Alan Gardner, who had recommended Vancouver to command the Northwest Coast exploration.

MAP 50 (*above*), in two parts.

From William Broughton's October 1792 survey of the first hundred miles of the Columbia River come these two details. The part at *left* shows the eastern extremity of the exploration, which Broughton named *Point Vancouver* after his commander; this was close to where the Hudson's Bay Company would found Fort Vancouver in 1822. Also named is *Point Possession*, on the south side of the river, approximately in the position of today's Troutdale, on the east side of Portland. *The River Baring* is the Willamette. The part at *right* shows *Grey's Bay*, near the river's mouth, which Broughton named in honor of Robert Gray, spelling the name in the English fashion. Note also the *Village Chenoke* (Chinook), on the north side, a Native village. A few years later, Fort Astoria would be founded on the south side, between *Point Adams* and *Tongue Point*.

MAP 51 (*below*).

Vancouver's original survey of Grays Harbor, which he named for Robert Gray.

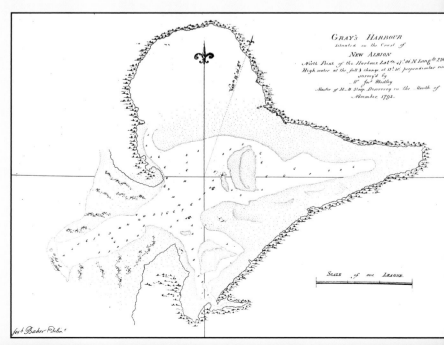

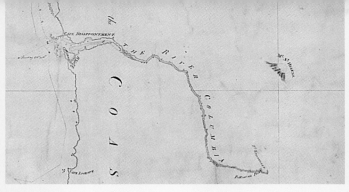

MAP 52 (*above*).
The River Columbia, shown on a pen-and-ink map drawn by Joseph Baker "under the immediate inspection of Captn. Vancouver" at the end of 1792. The map now incorporates William Broughton's survey of the river. Note *Mt. St. Helens*, named by Vancouver after Baron St. Helens, British ambassador to Spain and negotiator of the 1790 Nootka Convention (see page 23).

GRAY'S COLUMBIA

Vancouver missed the Columbia River, as had others before him. The first EuroAmerican to enter it was Robert Gray, on 11 May 1792. Gray was a Massachusetts fur trader who had sailed first to the Northwest Coast in 1788, when he had landed at Tillamook Bay. Then he did not attempt to enter the Columbia, but now, in his ship the *Columbia Rediviva*, Gray successfully crossed the bar at the mouth of the river and sailed thirteen miles upstream. A few weeks earlier he had again found the river but had been unable to get his ship over the bar. "We found this to be a large river of fresh water," he wrote matter-of-factly in his journal. "Vast numbers of natives came alongside," he added, which was what he wanted, for he was here to trade for furs, and many were purchased. And, of course, he named the river after his ship, christening it Columbia's River (MAP 54, *right*).

Before his second attempt to enter the Columbia, Gray had met George Vancouver and told him about the river; Vancouver investigated but was unable to get his own ship over the bar, instead sending the smaller *Chatham*, commanded by William Broughton, upriver. Broughton surveyed as far as the present site of Troutdale, on the east side of Portland, naming it Point Vancouver (MAP 50, *left, top*).

Below. The ship for which the river is named. This is a rare depiction of Gray's ship *Columbia Rediviva*, painted by one of his crew, George Davidson. It shows the *Columbia in Destress* in a December 1790 storm but serves to tell us what the eighty-three-foot-long ship looked like. In 1790 Gray, in the *Columbia*, became the first American to circumnavigate the globe.

COLUMBIA in DESTRESS.
In Latitude 44 South Long. 27 West December 1790

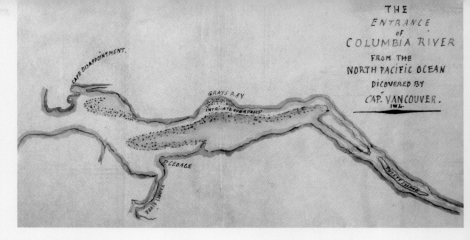

MAP 53 (*above*).
This delightful little map was drawn in 1805 by James Winter Lake, governor of the Hudson's Bay Company, as the company assessed its policies relating to the region west of the Rockies. Lake attributes the discovery of the mouth of the Columbia only to George Vancouver but shows *Grays Bay*. It is interesting to note that, as this map was being drawn, Meriwether Lewis and William Clark were on their way to the Columbia and would draw their own maps of the same area (see page 29).

MAP 54 (*below*).
Robert Gray's map of the first thirteen miles of his *Columbia's River*, shown here on a map given to or copied for George Vancouver. Note that north is to the left. Grays Bay is at top left.

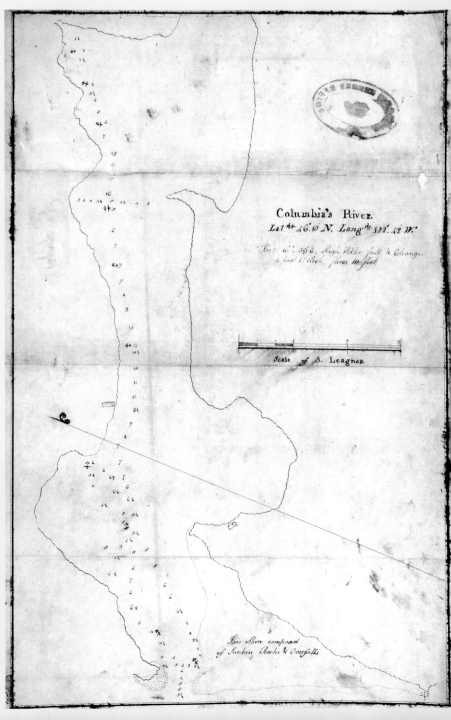

OREGON OVERLAND

Fur traders had coveted an overland route to the Pacific Coast—from where furs could be easily shipped to China—for a century or more, when Alexander Mackenzie, a trader with the British North West Company, became the first to reach the Pacific overland in 1793. That same year, Thomas Jefferson—the forward-thinking principal author of the Declaration of Independence—drew up instructions for an attempt sponsored by the American Philosophical Society that came to naught. In 1802, with the publication of Mackenzie's book, the commercial designs of the North West Company on the Northwest became known. The next year the Louisiana Purchase created a national imperative that the United States find its own route to the Pacific.

Since James Cook's visit to the Northwest Coast in 1778—with a newly invented chronometer for measuring longitude—the position of the West Coast (and hence the width of North America) had been known with some accuracy. What was not known was how to reach it by a land route.

MAP 56 (right).

The *Oregon, or R. of the West* flows to the entrance *Discovered by d'Aguillar* (Martín de Aguilar) south of that *Discover'd by Jn. de Foca* (Juan de Fuca) on this map drawn by French surveyor general of Louisiana, Antoine Soulard, about 1795 and copied in English, with a few additions, in 1802. The map was carried by Lewis and Clark and has a hole in the middle, part of which is shown here, worn by being carried in the explorers' pockets. Oregon is here *Unknown Country*.

MAP 57 (below).

Partly based on Aaron Arrowsmith's published map of 1802, this map by Nicholas King was also carried west with Lewis and Clark. The Missouri rises from many branches in a single line of mountains, while the Columbia (here the *River Oregan*), rising nearby, flows west. The river is labeled *The Indians say they sleep 8 Nights in descending this River to the Sea*. The map follows the theory that there was an easy portage across a single range of the Rocky Mountains, a notion that Lewis and Clark would soon disprove.

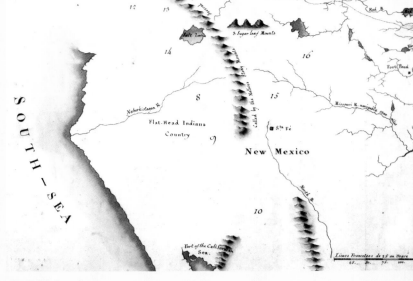

MAP 55 (above).

The North West Company's Peter Pond, a partner of Alexander Mackenzie, drew this map in 1784 and presented it to Congress the following year, urging an American interest in the Northwest. His *Naberkistagon R.* balances the *Missouri* flowing from a height of land *Called by the Natives Stony Mounts*—the Rockies. The *North R.* (Rio Grande) flows south. Contemporary theory called for the rivers of the continent to be balanced—that is, flow in all directions from a height of land.

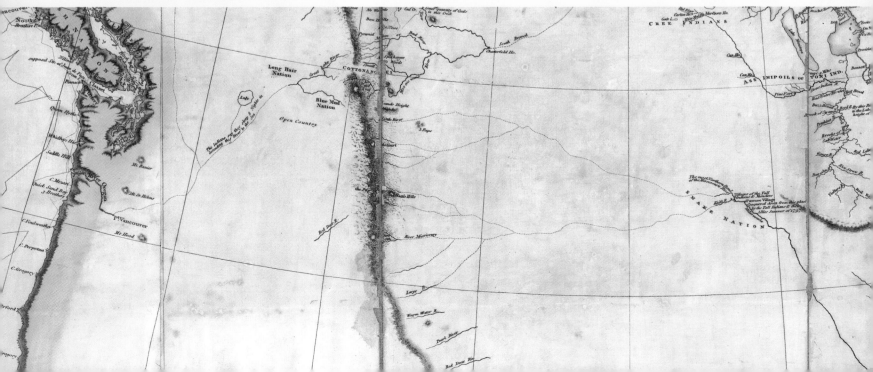

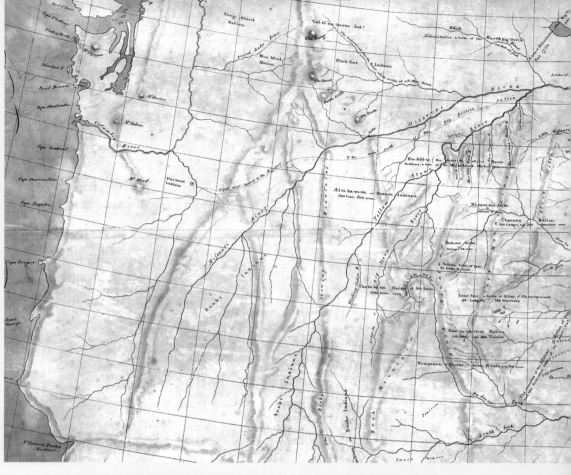

MAP 58 (*above*).
An 1802 map by James Pitot, a New Orleans merchant and sometime mayor of that city, depicted the Missouri flowing from British Columbia and the Columbia River as a much shorter stream within Oregon, a guess at how rivers were configured in the Northwest.

MAP 59 (*right*).
A composite copy of all of William Clark's maps made by Nicholas King before the Corps of Discovery set off. It represents the perception of what the expedition would find; there are now multiple ranges of mountains, but the map still depicts hoped-for easy passages between the river headwaters.

MAP 60 (*below*).
One of the famous maps of American history is this summary map of the Lewis and Clark expedition, engraved by Samuel Lewis and published in the book edited by Nicholas Biddle in 1814. The geography of the Northwest now approaches reality. The *Columbia* flows in from the north, *Lewis's River,* the Snake, flows from the southwest, and *Clark's River* (the Pend Oreille–Clark Fork and Bitterroot rivers), the major tributaries of the Columbia, are now shown. The Willamette is shown as the *Mult-no-mah R.,* mapped and named using information from Indians living near the river's mouth, but it is incorrectly shown flowing from too far south and east.

Jefferson, now president, asked Meriwether Lewis to organize and lead a new expedition west. Lewis soon enlisted the help of his friend William Clark, and in 1804 the Corps of Discovery, some twenty-seven strong, left St. Louis for the Pacific.

They overwintered at Mandan villages on the Missouri in today's South Dakota. They made an arduous crossing of the multiple ranges of the Rocky Mountains, not easily portaged, only after negotiating the purchase of horses from Shoshone, a deal made with the help of Sacagwea, the Shoshone wife of trader Toussaint Charbonneau, who had accompanied the expedition from the Mandan villages.

After crossing what Clark described as "emence mountains," the Corps found the Clearwater River, which drains westward into the Snake and then into the Columbia, reached on 16 October 1805. Beacon Rock, on the Columbia at the beginning of Pacific tidewater, was reached on 3 November. Four days later Clark recorded in his field book his famous line: "Ocian in view! O! The joy!"

It took another twenty days to reach the coast. After exploring the lower Columbia (MAP 62 and MAP 64, *overleaf*), both Lewis and Clark carved inscriptions on trees. Clark emulated Alexander Mackenzie when he carved "Capt. William Clark December 3rd 1805. By Land.

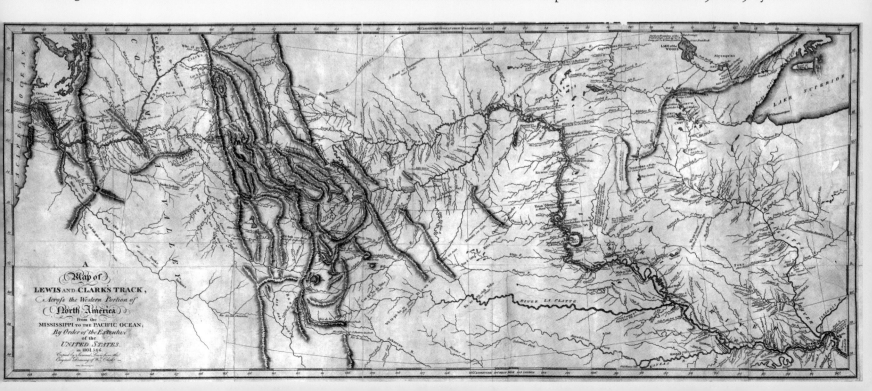

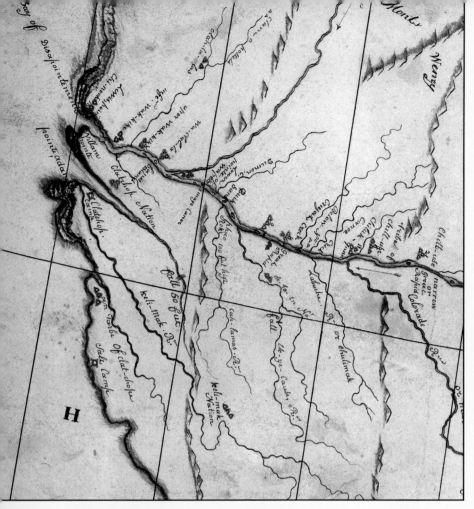

U. States in 1804–1805." The Corps built a stockaded winter camp just south of the Columbia (MAP 62, *below*) that they named Fort Clatsop after the local Indian tribe. Here they stayed from December 1805 to March 1806, with Lewis and Clark drawing maps and writing about natural history in their journals. Some forays were made into the surrounding countryside, including to the coast to set up a camp and boil seawater to make salt, so critical for the preservation of their food supply. The camp was at the location of today's Seaside (MAP 61, *left*).

Lewis and Clark were in the Northwest for only a few months, but the effect their visit lasted far longer. The dream of an easy route to the Orient was laid to rest, but by dint of first exploration the expedition established a firm American claim to the interior Northwest, and with the publication of a book about the expedition, which was delayed until 1814, inspired a dream of a transcontinental United States, setting in motion a train of events that would eventually lead to an American Oregon.

MAP 62 (*below*).
William Clark's map, drawn in early 1806, of the mouth of the Columbia and the coast south to today's Cannon Beach. Fort Clatsop, not named, is the black rectangle on the *Ne-tul River*, renamed the Lewis and Clark River in 1925. A number of encampment locations are shown, together with their dates. The salt camp is also indicated.

MAP 61 (*above*).
Robert Frazer (or Frazier), one of the Corps of Discovery, intended to publish a book about the expedition, but the book never saw the light of day. A map that he drew to illustrate it has survived, though it is rarely reproduced because of numerous inaccuracies. The map, drawn about 1807, shows the location of Fort *Clatsop* near the mouth of the Columbia as well as many Native villages (the red triangles). Also shown is the *Salt Camp* on the coast (today's Seaside), where salt was made from seawater between December 1805 and February 1806.

Below.
A modern replica of Lewis and Clark's overwintering post, Fort Clatsop, on what is today the Lewis and Clark River, near Astoria. This new structure was built in 2006 following a fire that destroyed a 1955-built replica the year before.

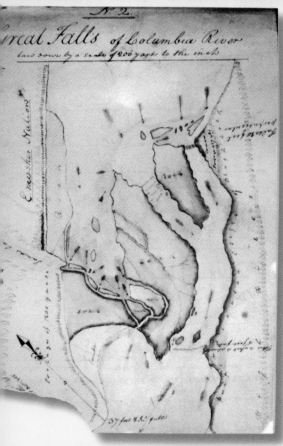

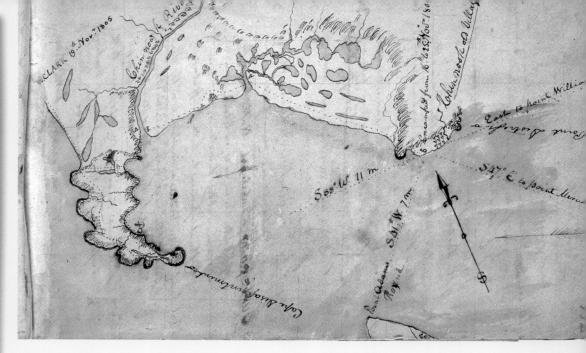

Map 64 (*above*).
Cape Disappointment and the northern side of the mouth of the Columbia on a map drawn by William Clark in his journal on 19 November 1805. Cape Disappointment was named not by Lewis and Clark but by British fur trader John Meares in 1788, reflecting his disappointment at not finding the mouth of the river.

Map 63 (*above*).
The *Great Falls of Columbia River*, a map drawn by William Clark in his journal. The river is flowing top to bottom on this map. The map shows Celilo Falls, which was a favorite Native fishing place until the falls were drowned by the rising waters behind The Dalles Dam in 1957.

Map 65 (*below, left*).
The northwest part of a large manuscript summary map of the Lewis and Clark expedition drawn by Clark while at Fort Clatsop. Until 1810 the map hung in Clark's St. Louis office, where he was superintendent of Indian affairs, and was updated from time to time, sometimes inaccurately. The *Multnomah River*—the Willamette—is shown extending south to the latitude of San Francisco Bay.

Map 66 (*below, right*).
Lewis's copy of Clark's map of the *Multnomah R,* the Willamette, the course of which was conveyed to Clark by a person Clark described as "an old and inteligent Indian." Note that the map is oriented with south at the top. The river's name came from the Native group living at the river's confluence with the Columbia. Lewis and Clark gained a great deal of information from Natives. They missed the mouth of the Willamette, finding only the much smaller *Quicksand River* (Sandy River). Natives told Lewis and Clark of the larger river, and Clark returned to investigate. Several maps appear to have been drawn by Natives for Clark using lumps of coal and a grass mat or simply with a finger in dust on the ground. This map was Clark's composite—this time drawn on paper.

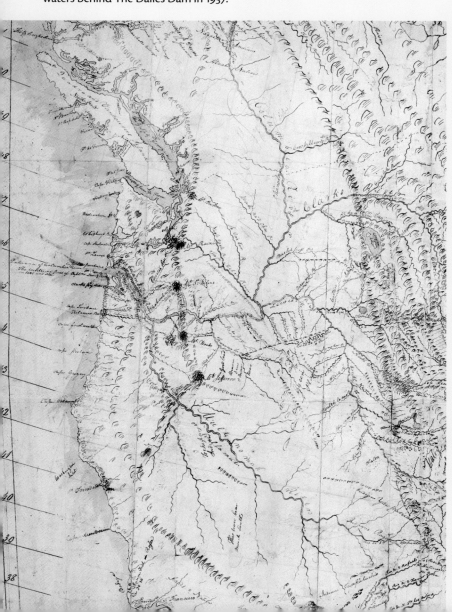

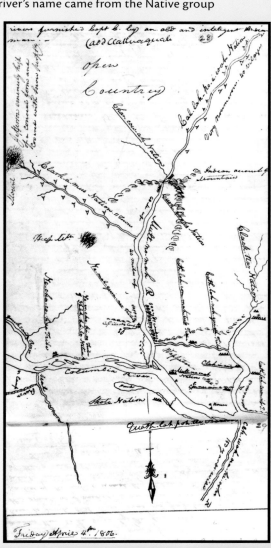

THE FUR TRADERS

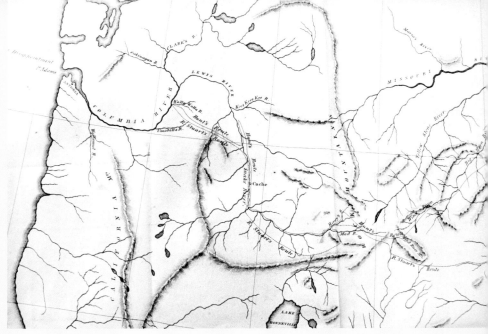

Once Lewis and Clark had revealed the Pacific Northwest to the world, the fur traders moved in and there followed a period of intense competition between British and American traders for the rich fur country. The first order of business was to set up a post on the coast that would collect and ship furs to China, a strategy formulated by Alexander Mackenzie and published in his book in 1801.

In 1808 John Jacob Astor created the American Fur Company to compete against the British North West Company and Hudson's Bay Company. A subsidiary, the Pacific Fur Company, was to concentrate on Oregon.

In 1810 a ship, the *Tonquin,* was dispatched from New York to the mouth of the Columbia. On board was Alexander Mackay—who had been Mackenzie's second-in-command on the Pacific Coast in 1793—with instructions to set up a trading fort. At the same time a land expedition, led by Wilson Price Hunt, left St. Louis. Mackay and his men reached the Columbia in March 1811 and, after a harrowing time trying to cross the bar at the river's entrance, built Fort Astoria, naming it after their sponsor. Hunt's party had a worse time, nearly starving en route, and, split into two parties, did not make it to Fort Astoria until January and February 1812. A return overland journey was made in 1812–13, led by Robert Stuart, who discovered South Pass at the south end of the Wind River Mountains in today's Wyoming. Apparently concealed at the time and rediscovered by Jedediah Smith in 1824, the pass was to become critical for settlers traveling to Oregon along the Oregon Trail, as it would accommodate wagons (see page 52). Smith sent this information to the American government.

MAP 67 (*above*).
This map shows the routes of the Pacific Fur Company traders: Wilson Price *Hunt's Route* going west in 1811–12 and Robert *Stuart's Route,* going east in 1812–13; the latter uses South Pass, not named on the map but just south of the *Wind River M^{ts.},* the critical pass across the Continental Divide that was suitable for wagons. Astor seems to have kept this route secret until he commissioned the book *Astoria,* by Washington Irving, in a late attempt to protect his historical legacy. This map comes from that book. *Astoria* is shown at the mouth of the *Columbia River.*

MAP 68 (*below*).
Part of David Thompson's original large map of the Northwest, now in poor condition, shows Oregon, including the course of the Columbia River, which he followed to the sea in 1811 only to find Fort Astoria already built—by traders arriving by sea. The North West Company's Thompson was the second to reach the mouth of the Columbia overland, after Lewis and Clark.

The North West Company also had its eyes on the West Coast. In 1808 Simon Fraser descended the river that now bears his name in British Columbia, and the company's master surveyor and

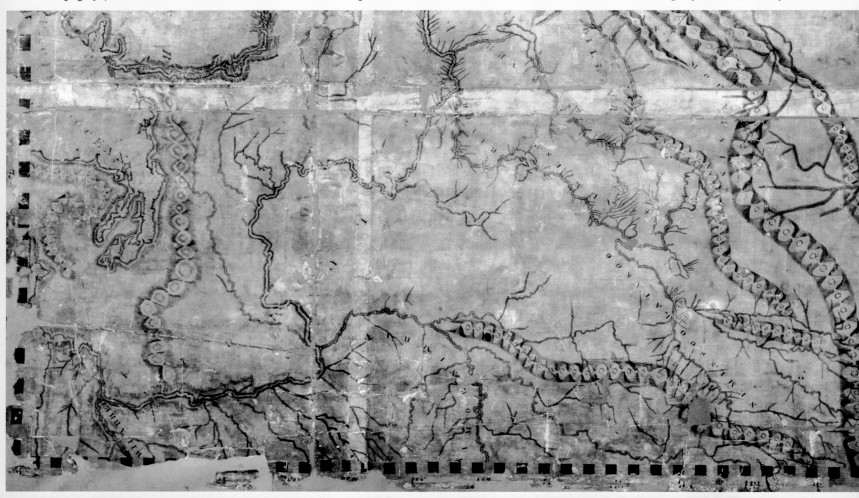

mapmaker, David Thompson, likewise descended the Columbia. He reached the mouth of the Columbia in July 1811 only to find the Astorians already ensconced there. He arrived well before Hunt's overland expedition but was beaten by the men in Astor's ship.

Astor had won—for the time being. But his victory was short-lived. From 1812 to 1814 the British and the Americans fought each other in the War of 1812. In September 1813 John McTavish of the North West Company arrived at Fort Astoria and informed Astor's men that a British supply ship and warship was en route to the Columbia with orders to seize Fort Astoria. In light of this information, and expecting to lose the fort by force, the Astorians, many of whom were British in any case, in October accepted McTavish's offer to buy the fort and all of its supplies. Hence the North West Company acquired a foothold in Oregon.

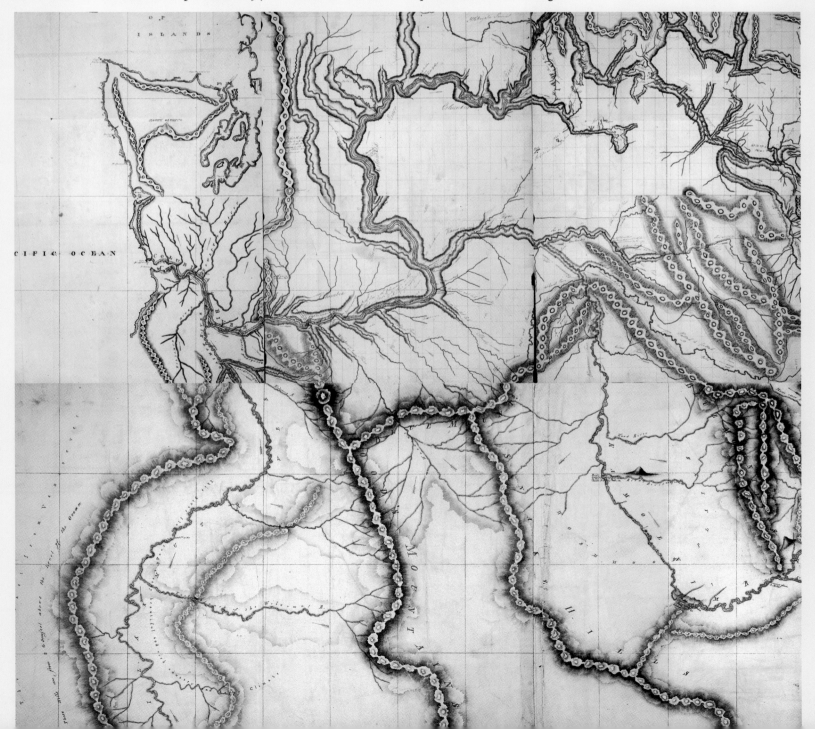

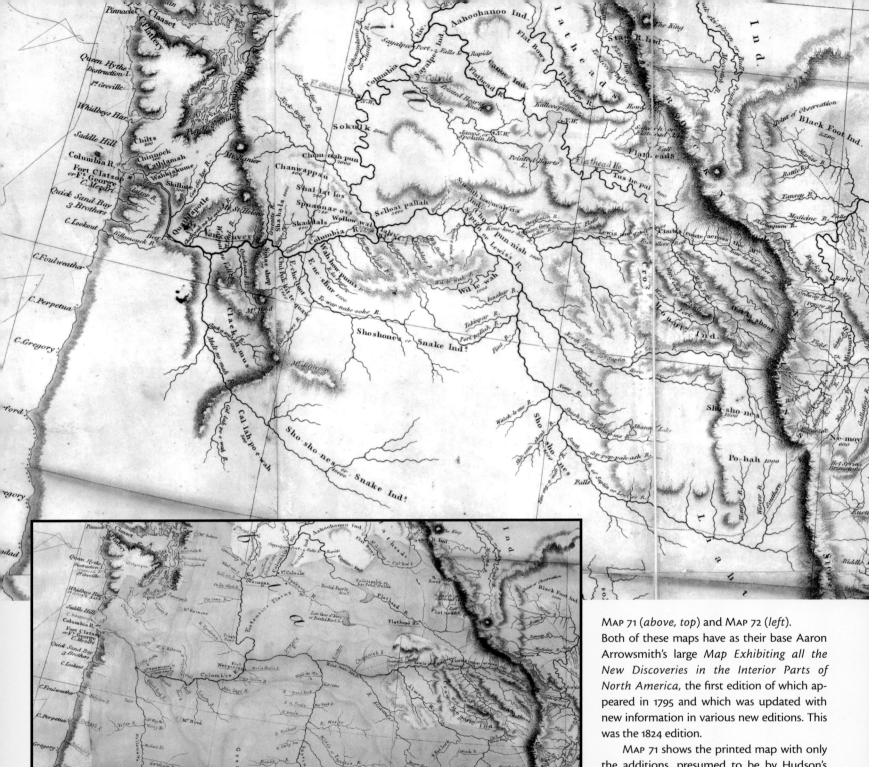

Map 71 (*above, top*) and Map 72 (*left*).
Both of these maps have as their base Aaron Arrowsmith's large *Map Exhibiting all the New Discoveries in the Interior Parts of North America*, the first edition of which appeared in 1795 and which was updated with new information in various new editions. This was the 1824 edition.

Map 71 shows the printed map with only the additions, presumed to be by Hudson's Bay Company employees, of company trading posts, shown in red ink. Note that the fictitious Caledonia River has been erased (see Map 96, *page 42*), except for its mouth, which likely represents the Skagit River.

Map 72 is very interesting, and a much larger reproduction of this map is shown on the title page of this book. The Hudson's Bay Company had a habit of updating printed maps in the field by pasting paper patches on them showing new information. Here most of what is now Washington and Oregon has been overlaid with a large paper patch, and new information, especially relating to the revised courses of rivers, has been drawn in red ink. The Snake River (*Saptin or Lewis's or Great Snake River*) is perhaps the most extensively revised, with the river's course now being shown much farther west than previously; this is a considerable increase in accuracy. The information was obtained as the result of the company's forays into the Snake Country starting in 1823, the beginning of the plan to turn the region into a "fur desert" so as to keep out American competition. The course of the middle Columbia has been extensively revised and is again correctly shown farther west. In addition the course of the *Willamette R.*—up to 1824 shown by most mapmakers much as Lewis and Clark had shown it, based in Indian reports—is now much more accurately depicted, and its modern name appears. The locations of the company forts are again shown: *Fᵗ· Vancouver*; *Fort Nezperces* (later Fort Walla Walla); *Fort Okanagan*; *Fᵗ· Colvile*; *Flathead Ho[use]*; and Spokane House, indicated just by *Ho.* on the *Spokane R.*

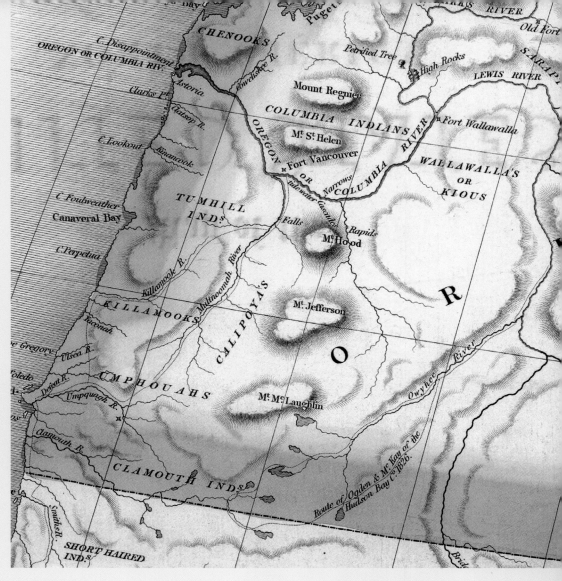

The British corvette *Racoon* arrived six weeks later, and its captain, William Black, intent on military glory, insisted on ceremonially seizing the fort for Britain, despite its sale to the North West Company. This seizure was used in 1846 boundary negotiations to reinforce American claims to the Columbia. Black renamed Astoria Fort George.

The North West Company merged with the Hudson's Bay Company in 1821 and remained a force in Oregon until well after 1846.

A new field governor for the Hudson's Bay Company, George Simpson, arrived on the scene, visiting Oregon in late 1824. He and John McLoughlin, the new chief of the company's Columbia Department (the region west of the Rockies), formulated a strategy for overtrapping the Snake River country to try to discourage American encroachment, sending the aggressive Peter Skene Ogden on the first of a series of expeditions south (MAP 74, *right, bottom*). The company developed a keen sense of the geography of Oregon in the process of its operations, sometimes revising maps considerably (MAP 72, *left, bottom,* and enlarged on the title page).

The threats of encroachment came from American fur traders recruited by General William Henry Ashley and his partner Andrew Henry. They included the likes of James Clyman, William Sublette, Jim Bridger, and, most famously, Jedediah Strong Smith. These mountain men, as they came to be called, crisscrossed the West in search of fur and also added much to the map in the process. Smith came to Oregon in 1828 only to have many of his men massacred as they camped along the Umpqua River. His route, and that of the survivors after the attack, is shown on MAP 73, *right, top,* a commercial map that is all we have, as Smith's own maps were lost. Smith and the other survivors made it to the newly established Fort Vancouver, where John McLoughlin made him welcome and even sent out a party to try to recover Smith's furs and equipment. Smith gathered information on the settlement possibilities of Oregon and later sent it to the U.S. secretary of war.

By the mid 1830s the halcyon days of the fur trade were drawing to a close, and in 1843, seeing the writing on the wall, the company established a new fort on Vancouver Island (Fort Victoria), which six years later would become its new western headquarters.

MAP 73 (*above*).
Official cartographer of the House of Representatives David Burr showed the travels of Jedediah Smith on this map published in 1839. Although not named on this extract, Smith's route in 1828 is the dotted line along the *Umpquagh R.* (Umpqua River) and, after the massacre on the *Defeat R.* (now Smith River), north to *Fort Vancouver,* with the survivors using two routes, down the *Multnoomah River* (Willamette) and the *Killamook R.* (the Tillamook or another of the coastal rivers, somewhat exaggerated). Also shown is the *Route of Ogden & Mc.Kay of the Hudson Bay Co. 1826,* on one of the several forays Ogden made into the Snake River country to exterminate beaver to discourage American competition. On another expedition in 1829 Ogden also traversed the Deschutes River in central Oregon, but this is not shown on this map, possibly attributable to the fact that Ogden's journal was lost to the river. *Lewis River* is the Snake.

MAP 74 (*right*).
Peter Skene Ogden's map of his fifth expedition, made in 1828–29, in which he reached south to California's Humboldt River (here called the *Unknown River,* Ogden's name for it). The map is oriented with north at the top. The *Snake River* is at right. The *Fort* at the top is Fort Nez Percés or Walla Walla. At left center, *Sylvailles Lake* is Malheur Lake, in south-central Oregon's Harney Basin.

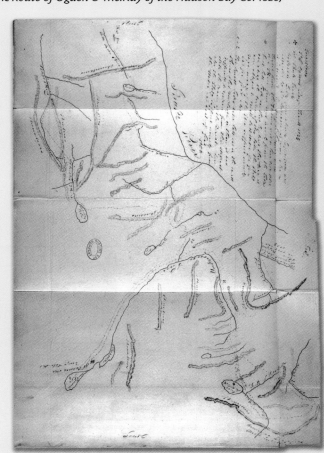

Alexander Ross's Map

Alexander Ross worked in the Oregon Country between 1811 and 1825, first for the Pacific Fur Company, then for the North West Company, and after 1821 for the Hudson's Bay Company. He traveled extensively in the Northwest and became one of those most familiar with the region's geography.

A Scot, he had immigrated to Canada in 1804 and worked as a schoolteacher before being recruited by John Jacob Astor's Pacific Fur Company in 1810. He arrived in Oregon on the *Tonquin* in 1811 and helped build Fort Astoria. When the North West Company took over the fort in 1813, Ross was one of those who elected to stay and work for the Canadian company. He was put in charge of the northern interior posts and was stationed at Fort Okanagan.

In 1823 the Hudson's Bay Company began exploring the Snake River country. In 1824 Ross was in charge of one of these expeditions, but the traveling governor of the company, George Simpson, implementing the policy of overtrapping the Snake Country to drive out American competition, decided that Ross was not suitable for this job and replaced him in 1825 with the more ruthless Peter Skene Ogden (see previous page).

Ross retired back to Ontario, where he later continued work on a massive and comprehensive map of the Columbia River basin he had begun in 1821 (MAP 75, *this page*).

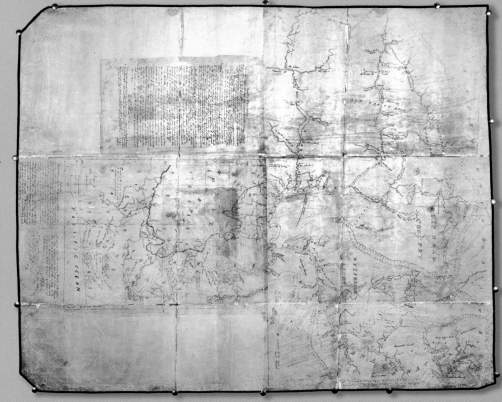

MAP 75 (*above*), complete, and six details (*below*).
Alexander Ross's detailed map of the Columbia River basin provides us with a wealth of information about the region as it was in 1821–25. The western half of the map was drawn in 1821, but the eastern half, though reflecting the 1821–25 period until Ross left Oregon, was drawn in his retirement, in 1849. The complete map shows the mouth of the *Columbia* at left and its confluence with the Snake in the center. Drawn in the thin lines of a quill pen, details are hard to read at this scale.

Enlarged details, counterclockwise from left, are the mouth of the Columbia, with *Fort Astoria in 1811* shown; the confluence of the Snake (*Lewis's River*) and the

Walla Walla River with an *Establishment in 1806*—Fort Nez Percés, later renamed Fort Walla Walla; the Snake River, with several annotations about killings and murders; the confluence of the *Wallamette* and the Columbia, with *Ft. Vancouver*; The *Oakinacken R.* with Fort Okanagan, *Built in 1811*; and (in center) *Fort Colville 1825* and *Spokane House built in 1809.*

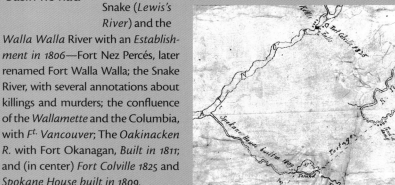

Hudson's Bay Company Forts

The British Hudson's Bay Company, its competitor and then merged constituent the North West Company, and its competitor the American Pacific Fur Company were the first EuroAmerican colonizing influences in the Oregon Country. Oregon was at the southern limit of the British fur trade influence west of the Rocky Mountains, but the whole, from New Caledonia in what is now northern British Columbia south to Oregon, formed the company's Columbia Department.

The Winship brothers of Boston, Nathaniel, Jonathan, and Abiel, arriving by ship in June 1810, were the first to attempt to build a post (intended to be permanent) in Oregon. Their fortified two-story house was built at Oak Point on the Columbia, twelve miles downstream of the present-day bridge at Longview. Within weeks the Winships were threatened by Native people, forcing them to abandon their venture.

Astor's Pacific Fur Company built Fort Astoria the next year; Fort Okanagan, at the confluence of the Okanogan River with the Columbia, later in 1811; and Fort Spokane, on the Spokane River, in 1812. All three were taken over by the North West Company following the forced sale of Astor's company during the War of 1812. Fort Astoria was renamed Fort George.

The North West Company's David Thompson built Spokane House in 1810, and the Pacific Fur Company established a short-lived rival post next to it two years later. It was not unusual for rival companies to set up posts adjacent to each other as they were trying to compete for the same furs, brought to the posts by local Native people.

In July 1811, on his way down the Columbia, David Thompson had marked a site near where the Snake River enters the Columbia with a pole and a notice claiming the region for Britain and announcing the intention of the North West Company to build a trading post at this strategic location. Fort Nez Percés was built seven years later, in 1818, a few miles downstream from the Snake, near the Walla Walla River, which was often used as a shortcut to reach the Columbia from the Snake.

In 1824 the regional headquarters of the Hudson's Bay Company were moved from Fort George to a site on the north bank of the Columbia close to its confluence with the Willamette. The decision to move farther upstream was simply to make it easier for traders to reach, but the decision to move to the north bank was a deliberate one, for the company was already coming to the conclusion that one day the region south of the river

MAP 76 (*below*) and MAP 77 (*below, center*).
The location of the North West Company's Fort George is noted on David Thompson's redrawn 1814 map as *N.W.Co.* (MAP 76) and later as the Hudson's Bay Company's *Fort George* (MAP 77), the latter drawn in 1845 by British military spy Mervin Vavasour (see page 48). Also shown is an engraving of the fort as Fort Astoria in 1810 (*above*) and as Fort George in 1845 (*below, bottom*), in a painting by fellow spy Henry Warre.

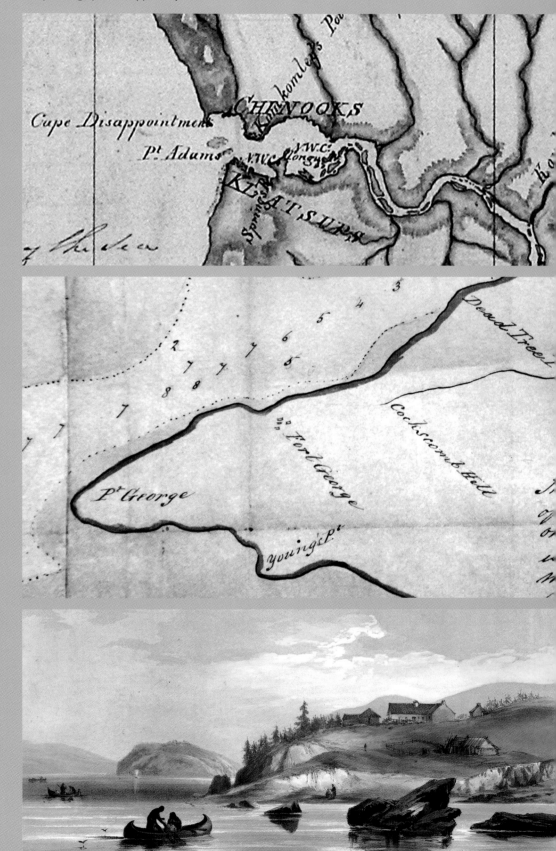

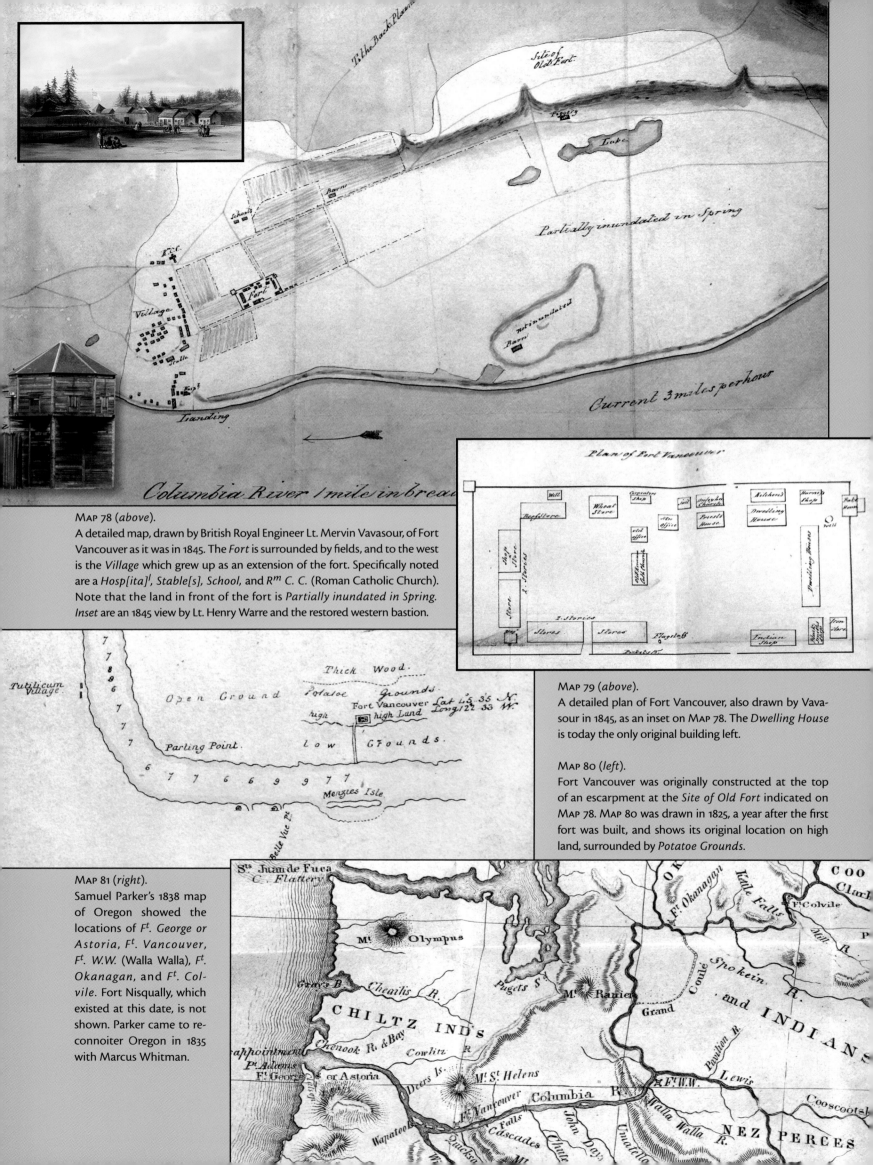

To the Back Plain

Site of Old Fort.

Lake

Partially inundated in Spring

not inundated Barn

Current 3 miles per hour

Barn

School

H.B.C.

Village

Fort

Stable

Hosp.

Landing

Columbia River 1 mile in breadth

Plan of Fort Vancouver

MAP 78 (*above*).
A detailed map, drawn by British Royal Engineer Lt. Mervin Vavasour, of Fort Vancouver as it was in 1845. The *Fort* is surrounded by fields, and to the west is the *Village* which grew up as an extension of the fort. Specifically noted are a *Hosp[ita]^l, Stable[s], School,* and *R^m C. C.* (Roman Catholic Church). Note that the land in front of the fort is *Partially inundated in Spring*. *Inset* are an 1845 view by Lt. Henry Warre and the restored western bastion.

Tutilicum Village.

Thick Wood.

Open Ground

Potatoe Grounds.

Fort Vancouver Lat 45.35 N.
Long 122 33 W.

high high Land

Parting Point.

Low Grounds.

Menzies Isle

Belle Vue Pt.

MAP 79 (*above*).
A detailed plan of Fort Vancouver, also drawn by Vavasour in 1845, as an inset on MAP 78. The *Dwelling House* is today the only original building left.

MAP 80 (*left*).
Fort Vancouver was originally constructed at the top of an escarpment at the *Site of Old Fort* indicated on MAP 78. MAP 80 was drawn in 1825, a year after the first fort was built, and shows its original location on high land, surrounded by *Potatoe Grounds*.

MAP 81 (*right*).
Samuel Parker's 1838 map of Oregon showed the locations of *F^t. George or Astoria, F^t. Vancouver, F^t. W.W.* (Walla Walla), *F^t. Okanagan,* and *F^t. Colvile.* Fort Nisqually, which existed at this date, is not shown. Parker came to reconnoiter Oregon in 1835 with Marcus Whitman.

St^s Juan de Fuca C. Flattery

Mt. Olympus

Grays B.

Chehalis R.

Pugets S^t

Mt. Ranier

Ft. Okanagan

Kettle Falls

F^t Colvile

COO

Clark

R

P

Grand Coule

Spokein R.

and INDIANS

CHILTZ IND^s

Chenook R. & Bay

Cowlitz R

disappointment Pt Adams

F^t George or Astoria

Deers Is.

Mt St Helens

Ft Vancouver

Columbia R.

AF^{t}W.W.

Wapatoo

Falls

Cascades

John Day

Lewis

Walla Walla R.

Cooscootes

Unatella R.

NEZ PERCES

Mud walls 12 feet high

Dwelling House

Stores

Stores

Columbia River
3/4 mile wide

Map 82 (*left*).
Fort Walla Walla, on the *Columbia*, shown in an 1845 map by Mervin Vavasour. It had *Mud walls 12 feet high* and mud bastions because of the lack of available timber. The fort was originally built by the North West Company in 1818 and named Fort Nez Percés. The first factor at the fort was Alexander Ross (see page 34). The engraving *inset left, top*, shows the fort in 1818. The modern photo (*inset left, bottom*) is all that is left today—a granite marker by the roadside at Wallula, Washington, with the Union Pacific tracks between it and the river.

Map 84 (*below*).
Nisqually, with the *Old Fort*, built in 1833, and *Nisqually House*. The Nisqually River is at bottom. Intended to provide food for both Company and Russian fur trade posts, Nisqually was principally an agricultural settlement, the first on Puget Sound. Note that *settlers from Red River* (near Winnipeg, Manitoba) were planned—farmers to help boost agricultural production. In the 1930s the fort was reconstructed twenty miles to the north, in Point Defiance Park, Tacoma. The site of the original Fort Nisqually (as shown on this map) is at Dupont, Washington.

Map 83 (*below*).
Fort Colvile HB.C. was considered by the company to be second only to Fort Vancouver in importance. It was a way station on the well-traveled route from the coast to Hudson Bay, used every year for supply and fur export. This map was drawn in 1867 by Alexander Caulfield Anderson, chief factor at Fort Colvile from 1848 to 1851. *Fort Shepherd,* just north of the forty-ninth parallel and also on the *Columbia*, was built as an alternative fort in 1847 after the boundary settlement placed Fort Colvile in American territory. Note the *Horse Trail*s to the fort.

might be lost to the United States. The company was also concerned about possible Russian incursions from its base in Alaska, and at the same time reconnoitered a location for a new fort on the lower Fraser River, nearly three hundred miles to the north. Fort Langley was built there three years later.

Spokane House, which was away from the main route up the Columbia used by traders traveling between Oregon and Hudson Bay (from where furs were shipped to Britain), was replaced by a new post, Fort Colvile, at Kettle Falls, on the Columbia, in 1825.

In 1833 the Hudson's Bay Company came to an arrangement with the Russian American Company to supply the latter's posts to the north and, as part of this plan, built Fort Nisqually, with extensive farmlands around it. The fort was the first EuroAmerican settlement on Puget Sound.

Fearing the loss of the Oregon Country to the Americans, the Hudson's Bay Company built Fort Victoria at the southern tip of Vancouver Island in 1843, and after the 1846 boundary decision (see page 48), moved its headquarters there in 1849.

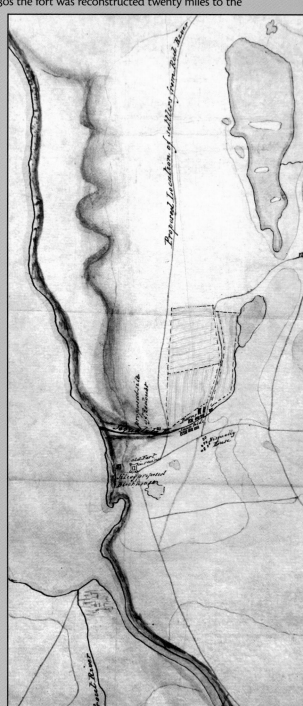

PROMOTERS AND MISSIONARIES

Not surprisingly, the Hudson's Bay Company wanted to discourage any settlement in its domain, since settlement, it felt, would lead to a reduction in fur-bearing animals and hence be injurious to the company's trade. The region was declared to be one of joint occupancy in 1818, open to citizens of both Britain and the United States. It was not long before Americans seeking new lives began to cast their gaze westward.

A New England schoolteacher, Hall J. Kelley, began to take an interest in Oregon right away and became the first to promote a permanent American presence in Oregon. In 1828 Kelley petitioned Congress to support a colonizing company he intended to organize, but it was modeled on the exclusive grant the Hudson's Bay Company had from the British government and Congress would not support that. The following year Kelley organized the American Society for Encouraging the Settlement of Oregon and began writing a series of pamphlets. A book, published in 1830, contained what is thought to be the first map solely depicting Oregon (MAP 86, *right*).

Kelley finally visited Oregon himself in 1834, arriving by horse northward from California with Ewing Young, a fur trader who stayed in Oregon, becoming one of its pioneer settlers. By the time Kelley arrived, he was ill, and he was not made welcome by John McLoughlin, factor at Fort Vancouver. So, the following year, Kelley sailed back to Boston via Hawaii. Despite

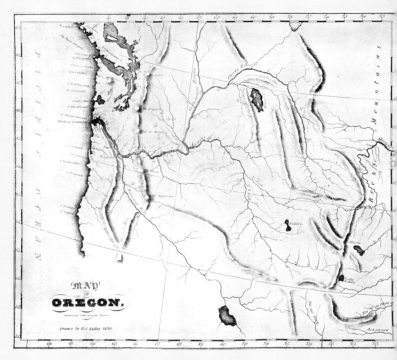

MAP 85 (*below*).
Hall Kelley's proposal for a town at the confluence of the *Multnomah R.* (Willamette) and the *Columbia River*, an area now within the boundaries of Portland and the site of Kelley Point Park.

MAP 86 (*above, right*).
The map of Oregon published in Hall Kelley's book *A Geographical Sketch of That Part of America Called Oregon*, published in 1830.

MAP 87 (*right*).
In 1839 Kelley sent this map to Joel Poinsett, secretary of war, complete with proposed names of Northwest mountains, naming them after U.S. presidents. Kelley for a while tried to get the Cascade Range renamed the *Presidents Range*. Only the name of Mount Adams remains. The mountain that Kelley had intended to be called *Mt Adams*, as shown here, was Mount Hood, but a later mapmaker (MAP 178, *page 80*) placed the name farther north, where, purely coincidentally, there was an as-yet-unnamed mountain—and one that Kelley did not know was there—that thus assumed the name.

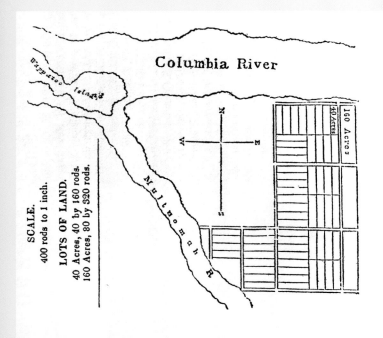

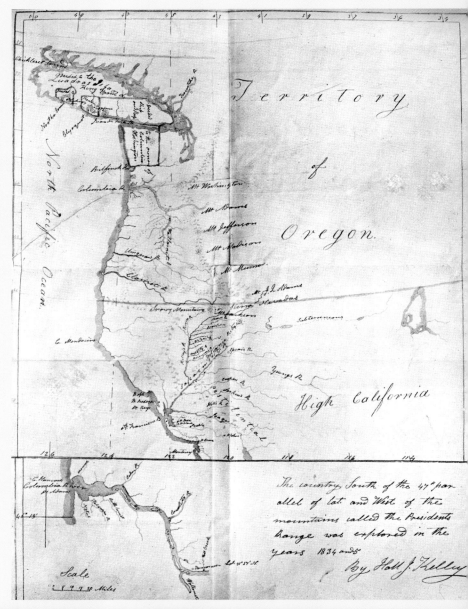

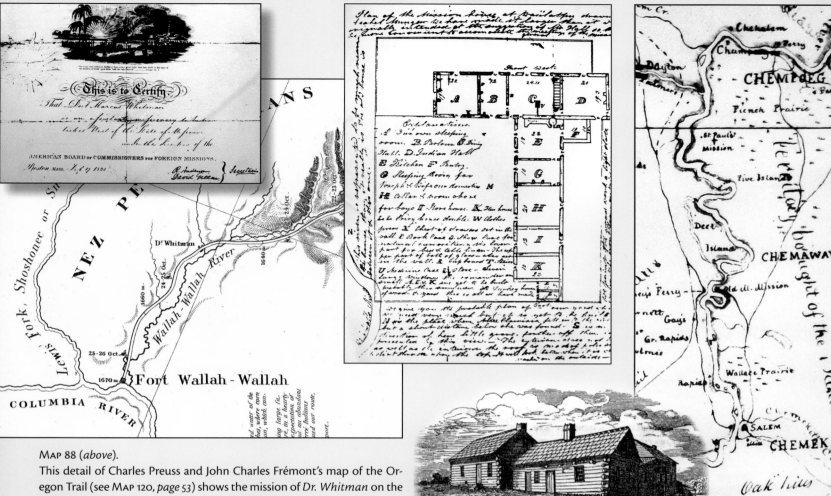

MAP 88 (*above*).
This detail of Charles Preuss and John Charles Frémont's map of the Oregon Trail (see MAP 120, *page 53*) shows the mission of *Dr. Whitman* on the *Wallah Wallah River*, twenty-five miles east of the Hudson Bay Company's *Fort Wallah-Wallah*, marked as *1670 m[iles]* from the Missouri River. *Inset* is Whitman's certificate from the American Board of Commissioners for Foreign Missions, dated February 1835. Whitman is certified *an assistant missionary to Indian tribes West of the State of Missouri.*

MAP 89 (*above, center*).
A page from Marcus Whitman's journal shows his plan of the mission together with a key. The *Columbia River* is at bottom left.

MAP 90 (*above, right*).
This extract from an 1851 Indian Affairs map shows *St. Paul's Mission*, near the Willamette at *French Prairie*, established in 1836, and Jason Lee's *Old M. [Methodist] Mission*, established two years before, a few miles north of *Salem*; Lee's mission is also illustrated, *inset*.

MAP 91 (*below*).
The *Catholic Mission*, established in 1836, *St. Paul's Church*, built in 1846, and a *Nunnery* are shown on this 1852 General Land Office survey of French Prairie (see MAP 124, *page 56*). Note also the *Catholic Fathers' Claim*, land claimed under the Donation Land Claim Act (see page 57).

this setback Kelley continued to promote Oregon and was responsible for a considerable upwelling of public interest in the Northwest.

Kelley's neighbor in Boston had been businessman Nathaniel J. Wyeth, and he was a convert to Kelley's ideas. In 1832 and 1834 Wyeth led pioneering overland expeditions to the lower Columbia, intending to found a colony based on international trade in lumber, salmon, fur, and agricultural produce. He set up a trading post called Fort William on Sauvie Island, at the mouth of the Willamette. In 1832 his supply ship was wrecked on the Columbia bar, and his second attempt, when he tried to set up a commercial salmon fishery, failed largely because of the Hudson's Bay Company's lack of support.

With Wyeth in 1834 was Thomas Nuttall, the famous British naturalist. Also in the party was a new breed of settler, traveling to Oregon not for personal gain but to convert the Indians to Christianity. Jason Lee, a Methodist who had been sent to Oregon to work with the Indians of the interior, instead chose to found a mission in the Willamette Valley (MAP 90, *right*). The prevailing thought was that Indians had to be persuaded to convert to a sedentary lifestyle and become farmers, and so missions became agricultural centers as well.

Lee was followed to Oregon by Marcus Whitman. In 1835 the American Board of Commissioners for Foreign Missions sent Whitman

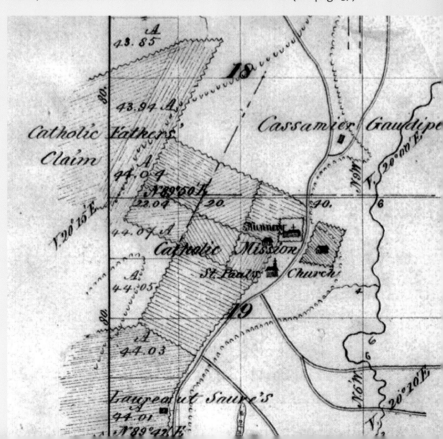

Pays des Parteurs
deux stations.
436 baptisés en 1842-43
par M. De Mers.

P.D.S.

MAP 92 (left).
Rivers were the highways of the Pacific Northwest, and this simple fact is evident in this map of the area that is now north-east Washington. The map was drawn in 1845 by perhaps the most famous of western missionaries, the Jesuit Pierre-Jean De Smet, known as Blackrobe. The map is signed *P.D.S.* He traveled 200,000 miles in the West between 1843 and 1868. There are a number of missions shown on this map, which stretches into modern-day north-west Montana. The map shows another *St. Paul*'s Mission, this one at Kettle Falls on the *Fleuve de la Colombie* close to *Fort Colville*. On the *Flv Spokane* De Smet has shown a *Mission protestante*—the Methodist Tshimakain Mission, built in 1839. At right is *Ste Marie*, a mission founded by De Smet in 1841, near present-day Missoula, Montana. The map demonstrates an excellent working knowledge of the geography of the region, essential for a traveling missionary like De Smet.

MAP 93 (right).
This important map of the *Territory of Oregon*—then, of course, not yet a territory in the legal sense—accompanied Senator Lewis Linn's first bill proposing the annexation of Oregon in 1838. The map was drawn by Washington Hood, a captain in the U.S. Corps of Topographical Engineers, but was more or less a copy of an earlier commercial British map published by John Arrowsmith, which incorporated reasonably up-to-date information from the Hudson's Bay Company and from American sources, notably the coastal rivers from Jedediah Smith. Linn's 1838 bill incorporated an international boundary along the forty-ninth parallel, including through Vancouver Island; the 1818 convention between the United States and Britain had agreed on this boundary only as far west as the Rocky Mountains. Hood's justification, spelled out in the text at bottom left, comes from previous negotiations in 1826.

and Samuel Parker (see MAP 81, *page 36*) to Oregon to assess its suitability for missions. Whitman returned to Oregon two years later—now married, as was required by the board—with his new wife, Narcissa, and accompanied by Henry and Eliza Spalding, who founded a mission at Lapwai, now in Idaho. The Whitmans founded their mission adjacent to what would become the Oregon Trail at Waiilatpu, in the Walla Walla River valley (see page 52). Their mission thrived until 1847, when Cayuse and Umatilla massacred Whitman, his wife, and twelve others. The Indians had noted that their people were being decimated by disease—measles—and yet the EuroAmerican travelers on the trail were not (owing to immunity due to previous exposure) and may also have been persuaded by a disgruntled immigrant that Whitman was deliberately poisoning them to clear them off the land. Forty-seven others were taken captive, and three of these, all children, died while in captivity. The Hudson's Bay Company arranged for ransom to be paid to free the hostages a month later. The incident initiated a period of unrest called the Cayuse War (see page 64) and made Congress realize the need for proper government; the Oregon Territory was created the following year.

The Methodists were soon joined in Oregon by Catholic missionaries, initially subsidized by the Hudson's Bay Company for their farmers in the Cowlitz Valley, working for the Puget's Sound Agricultural Company, set up in 1838. Missionaries François Norbert Blanchet and Modeste Demers were also given the task of converting the Indians to the Catholic faith. A mission called St. Francis Xavier was established on the Cowlitz River, and a second, called St. Paul's, near the settlements on the Willamette (MAP 90, *previous page*) was opened in 1839. By this time the Hudson's Bay Company had been persuaded of the wisdom of allowing its retiring employees to remain in Oregon if they wished, rather than insisting they return to where they were hired. The retired former employees began to congregate on one of the fertile prairies of the Willamette Valley, an area that became known, because most employees were of French Canadian origin, as French Prairie.

One tireless promoter of an American Oregon was the U.S. senator from Missouri, Lewis Fields Linn. In 1838 he introduced a bill that would have authorized the United States to occupy "the Columbia or Oregon River," and create "a territory north of latitude 42 degrees, and west of the Rocky Mountains, to be called Oregon Territory." The bill called for the establishment of a military fort and occupation by the military forces of the United States. Accompanying Linn's bill was an influential map by Topographical Engineer Washington Hood (MAP 93, *right*). The bill failed to pass, but Linn was not one to give up easily, and in 1841, 1842, and 1843 Linn introduced a similar bill each year. These included provision for free land for settlers, an idea that was finally passed in 1850 as the Oregon Donation Land Claim law (see page 57). In 1843 Linn's bill passed the Senate but failed in the House, nevertheless stirring up interest in Oregon—people were becoming convinced that Oregon would end up in American hands—and that year saw the beginning of mass migration west along the Oregon Trail.

MAP
of the
UNITED STATES
TERRITORY OF OREGON
West of the Rocky Mountains,

Exhibiting the various Trading Depots or Forts
occupied by the British Hudson Bay Company, con-
-nected with the Western and northwestern Fur Trade.

Compiled in the Bureau of Topographical
Engineers, from the latest authorities, under
the direction of Col. J.J. Abert, by
Wash: Hood.
1838.
M.H.Stansbury del.

The prolongation of the 49th parallel of latitude from the Rocky
Mountains to the Pacific has been assumed as the Northern Boundary
of the U.States possessions on the N.W. coast, in consequence of the
following extract from the Hon. H.Clay's letter to Mr Gallatin
dated June 19th 1826. (see Doc.199. 20th Cong.1 sess.Ho: of R.) "You are"
"then authorised to propose the annulment of the third article of the"
"Convention of 1818, and the extension of the line on the parallel of"
"49, from the eastern side of the Stony Mountains, where it now"
"terminates, to the Pacific Ocean as the permanent boundary"
"between the territories of the two powers in that quarter. This is"
"our ultimatum: and so you may announce it."

The Posts of the British Hudson's Bay Company are marked thus. ○

THE OREGON QUESTION

The Oregon Question, as it became known in governmental and popular language between 1818 and 1846, asked to whom Oregon belonged, and, if it was to be divided between the United States and Britain, where was a boundary to be drawn?

A convention in 1818, a tying up of loose ends left over from the 1814 Treaty of Ghent, which ended the War of 1812, established Oregon as a region of joint occupancy, shared between the two nations. During these negotiations a map produced by Britain's North West Company was studied (MAP 96, *right*), and because it showed a river claimed to be a fur trade route emptying in two places into Puget Sound, the United States offered a boundary just to its south, an offer Britain turned down. Had it been accepted, the international boundary might today be at approximately the latitude of Everett, Washington.

After 1819 there were but two territorial contenders for the Northwest: Britain and the United States. For that year Spain conceded all her claims beyond 42°N to the United States by the Transcontinental Treaty, skillfully negotiated by secretary of state John Quincy Adams. This treaty established the 42°N southern boundary of Oregon.

From 1821 to 1824, however, a third territorial contender of sorts entered the arena—Russia. The tsar of Russia, encouraged by those with interests in the Russian American Company, issued a *ukase,* or edict, in 1821, claiming Russian sovereignty south to 51°N (MAP 97, *right*). This compromised the American claim to Oregon north to 54°40′N inherited from Spain and was the catalyst for the Monroe Doctrine, announced in late 1823 by President James Monroe (but authored by Adams), that the western hemisphere was not to be further colonized by European powers. After this, and other American and British protests, Russia agreed to the 54°40′N line as the southern limit of its interests by treaty with the United States in 1824 and with Britain the following year.

Boundary negotiations between the United States and Britain took place again in 1823 and 1826–27, and the joint occupancy agreement was maintained, although the British offered a boundary following the most northeasterly branch of the Columbia, the Kootenai River, then known as McGillivray's River and then following the Columbia to the sea. This boundary and variants of it showed up on maps for many years

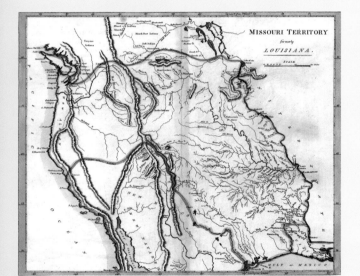

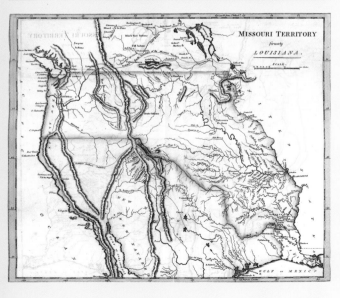

MAP 94 (*left*) and MAP 95 (*left, bottom*).
Two editions of the same map show the extent of the Louisiana Purchase—and hence American territory as interpreted in 1814 (MAP 94) and 1818 (MAP 95) by American mapmaker Mathew Carey. Missouri Territory was the name given to the remainder of Louisiana once the state of Louisiana was created. The straightening of the boundaries of Oregon is the result of the 1818 convention with Britain, which extended the forty-ninth-parallel boundary to the Rocky Mountains, and the 1819 Transcontinental Treaty with Spain, in which Spanish claims north of 42° were conceded to the United States.

MAP 96 (*above, top*).
This 1817 map from the North West Company—the cartographic authority for the Northwest at this time—showed an international boundary at about 46° 40′N, taken from an 1816 map by John Melish. The *Caledonia Riv.* has been added, a totally fictitious river with two outlets in Puget Sound purported to be used by fur traders. American negotiator Albert Gallatin offered an international boundary south of these outlets, but it was rejected by Lord Castlereagh, the British negotiator. Twenty-eight years later Britain would settle for a boundary eighty miles farther north.

MAP 97 (*above*).
The Russian attempt to extend its influence southward is depicted on this 1822 map by American mapmaker Henry Tanner. The *Boundary as claimed by Russia* reflects the *ukase* of 1821.

(MAP 99, *far right*). Another variant, which appears to have originated with Hudson's Bay Company director George Simpson, was to drop the forty-ninth-parallel boundary to 46°20′N at the Rockies and then run it to the Snake River, which it would then follow to the Columbia and thence to the Pacific.

But American negotiators maintained their contention that the whole of the Oregon Country belonged to the United States, and this claim was reflected in maps such as MAP 98, *right.*

During the 1826–27 negotiations Britain, recognizing that the United States' position was partly due to a desire for the deep-water harbors of Puget Sound, offered an additional enclave of territory roughly corresponding to the Olympic Peninsula. This offer was turned down, but the proposal showed up on maps for many years afterward (MAPS 105 and 107, *pages 46–47*).

There were no further boundary negotiations in the 1830s, but the American hold on Oregon tightened when public and political interest was stirred by advocates such as Hall Kelley and Samuel Parker (who published an influential book in 1838), and by Senator Linn's efforts in government; and settlers had begun to move in (see page 54).

In 1842 the international boundary in the Northeast was settled, and Britain reoffered a solution similar to the Simpson-derived boundary, but it was again rejected by the United States. Further offers two years later were also turned down.

That year the report of Charles Wilkes's U.S. Exploring Expedition was published (see overleaf), further increasing American desire for Oregon. Settlers were now pouring

[Continued on page 48.]

MAP 100 (*right*).
Oregon reaches from 54°40′ in the north to 42° in the south and from the Rocky Mountains to the Pacific on this map of *Oregon and Upper California*, published in 1846. Several boundaries are shown: the forty-ninth parallel extended from the Rockies and through Vancouver Island to the Pacific; the detached area of the Olympic Peninsula, offered by Britain to the United States in an 1826–27 proposal; and the boundary favored by the British following the *Columbia R.* and *McGillivray's R.* (Kootenai River in Washington and Kootenay River in British Columbia).

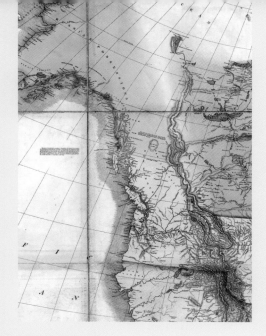

MAP 98 (*above*).
This 1825 map by Henry Tanner seems to have been the first to suggest that American territory might cover the Northwest north to *Russian Possessions* at 54°40′N. This was wishful thinking, as the treaty allowing joint occupancy of Oregon with Britain was still in force.

MAP 99 (*right*).
The (British) Society for the Diffusion of Useful Knowledge published this map in 1834. The map depicts British territory as extending west to what is now the Alaska Panhandle and to 141°W, the boundary line with Russian Territory agreed in 1825. In the south the boundary line extends the forty-ninth parallel east of the Rockies west to meet the *R. Columbia or Oregon*, which it then follows to the Pacific. This was the British version of wishful thinking in the 1818–46 period.

MAP OF THAT PART OF THE WEST COAST OF NORTH AMERICA COMPRISING OREGON AND UPPER CALIFORNIA

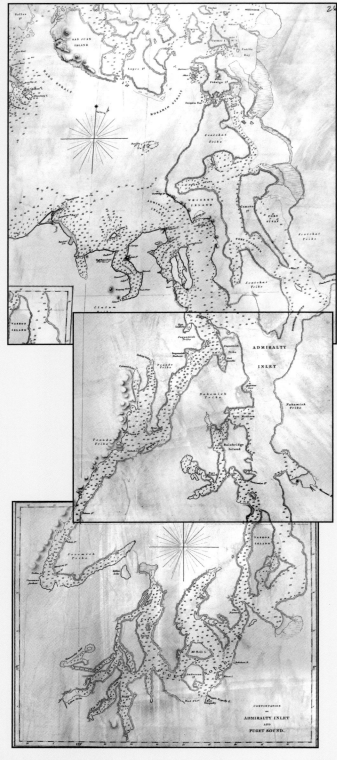

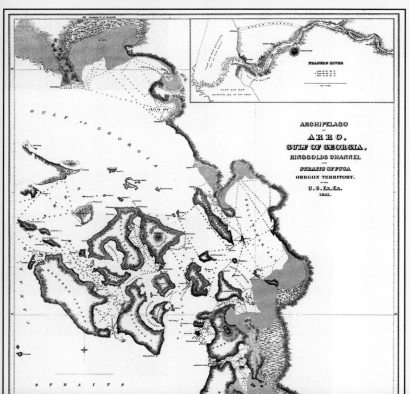

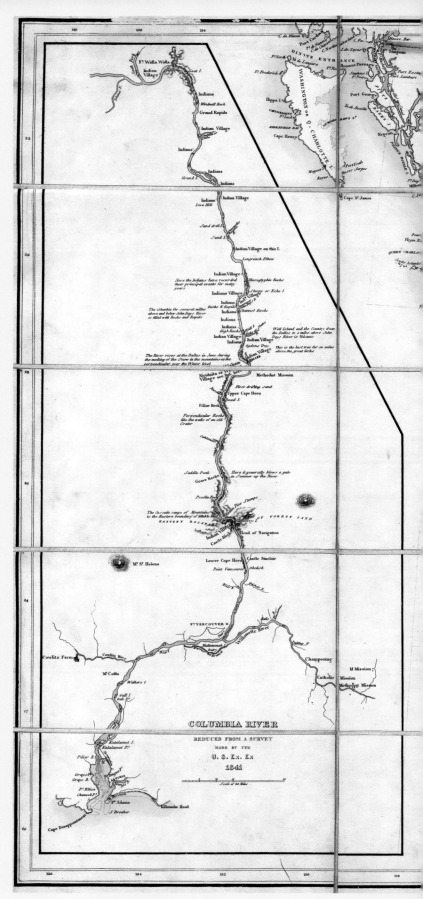

Map 101 (*left, top*), in three sections, Map 102 (*left, bottom*),
Map 103 (*above*), and Map 104 (*right, bottom*).
Maps from the United States Exploring Expedition, a round-the-world
scientific expedition that reached Oregon in 1841. Charles Wilkes and the
U.S. Ex. Ex. report were enormously influential in the United States' deter-
mination not to settle for a solution that did not include Puget Sound.
The wreck of one of the expedition's ships, the *Peacock*, on the sandbanks
at the entrance to the Columbia River (Map 104) emphasized to Wilkes
the importance of a safe harbor on a coast poorly endowed with deep-
water shelter. Map 101 is a composite of three sheets showing in detail the

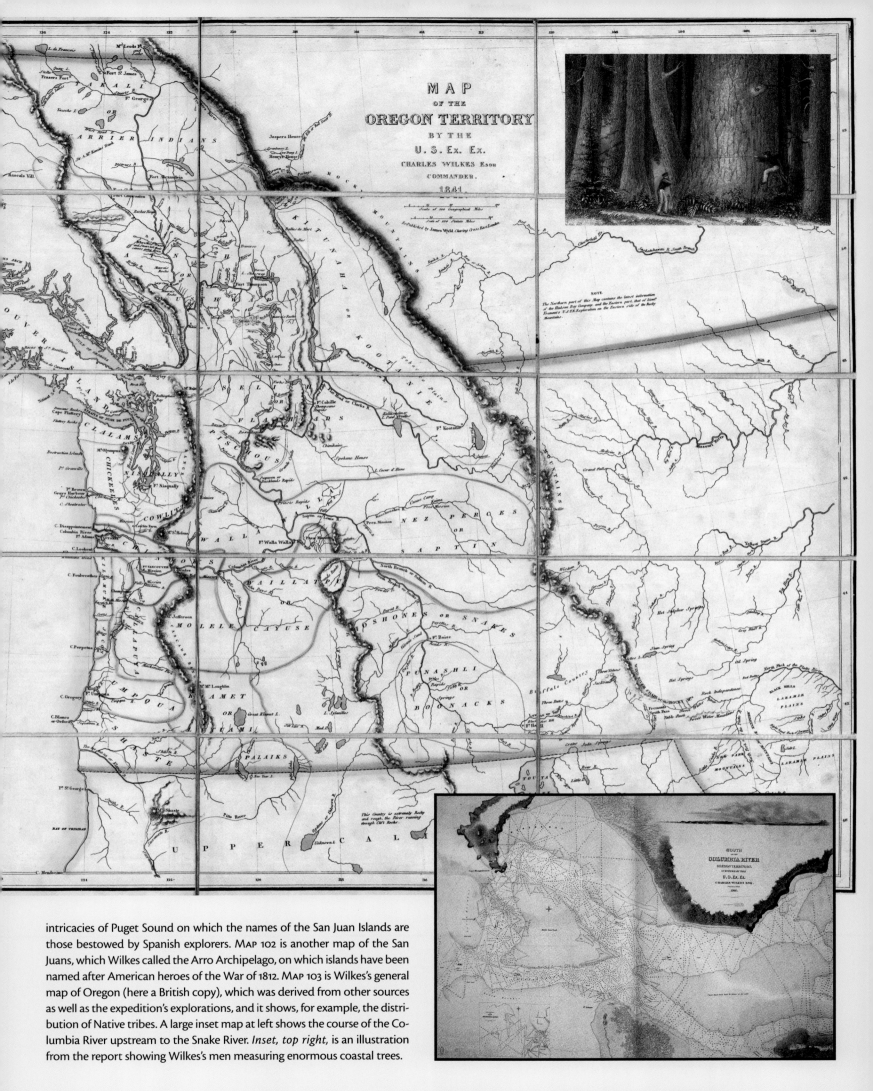

MAP
OF THE
OREGON TERRITORY
BY THE
U. S. Ex. Ex.
CHARLES WILKES Esqr.
COMMANDER.
1841

intricacies of Puget Sound on which the names of the San Juan Islands are those bestowed by Spanish explorers. MAP 102 is another map of the San Juans, which Wilkes called the Arro Archipelago, on which islands have been named after American heroes of the War of 1812. MAP 103 is Wilkes's general map of Oregon (here a British copy), which was derived from other sources as well as the expedition's explorations, and it shows, for example, the distribution of Native tribes. A large inset map at left shows the course of the Columbia River upstream to the Snake River. *Inset, top right,* is an illustration from the report showing Wilkes's men measuring enormous coastal trees.

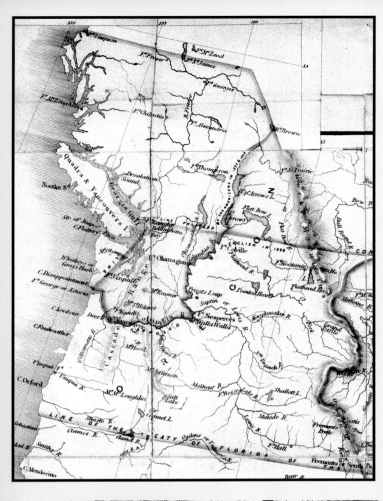

MAP 105 (*left*).

A map produced in 1845 by British mapmaker James Wyld to show the history of claims to the Oregon Country. Wyld's British bias shows in an accompanying leaflet, where he describes the United States' "claims," comparing them with British "rights." The purple line is the supposed boundary of Louisiana, inherited by the United States—which excludes Oregon, of course. The green line is the *Line Proposed by the Americans in 1824 & 1825,* and the red line is that *Proposed by English in 1826.*

MAP 106 (*below*).

Red lines depict various boundary proposals in this map by Eugène Duflot de Mofras in 1844. He was a trade official with the French legation in Mexico City who traveled the West Coast in 1841 looking for trade opportunities and in the process collected information that resulted in the most accurate map of the West for its time, one copied by many others. Other red lines indicate routes to the West Coast; some depict those of immigrant settlers, but Alexander Mackenzie's track in 1793 is shown to the north, and Lewis and Clark's route is also shown. The northern boundary line (in red) is that of the Russia–United States treaty of 1824.

MAP 107 (*right*).

This is the British Foreign Office copy of a commercial map published by W. & A.K. Johnston, dated 1 January 1846, on which there are many manuscript additions. The forty-ninth-parallel boundary line has the notation *Boundary settled by the Treaty of 1846,* and its previous extension through Vancouver Island has been erased, leaving a white band in its place. The map as published, part of which is shown in the *inset,* shows this line as the *Boundary proposed by the United States 1824–36.* The boundary Britain wanted at this time is shown by the red line following the *Columbia R.* The Olympic Peninsula is the *Detached Territory Offered by G.t Britain 1826.* In the valley of the *Willamet R.,* *American Settlements* are noted, and also *Settlements of French Canadians,* carefully and erroneously placed to the south of the American, no doubt a negotiating point for the British, as were the locations of the Hudson's Bay Company posts, underlined in black. Note also in southern Oregon the *Waggon Route from the United States to the Willamet R.* The map was based on that of Duflot de Mofras (**MAP 106,** *below*).

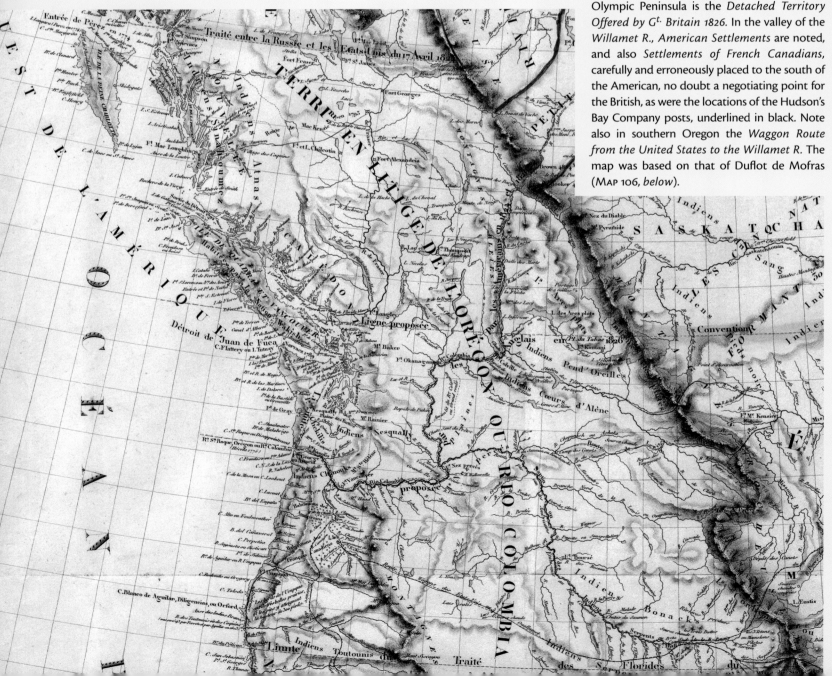

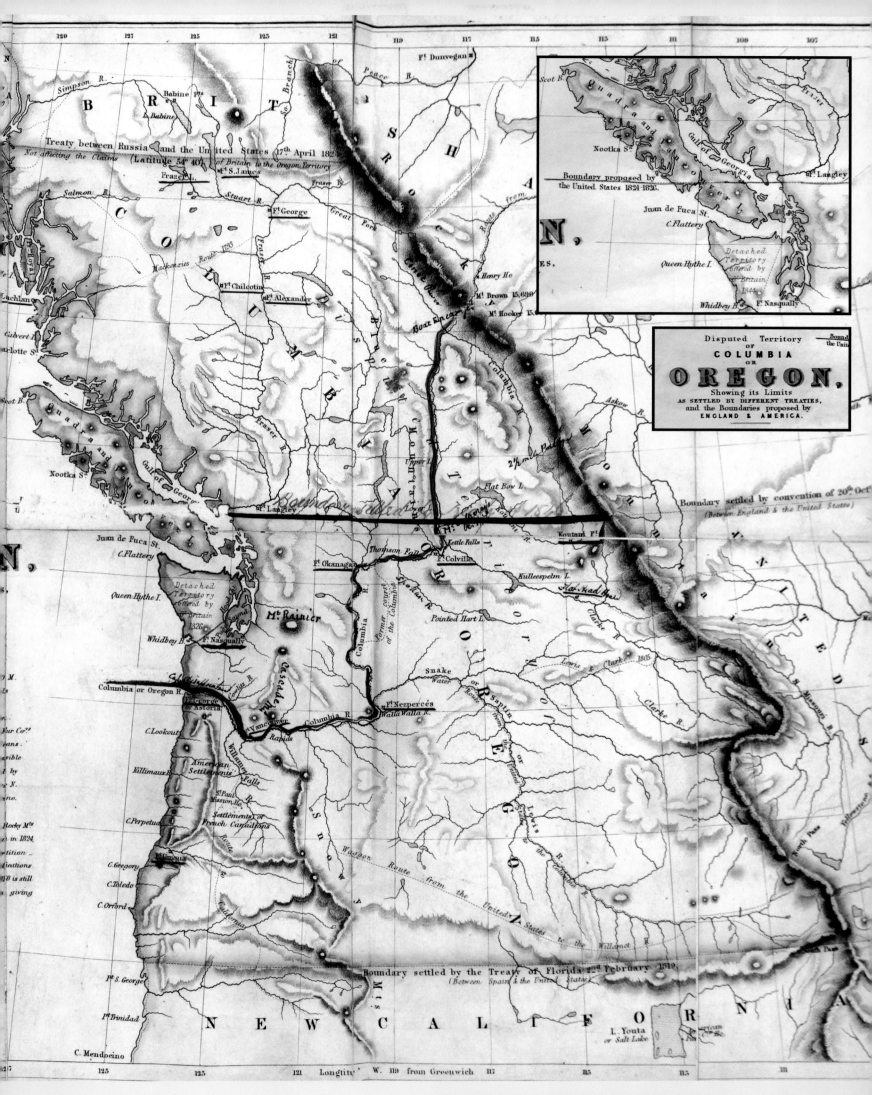

Disputed Territory
OF
COLUMBIA
OR
OREGON,
Showing its Limits
AS SETTLED BY DIFFERENT TREATIES,
and the Boundaries proposed by
ENGLAND & AMERICA.

Treaty between Russia and the United States 17th April 1824
Not affecting the Claims (Latitude 54° 40') of Britain to the Oregon Territory

Boundary proposed by
the United States 1824-1826.

Boundary settled by convention of 20th Oct

Boundary settled by the Treaty of Florida 22nd February 1819
(Between Spain & the United States)

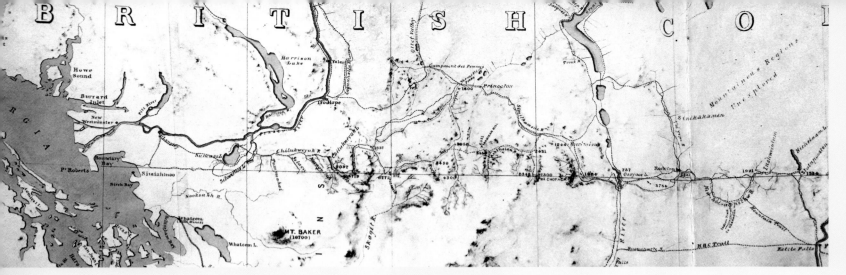

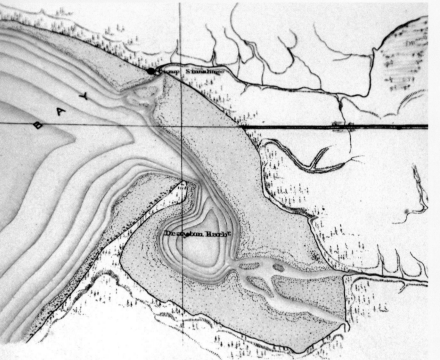

MAP 108 (*above*).
The westernmost part of the continental boundary line agreed to in 1846, shown on a British government map documenting the boundary survey of 1858–59; the map was created in 1869 as part of a large boundary atlas. Detail is enhanced along the zone of the boundary.

MAP 109 (*left*).
From the same atlas comes this detail of the point where the forty-ninth-parallel boundary line meets the ocean for the first time, at today's Blaine, Washington. The line was later shown to be incorrectly located, being a few hundred feet south of where it ought to have been, an error later attributed to the gravitational pull of the mountains a few miles to the north. An agreement to leave the boundary where it was surveyed, even if not exactly correct, was signed by Britain and the United States in 1912.

into Oregon in ever-increasing numbers, and a provisional government had been set up.

Information on American defenses was gathered in 1845 by British spies, Lieutenant Henry Warre and Lieutenant Mervin Vavasour (see, for example, MAP 83, *page 37*, and many others) in case the two nations went to war over the boundary issue.

The expansionist James Polk became president in 1845, riding a wave of nationalism over a newly coined term—Manifest Destiny—the idea that the United States was ultimately destined to control the entire North American continent. Polk maintained that the United States had a "clear and unquestionable" right to the whole of the Oregon Country. Polk rescinded the joint occupancy agreement. The rallying cry "Fifty-four-forty or Fight!" became popular. Several popular senators urged Polk to annex all of Oregon to 54°40´N. The stage was set for confrontation, yet neither side wanted it. The British government of prime minister Robert Peel had domestic problems and would soon fall—there was famine in Ireland and Peel determined against much opposition to repeal the Corn Laws, which maintained import duties; Polk was set to go to war with Mexico and did not want to have to fight on multiple fronts.

Peel instructed his foreign minister, the Earl of Aberdeen, to settle what, to the British government, had become not the Oregon

Question but "the Oregon problem." The Columbia boundary seemingly unattainable, Aberdeen was willing to settle for any reasonable compromise and found it in a proposal that the boundary follow the forty-ninth parallel to the sea but then retain all of Vancouver Island as British territory, as the Hudson's Bay Company had built a new fort there—Fort Victoria—in 1843, which could replace Fort Vancouver as its Columbia Department headquarters (and which it did in 1849).

The agreement, negotiated in Washington, D.C., by foreign secretary James Buchanan and British ambassador Richard Pakenham, was signed on 15 June 1846; the Senate ratified it three days later and Polk signed it the following day. Ten days later Peel, having repealed the Corn Laws, resigned as prime minister—over issues relating to the government of Ireland, not the negotiations with the United States.

Known as the Oregon Treaty in the United States and the Treaty of Washington in Britain, the agreement provided that, boundary line notwithstanding, navigation in the Strait of Georgia and the Strait of Juan de Fuca was to be free and open to both nations. Navigation on the Columbia was to be open to the Hudson's Bay Company and those trading with it. War was averted, and the United States assumed uncontested sovereignty over the continental Northwest south of 49°N.

The British government wanted nothing more to do with American Oregon. During the Civil War, when the United States would have had great difficulty defending the Northwest, James Douglas, premier of the new British Colony of British Columbia, proposed that he use troops at his disposal to retake Puget Sound and, with more troops shipped in, to retake the country as far south as the Columbia. It could likely have been done given American military weakness at the time, but the British government turned his proposal down flat.

An Island Peninsula

The terms of the Oregon Treaty provided that the forty-ninth-parallel boundary line run westward "to the middle of the channel which separates the continent from Vancouver's Island." This wording, created in some haste to solve a problem at hand, was to nearly cause a war (see overleaf), but it also left Point Roberts as a cut-off peninsula of American territory. The negotiators would have been aware of this because the maps they were using were quite accurate enough to show it; they had been since George Vancouver mapped and named the headland after his friend Captain Henry Roberts.

Clearly, however, the issue of leaving a small isolated piece of territory completely surrounded by the other's was not considered important in the context of such a significant treaty and a boundary line that had been argued about for decades. And the government of which British negotiator the Earl of Aberdeen was part was about to fall in any case; more urgent issues were in play.

The issue was addressed later, when the boundary was being surveyed by both British and American surveyors in 1858. The British boundary commissioner, James Prevost, received instructions from his government that noted that it would be "more convenient" if Point Roberts were excluded from American territory, and Prevost was therefore to propose that it should "be left to Great Britain." If, however, the American boundary commissioner, Archibald Campbell, would not agree, then Prevost was authorized to offer in return some "equivalent compensation by a slight alteration of the Line of Boundary on the Mainland." This would, the instructions continued, "be attended with some convenience to this Country, while it would relieve the American Govt from the inconvenient appendage to the territory of the Union of a patch of ground of little value in itself & inaccessible by land to the other territories of the Confederation without passing through British Territory."

But Campbell was not amenable to any such arrangement, and the isolation of Point Roberts from the rest of Washington State continues to this day. At considerable expense three hundred students from grade four up are bused every day across two international borders to school in Blaine, and most residents have to travel likewise to Bellingham for medical and other services. Point Roberts remains inconvenient indeed.

MAP 110 (*above*).
This 1869 British official survey map, from the same source as **MAP 108** and **MAP 109**, *left*, shows the international boundary cutting off the Point Roberts peninsula. The area shown as sandbanks is normally under water at high tide.
Right. The initial boundary marker erected on the western side of Point Roberts in 1861. This is the *Obelisk* noted on the map.

MAP 111 (*below*).
Point Roberts with the forty-ninth-parallel boundary line extending across *Boundary Bay*, thus creating the cut-off peninsula of American territory. This was part of a hydrographic survey carried out by Captain George Henry Richards in the survey ship HMS *Plumper* in 1858, in the wake of the British Columbia gold rush of that year.

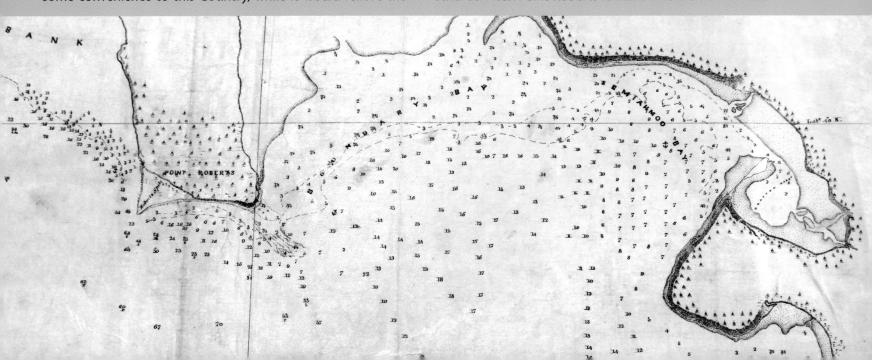

The Middle of the Channel

A much more serious issue than a piece of isolated territory also arose from the wording of the Oregon Treaty. The boundary was supposed to follow the "middle of the channel" south and around Vancouver Island, yet the "channel" was full of islands; where *was* the middle?

The British may have at first thought that Haro Strait was a fair interpretation of the "middle," but it soon became clear that San Juan Island could threaten the new settlement of Victoria, and to Britain, it became imperative that the island be in British hands.

In 1856 Britain and the United States agreed to carry out a joint boundary survey. While the survey was taking place, a dispute erupted over the ownership of San Juan Island. In June 1859 an American settler shot a pig that had wandered onto his land—a pig that belonged to the Hudson's Bay Company. The settler then refused to pay the $100 in damages demanded by the company. General William S. Harney, aggressive and anti-British military commander of the Department of Oregon, ordered

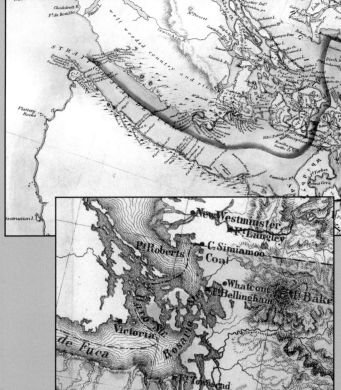

Map 112 (*left*).
This British map dated 3 December 1857 was copied from an American naval map and the potential boundary lines through the San Juan Islands added. The map was signed by George Henry Richards, captain of the *Plumper*, and James Provost, captain of the *Satellite*, both British survey ships. The black line passes through *Rosario Strait*, the red through the *Canal de Haro* (Haro Strait), and the green through the middle channel, which would have given *San Juan Island* to the British but the rest of the San Juan Islands to the United States. The blue line, according to a key, "passes through the centre of the whole space between the Continent and Vancouver's Island" at the expense of bisecting *Orcas Island* and *Shaws Island*.

Map 113 (*above, top*) and Map 114 (*above*).
An 1858 British map and an 1865 American map produced during the period of conflicting claims to the San Juans; each shows its respective nation's position as if it were a fait accompli.

troops to the island, and their commanding officer, Captain George E. Pickett (later famous for Pickett's Charge in the Civil War Battle of Gettysburg in July 1863) issued a proclamation claiming the island as United States territory.

In response, the governor of British Columbia, James Douglas, dispatched three ships and detachments of Marines and Royal Engineers. The resulting standoff was dubbed the Pig War. The British ensconced themselves at the northern end of the island, and the Americans established a camp at the southern end. The U.S. federal government was not pleased with Harney's actions and sent General Winfield Scott, veteran of the war with Mexico, to deal with the matter. Joint occupancy returned to Oregon when Scott and Douglas agreed that a company

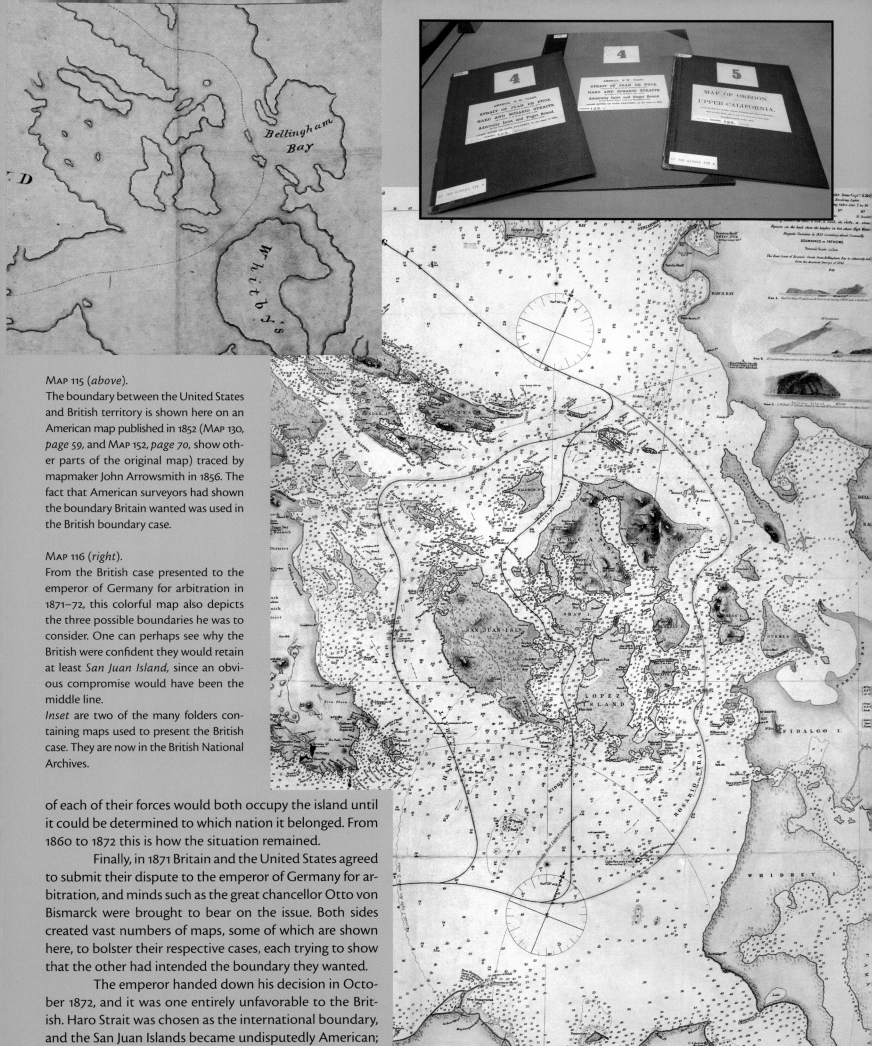

Map 115 (*above*).
The boundary between the United States and British territory is shown here on an American map published in 1852 (Map 130, *page 59*, and Map 152, *page 70*, show other parts of the original map) traced by mapmaker John Arrowsmith in 1856. The fact that American surveyors had shown the boundary Britain wanted was used in the British boundary case.

Map 116 (*right*).
From the British case presented to the emperor of Germany for arbitration in 1871–72, this colorful map also depicts the three possible boundaries he was to consider. One can perhaps see why the British were confident they would retain at least *San Juan Island,* since an obvious compromise would have been the middle line.
Inset are two of the many folders containing maps used to present the British case. They are now in the British National Archives.

of each of their forces would both occupy the island until it could be determined to which nation it belonged. From 1860 to 1872 this is how the situation remained.

Finally, in 1871 Britain and the United States agreed to submit their dispute to the emperor of Germany for arbitration, and minds such as the great chancellor Otto von Bismarck were brought to bear on the issue. Both sides created vast numbers of maps, some of which are shown here, to bolster their respective cases, each trying to show that the other had intended the boundary they wanted.

The emperor handed down his decision in October 1872, and it was one entirely unfavorable to the British. Haro Strait was chosen as the international boundary, and the San Juan Islands became undisputedly American; they were added to Washington Territory.

THE OREGON TRAIL

Left.
The Oregon Trail, doubtless looking a little more trimmed and cultivated than it did in its heyday, runs past the fence around the Whitman Mission in northeastern Oregon (see page 39). The original ruts in the trail are still evident, made by the passage of thousands of wagons. The covered wagon shown, a reproduction of bona fide design, has been placed on the trail, as today this is a National Historic Site.

those initially led by John Bartleson and John Bidwell, many of whom branched south along the California Trail, but about half arrived in Oregon, though without their wagons, which were left at Fort Hall.

In what has been called the "Great Migration of 1843," perhaps as many as a thousand emigrants left for Oregon that year along the Oregon Trail. When they arrived at Fort Hall, Hudson's Bay Company men told them they should abandon their wagons and proceed with packhorses. Missionary Marcus Whitman, who was traveling with the group, disagreed, and led the wagon

MAP 117 (below).
A map of the Oregon Trail and Oregon produced in 1851 by the Jesuits, in French, as a guide for missionaries. The locations of the various Jesuit missions are shown. Here the Oregon Trail looks more like a railroad!

No history of Oregon would be complete without mention of the Oregon Trail, for the majority of the first EuroAmericans to settle in the region arrived in wagons along the trail.

The route across the Continental Divide at the key South Pass was originally discovered by Astorian Robert Stuart in 1813, and rediscovered by Jedediah Smith in 1824 (see page 30). U.S. Army explorer Benjamin Louis Eulalie de Bonneville, with businessman Nathaniel Wyeth (see page 39), followed the trail across South Pass in 1832 and was probably the first to do it with wagons, a feat that would have enormous consequences in the following decade.

Between 1842 and 1844 Army topographical engineer John Charles Frémont, who was to play an important role in the American acquisition of California, explored and mapped routes west, including the Oregon Trail, earning the epithet "The Pathfinder." He and his mapmaker Charles Preuss produced a series of the most accurate maps available to emigrants, including a very influential and much-used seven-section, large-scale map of the entire trail from the confluence of the Kansas River with the Missouri to the Columbia (MAP 119, far right, center).

The first wagons to reach the Columbia arrived in 1840, a three-wagon train led by Joseph Meek and Robert Newell (who would settle near Champoeg and whose land is marked on MAP 124, page 56). They had driven wagons from Fort Hall (near today's Pocatello, Idaho). The first emigrants credited with using the whole of the Oregon Trail were

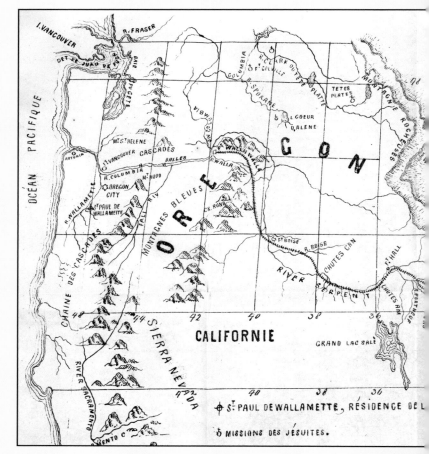

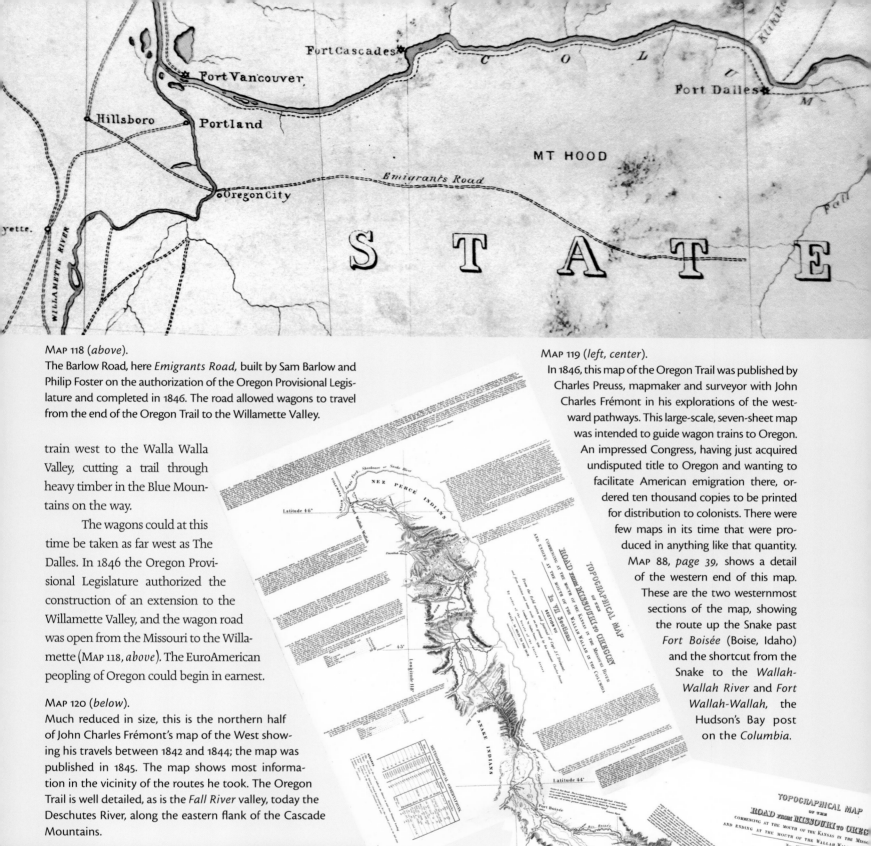

MAP 118 (above).
The Barlow Road, here *Emigrants Road,* built by Sam Barlow and Philip Foster on the authorization of the Oregon Provisional Legislature and completed in 1846. The road allowed wagons to travel from the end of the Oregon Trail to the Willamette Valley.

train west to the Walla Walla Valley, cutting a trail through heavy timber in the Blue Mountains on the way.

The wagons could at this time be taken as far west as The Dalles. In 1846 the Oregon Provisional Legislature authorized the construction of an extension to the Willamette Valley, and the wagon road was open from the Missouri to the Willamette (MAP 118, *above*). The EuroAmerican peopling of Oregon could begin in earnest.

MAP 120 (below).
Much reduced in size, this is the northern half of John Charles Frémont's map of the West showing his travels between 1842 and 1844; the map was published in 1845. The map shows most information in the vicinity of the routes he took. The Oregon Trail is well detailed, as is the *Fall River* valley, today the Deschutes River, along the eastern flank of the Cascade Mountains.

MAP 119 (left, center).
In 1846, this map of the Oregon Trail was published by Charles Preuss, mapmaker and surveyor with John Charles Frémont in his explorations of the westward pathways. This large-scale, seven-sheet map was intended to guide wagon trains to Oregon. An impressed Congress, having just acquired undisputed title to Oregon and wanting to facilitate American emigration there, ordered ten thousand copies to be printed for distribution to colonists. There were few maps in its time that were produced in anything like that quantity. MAP 88, *page 39,* shows a detail of the western end of this map. These are the two westernmost sections of the map, showing the route up the Snake past *Fort Boisée* (Boise, Idaho) and the shortcut from the Snake to the *Wallah-Wallah River* and *Fort Wallah-Wallah,* the Hudson's Bay post on the *Columbia.*

SETTLING THE WILLAMETTE

EuroAmerican settlement in Oregon—rather than just the staffing of fur trade posts—began with retired Hudson's Bay Company employees who wished to spend the rest of their lives on the West Coast rather than return to Eastern Canada or Scotland, where most came from, and where they were supposed to return according to their contracts. They were aided and encouraged in this decision by John McLoughlin, company factor at Fort Vancouver, even though it was against company policy. McLoughlin became a settler himself when he retired following the 1846 boundary decision. In 1829 he had claimed land at the falls on the Willamette, and two years later he built a water-powered sawmill at a place that came to be called Oregon City. This, along with a gristmill he built in 1844, are shown on MAP 121 (*above, right*). Oregon City was the end of the Oregon Trail (MAP 126, *page 57*), and in 1844 it became the first incorporated city west of the Rocky Mountains and was the first territorial capital between 1848 and 1851.

The early company settlers found a ready market for their produce in the fur trade posts and in Hawaii, though after the discovery of gold in California in 1849 it became more profitable to ship their produce there. Some of the first settlers were French-Canadian ex–Hudson's Bay Company men such as Joseph Gervais, Louis Labonte, and Étienne Lucier, who arrived in 1836. The names of some of the first settlers are found on the early surveys such as MAP 125, *overleaf*. Pioneers were attracted to the large patches of open prairie land found in the Willamette Valley, areas created by Indian burning. This land was not only fertile but also did not require labor-intensive

Above.
A sketch of Oregon City in 1845 from the west bank looking across the Willamette. It was painted by British military spy Lieutenant Henry Warre, whose reconnaissance of Oregon with Lieutenant Mervin Vavasour that year left a legacy of excellent maps and sketches (see also MAP 121, *above, right*).

Left.
John McLoughlin's house in Oregon City as it is today. It was moved in 1909 to its present site atop the escarpment, a site marked *Public Square* on MAP 122 (*right*). The house's original location can be seen on MAP 121 (*above, right*), where it as noted as D^r. M^cL^n.

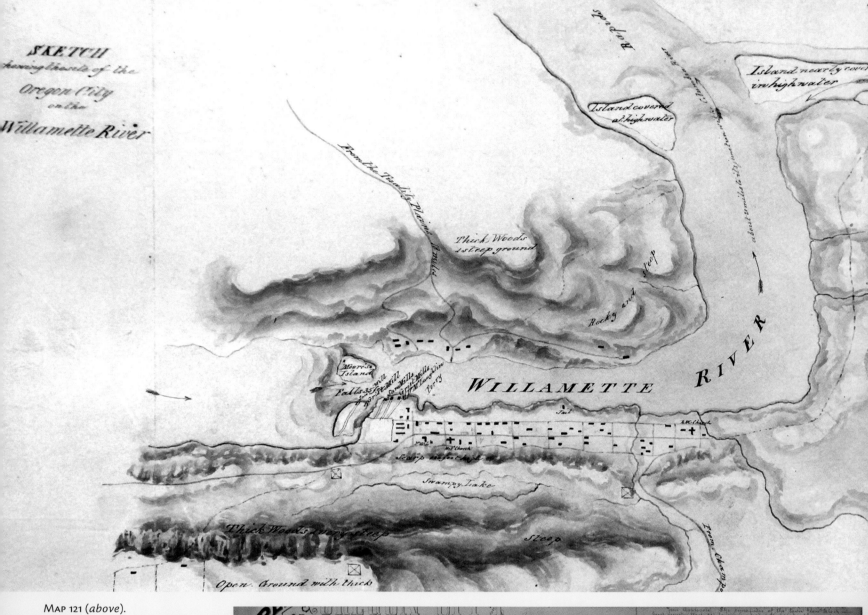

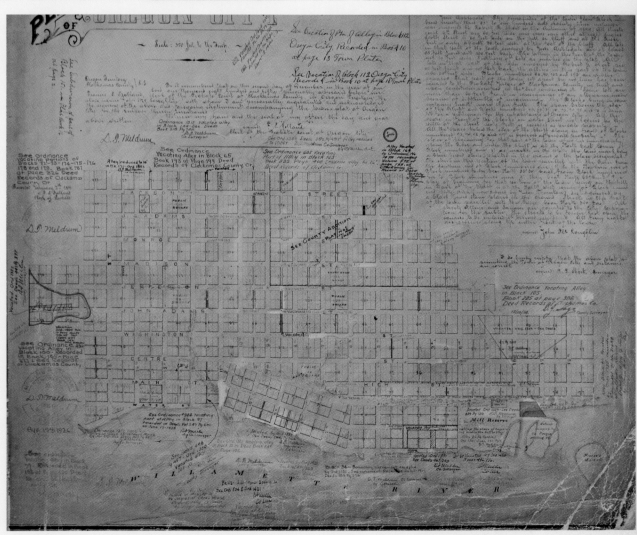

Map 121 (*above*).
Oregon City in 1845, depicted by Henry Warre and Mervin Vavasour. The first part of the city is platted below the river escarpment noted as *Scarp 120 feet high*. At right is the road *From Champoeg*, and the *Grist Mills of Dr McLoughlin* are noted at the *Falls*, along with several other *Saw Mills* and *Grist Mill[s]*.

Map 122 (*right*).
This detailed plan of *Oregon City*, complete with notes by and the signature of John McLoughlin, dates from 1850. The area at the top of the escarpment has now been platted, and at a different angle from the first part. The map has been updated by a number of surveyors since that time; it was a working document for almost a hundred years. As a result it has been trimmed and become a little dog-eared with use. It survives today proudly displayed in the Museum of the Oregon Territory in Oregon City.

MAP 123 (*left* and *above*).
The town of Champoeg is no more, but in its place is a unique map—and a full-scale one at that. Intersections of roads have been marked with stakes labeled with the road name, so that it is possible to walk the very streets that once carried the commerce of much of the Willamette Valley.

MAP 124 (*right*).
Champoeg is shown on the banks of the Willamette on the edge of French Prairie on this 1852 General Land Office survey map. The town grew up on the land claim of *Robert Newell*, who had arrived via the Oregon Trail in 1840 (see page 52). Note, near the bottom of the map, *James McKay's Flouring Mill* and *Saw Mill*, and the *Mill Pond*.

MAP 125 (*right, bottom*).
The confused pattern of land ownership staked under the Donation Land Claim Act is apparent on this map of the claims on French Prairie near Champoeg, drawn in 1860. The names of some of the earliest Oregon settlers are seen here.

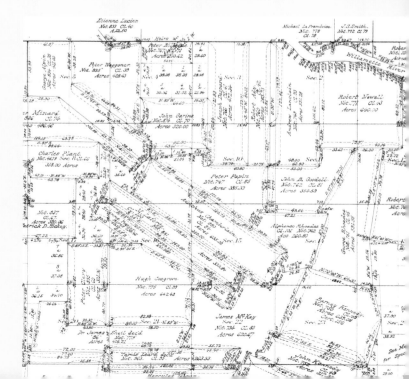

clearing of trees. Claims were permitted by both British and American subjects under the terms of the joint occupancy agreement of 1818.

In 1841 settlers began meeting at a place called Champoeg (pronounced Cham-poo-ey) to discuss how to deal with wolves that were killing livestock and how to handle estates, such as that of Ewing Young, who had died that year. In 1843 a vote was taken as to whether a provisional government should be established. The measure only narrowly passed, by fifty-two votes to fifty; many settlers thought government would simply mean regulation and taxation. A set of laws called the Organic Laws of Oregon were drawn up, based on the laws of Iowa Territory. A representative—William Gilpin, who had traveled with John Charles Frémont—was

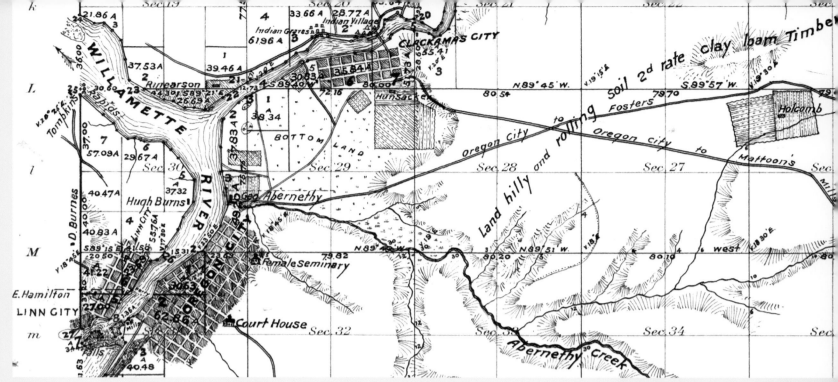

elected to go to Washington, D.C., to petition for American government, an issue resolved to the settlers' liking in 1846 (see page 48). Champoeg, which had become the de facto capital of Oregon, lost this position to Oregon City when territorial status was achieved in 1848. Champoeg in any case proved to be an unsuitable site for a capital, as it was subject to flooding, and one such flood essentially wiped out the town in December 1861; it was not rebuilt.

In 1848 the new Territory of Oregon sent Samuel R. Thurston to Congress as a delegate where he lobbied for a land claims law. This goal was achieved in September 1850 when Congress passed the Oregon Donation Land Claim Act. This validated and documented previous land claims—which were often irregular in shape (MAP 125, *left, bottom*)—and allowed land to be granted to settlers claiming it before 1855. The result was a massive increase in immigration to Oregon and the Willamette Valley. Some 25,000 to 30,000 new immigrants swelled the population by 300 percent, and 7,437 land grant patents were issued.

To facilitate the claiming of land, a surveyor general, John B. Preston, was appointed to begin the process of extending the rectangular public land survey system to Oregon. Far from the rough-and-ready ad hoc system used before, the Public Land Survey, or Township and Range System, required precise measurement

MAP 126 (*above*).
Oregon City, Clackamas City, and area, shown on an 1852 General Land Office map. The road marked *Oregon City to Fosters* is the end of the Barlow Road, the final link with the Oregon Trail to carry a wagon road to the Willamette.

MAP 127 (*below, left*).
Peoria was one of many small towns platted in the Willamette Valley. The town seems to have been platted about 1850, but before 1853 the settlement was moved a mile north to avoid constant flooding, which plagued the first townsite. This is the plat of the original Peoria (north is to the right), together with, *inset,* the 1853 survey map; the town's first location is at the most southerly bend of the river. By the *Ferry* is the new settlement.

MAP 128 (*below*).
The territorial capital, *Salem,* on an 1852 General Land Office map. The settlement, originally called Chemetka, was platted in 1850–51, after steamboats began plying the waters of the Willamette. In 1851 the territorial capital moved from Oregon City to Salem.

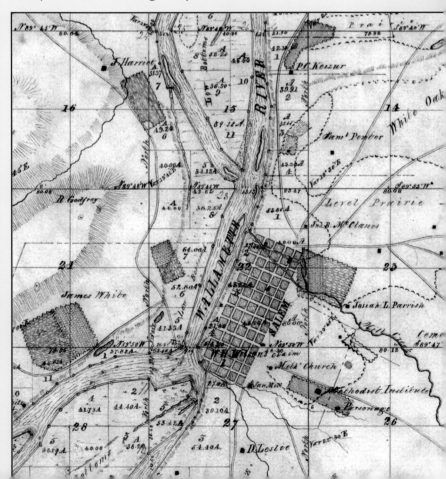

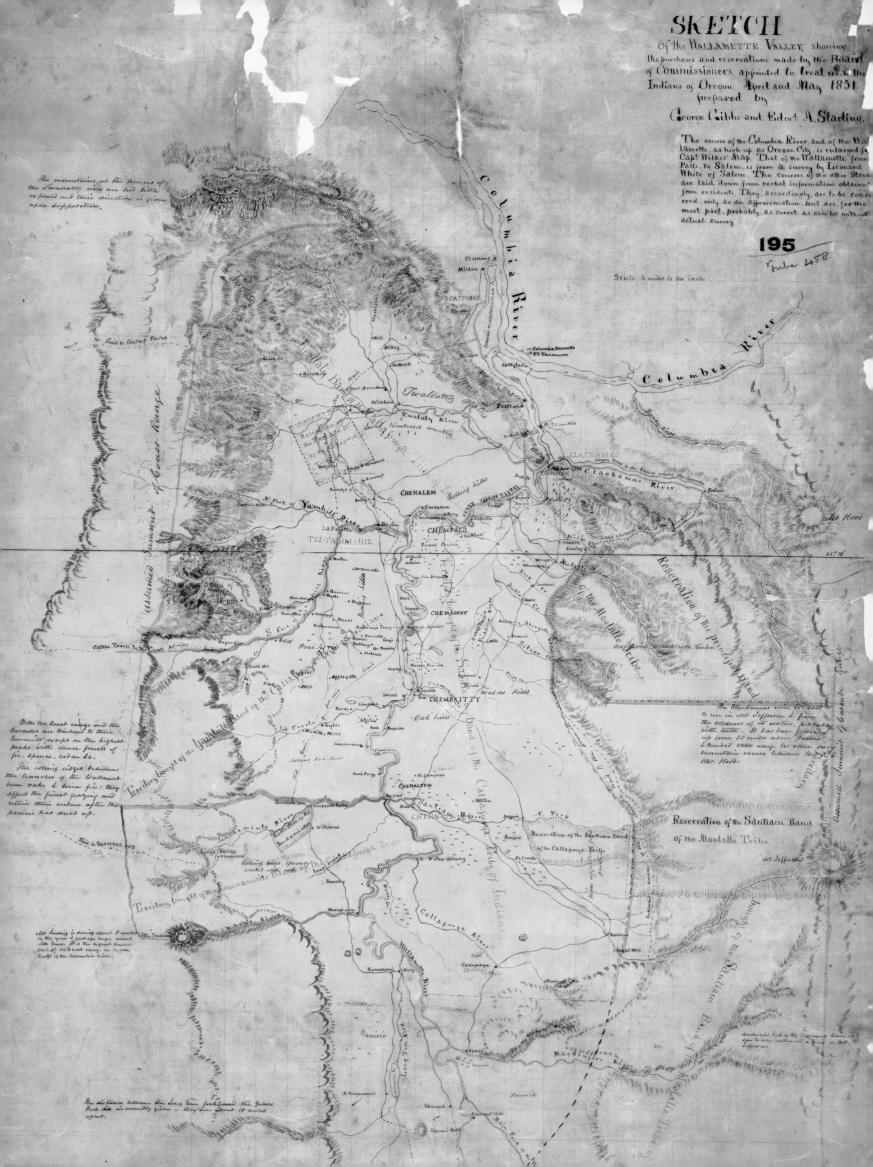

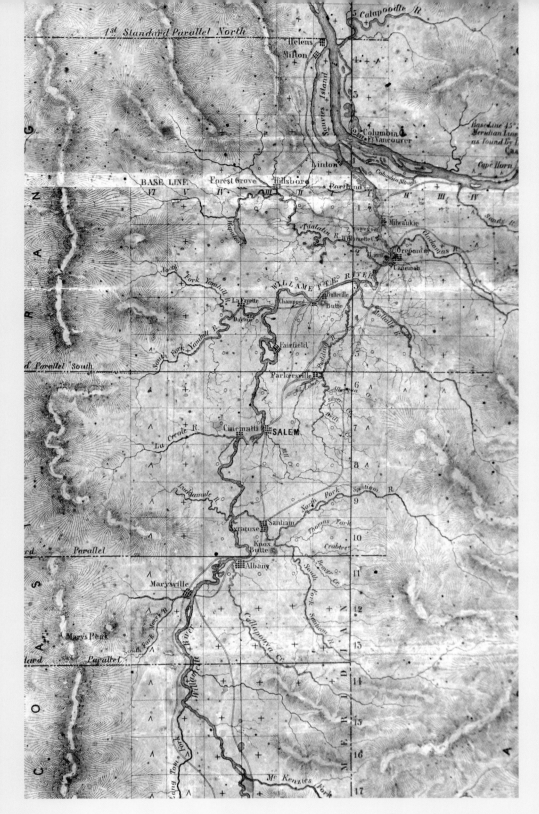

MAP 130.

The northern Willamette Valley part of an important map published by surveyor general John Preston in 1852. His *Diagram of a Portion of Oregon Territory* extended from the California boundary north to the international boundary and showed the extent of the Township and Range System at that date. The initial *Base Line* and *Meridian* Line cross just west of Portland at a monument called the Willamette Stone, placed on 4 June 1851. The squares are the six-mile-square survey blocks called townships. Many of the early centers of commerce in the Willamette Valley that have now disappeared are shown: *Fairfield, Parkersville, Champoeg, Canemah, Cincinatti,* and others. *Marysville* became Corvallis in 1853. Eugene City and Springfield (at *Mc. Kenzies Fork,* the McKenzie River, at the bottom of the part of the map shown), having been platted in 1853 and 1856, respectively, are not shown. Surprisingly, perhaps, for a map drawn by surveyors, the position of the Willamette is mispositioned in places, sometimes in error by as much as four miles when compared with detailed surveys done the same year. In addition, Barlow Road east of *Oregon City* seems to be incorrectly shown following the *Clackamas R.* Other portions of this map are MAP 149, *page 68,* and MAP 152, *page 70.*

of survey baselines (shown on MAP 130, *left*) and detailed division of the land into six-mile squares—townships—and then into thirty-six one-square-mile (640-acre) sections, which could be further subdivided. It was a regular and predictable system that defined the Euro-American colonization of most of the West.

The act creating Oregon Territory in 1848 had, in theory, acknowledged and guaranteed Indian land title, unless legally "extinguished." In practice, extinguishment was coerced. EuroAmericans wanted the land and made sure Indians were herded onto reservations. In 1850, just ahead of the Donation Land Claim Act, Congress passed a law known as the Indian Treaty Act, which specifically provided for "the extinguishment of their claims to lands lying west of the Cascade Mountains." The law allowed for the appointment of commissioners who were, if possible, to "remove all these small tribes" and "leave the whole of the most desirable portion . . . open to white settlers."

Although Indian resentment flared into resistance elsewhere, the Indians of the Willamette Valley proved no match for the flood of immigrants in search of their land—land that the government freely gave them. There were about 2,100 EuroAmericans (except for a few Indian spouses who were also counted) in Oregon south of the Columbia in 1842. This had grown to 8,779 persons by 1849, and the first federal census, conducted between September 1850 and March 1851, enumerated 11,873 persons, the majority settled in the northern Willamette Valley. At the same time the Indian population began a precipitous decline (see page 10).

MAP 129 (*left*).

This important *Sketch of the Wallamette Valley* was prepared in 1851 by Indian commissioners George Gibbs and Edmund A. Starling, "appointed to Treat with the Indians of Oregon," to show their progress in confining the Indians of the valley to reservations so that their land could be appropriated for use by settlers. The map relied on existing maps for some of its information; the course of the Columbia, for example, was taken from John Wilkes's U.S. Ex. Ex. map published in 1844 (MAP 103, *pages 44–45*). Then settlements, as they were in 1851, were added. *French Prairie* and other prairies are depicted. Locations of bands of the *Callapooya* (Kalapuya) are noted, and the recommended reservations are shown. These were not approved by Congress, which instead created the Grande Ronde reservation in 1857.

EARLY PORTLAND

A number of townsites jockeyed to rival Oregon City as the region's primary center of commerce. Linn City, opposite Oregon City on the Willamette; Milwaukie; and Multnomah City—until it was swept away by a flood in 1853—were contenders, but it was Portland that emerged the winner.

For Portland was ultimately found to be the highest safe point for navigation by oceangoing ships. William Overton, an emigrant from Tennessee, filed a land claim in 1843, giving his lawyer, Asa Lawrence Lovejoy, a half-interest in return for doing the filing and paying the fee. The fee was twenty-five cents, and that is how much Lovejoy paid for half of what is now downtown Portland. Overton sold his interest to a merchant, Francis W. Pettygrove, the following year for fifty dollars' worth of supplies—not, it would seem, a bad return for Overton on an investment of nothing.

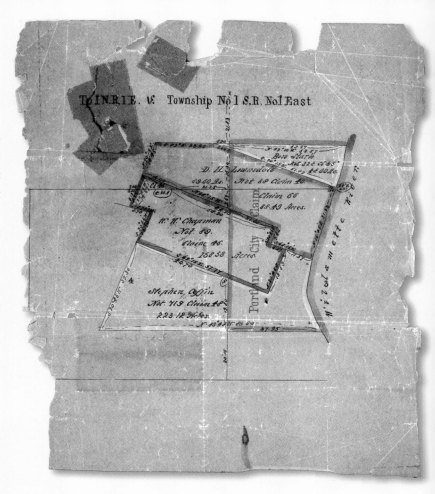

MAP 131 (*right*).
The ownership of Portland by 1850. The land ownership beyond Pettygrove and Lovejoy shown on this map is derived thus: Benjamin *Stark* purchased a half-interest from Lovejoy in 1845; his interest, though a smaller area, included most of the buildings. Pettygrove sold his half-interest in 1848, when he left for the gold fields of California, to Daniel H. *Lownsdale,* a tanner, for leather worth $5,000. Lownsdale in turn sold half of his interest to *Stephen Coffin,* an emigrant New Englander, for $6,000. In December 1849 Lowns- dale and Coffin sold part of their land to William W. *Chapman,* an Oregon City lawyer. As with countless other successful western townsites, the land was sequentially subdivided, increasing in value at each transaction.

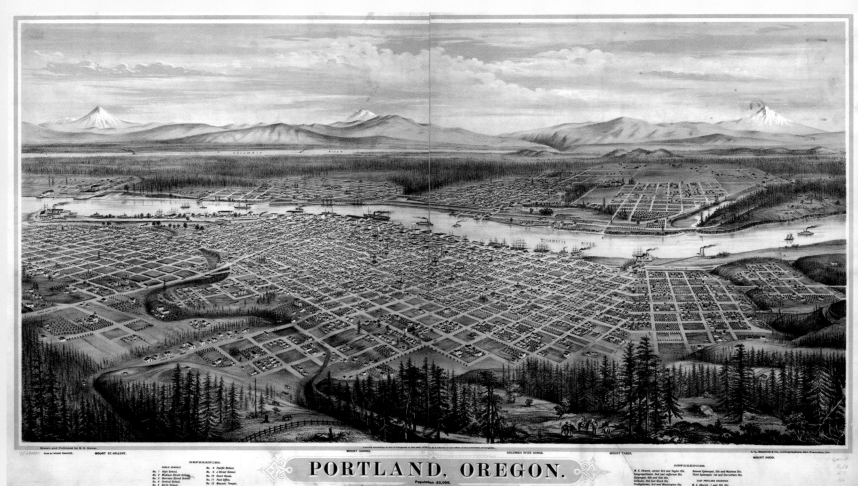

PORTLAND, OREGON.
LOOKING EAST TO THE CASCADE MOUNTAINS.

The story of how Portland was named is well known. Joint owners Pettygrove and Lovejoy argued about a name for the town they were going to create and agreed to flip a coin. Pettygrove, from Portland, Maine, won the toss, and Portland thus became Portland and not Boston, Lovejoy's hometown.

MAP 132 (*left, bottom*).
By 1879 *Portland* had grown into a fine city and boasted a population of 22,000, as this promotional bird's-eye map proclaims. East Portland, across the river, would merge with Portland in 1891 (see page 106).

MAP 133 (*right, top*) and MAP 134 (*right, bottom*).
Covering much of the area of modern Portland, these two survey maps were published by the surveyor general of Oregon's office in 1852. At the bottom left-hand corner of the top map and top left corner of the bottom one is the intersection of the survey base lines, the *Willamette Meridian* and the Portland *Base Line*. Portland is now a plat of many blocks on the land once owned by Pettygrove and Lovejoy. The lake on MAP 133 was Guild's Lake, the area that would be filled and reclaimed for the 1905 Lewis and Clark Exposition (see page 108). Modern sources of power are in evidence in Portland: a *Steam Grist* mill and two *Steam Saw Mill*s. On the east side of the river is the land claim of *James B. Stephens*, who had acquired the area that would become East Portland by purchase from John McLoughlin in 1846. A *Ferry* connects Stephens's claim with Portland. At bottom is *Milwaukie*, the plat of *Lot Whitcomb*, whose claim is also noted on the map. In 1850 Whitcomb launched the steamboat *Lot Whitcomb*, assembled from parts shipped from the eastern United States, in a bid to ensure the growth of Milwaukie. The steamboat, whose first captain was John C. Ainsworth (one of the founders of the Oregon Steam Navigation Company; see page 76), served Astoria and Oregon City and points between—including Milwaukie—until 1854.

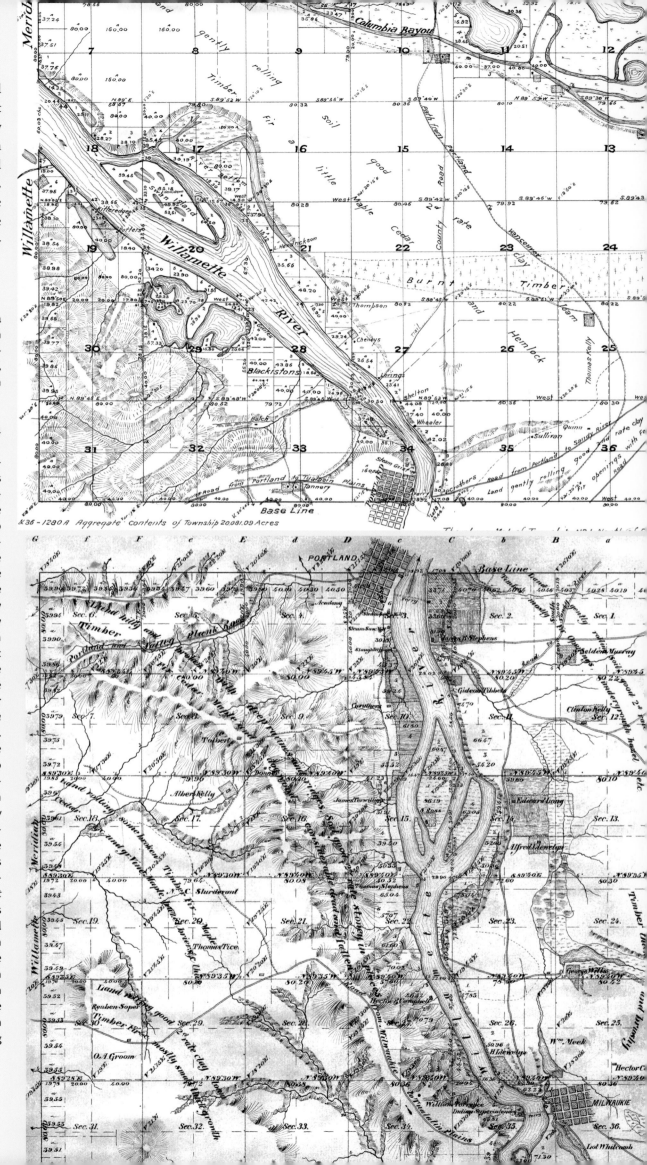

Paths to Statehood

The seals of the states of Oregon (*above*) and Washington (*above, right*), photographed on state edifices in Salem and Olympia.

MAP 135 (*below*).
Part of a map of the United States published in 1850 shows Oregon Territory extending east to the Rocky Mountains.

MAP 136 (*below, bottom*).
This 1859 map, published just before Oregon's admission to the Union, depicts the two territories still extending to the Rockies.

The region below the continental forty-ninth parallel had become American territory in 1846 (see page 48), and two years later demands for a formal government structure led to the creation Oregon Territory, embracing the country west of the Rockies (MAP 135, *below, left*).

It did not take long for the settlers north of the Columbia—far fewer in number than those to the south—to voice their discontent over distant representation, and the move of the territorial legislature from Oregon City to Salem in 1851 only heightened demands for a separate territory. Conventions at Cowlitz (see MAP 152, *page 70*) in August 1851 and Monticello (MAP 156, *page 73*) in November 1852 adopted a memorial to Congress requesting the creation of Columbia Territory. Joseph Lane, previously Oregon's first governor and then congressional delegate, had senatorial ambitions and had already introduced legislation to create Columbia Territory; his idea was to make Oregon smaller and thus more likely to be accepted as a state. Territorial status north of the river was achieved in 1853 when what had been called North Oregon became not Columbia but Washington Territory, and it too extended to the Rockies (MAP 136, *below*).

In 1857, as the national debate over slavery approached its climax, Oregonians held a convention at Salem and drafted a constitution. Almost everyone opposed slavery, and the Dred Scott case of 1856 had declared that territories had no right to exclude slavery—hence the urgent move toward statehood. The idea of including only the area west of the Cascades in a new state was discussed, but the final area agreed upon extended east to the Snake River. Oregon became a state on 14 February 1859, and Joseph Lane became one of its two first senators. The balance of what had been Oregon Territory was added to Washington Territory (MAP 137, *right, top*). The region east of 117°W and the Snake River was removed from Washington Territory in 1863 to become part of a new Idaho Territory, and Washington closely assumed its present-day boundaries (MAP 138, *right, bottom*).

It was to take another thirty years before Washington achieved statehood, largely because of partisan infighting. An enabling act was signed by lame-duck Democrat president Grover Cleveland on Washington's birthday, 22 February 1889. A convention met in Olympia in July and August to draft a constitution, which was approved by the electorate in October by only 77 percent—Democrats generally opposed admission—and new Republican president Benjamin Harrison proclaimed Washington a state of the Union on 11 November.

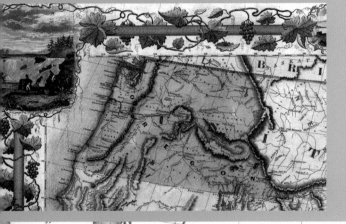

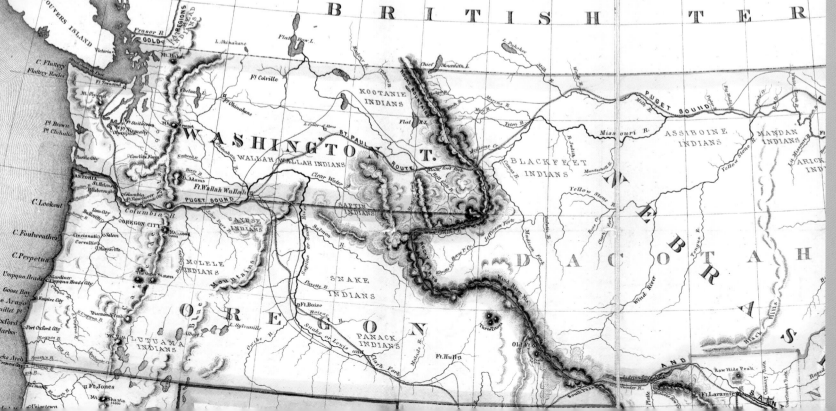

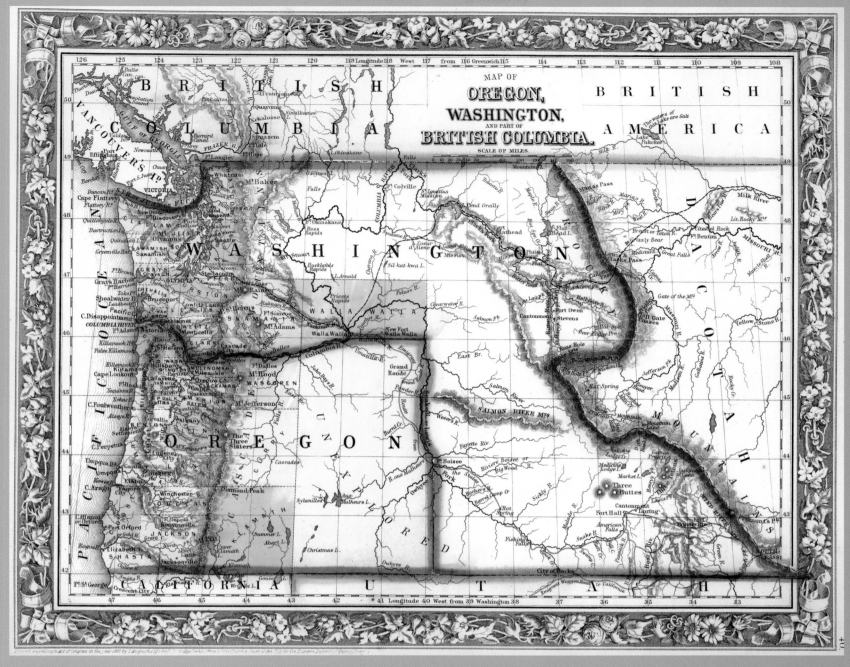

MAP 137 (*above*).
After Oregon became a state, Washington Territory embraced much of what is today the state of Idaho. This ornately bordered map was published in 1860.

MAP 138 (*left*).
Idaho Territory covers what was eastern Washington Territory on this 1864 map. Note that Washington excludes the San Juan Islands.

MAP 139 (*right*) and **MAP 140** (*right, bottom*). Coffee maker Arbuckles issued a series of illustrated postcard maps of American states and territories in 1889; these are the Washington and Oregon cards. The location of *Monticello* is shown on **MAP 139**. Note the population figures: Washington 75,115 and Oregon 174,768.

INDIAN WARS

With the surge in EuroAmerican immigration to Oregon following the resolution of the boundary issue in 1846, and particularly during the 1850s, the Native peoples were increasingly forced off their lands and onto reservations that were too small to support a hunting and gathering lifestyle. The predominant view of the settlers was that the Indians were incapable of fully utilizing the land and hence deserved to be removed from it or, worse, that they were a scourge—for sometimes they fought back—and should be exterminated.

The clash of cultures and values led to violence. The period of Indian–EuroAmerican turmoil lasted from about 1847 to 1879 in the Northwest but erupted into more widespread violence from time to time, and these periods have been named as various wars by later historians. But many so-called battles were little more than murders of Indians.

The settlers not only took Indian land but brought diseases against which the aboriginal peoples had no immunity, and it was this that led to the massacre at the mission of Marcus Whitman in November 1847 (see page 40). Retribution was swift. A rifle company called the Oregon Rifles led by Major Henry A.G. Lee, and a force of five hundred led by Colonel Cornelius Gilliam pursued various groups of

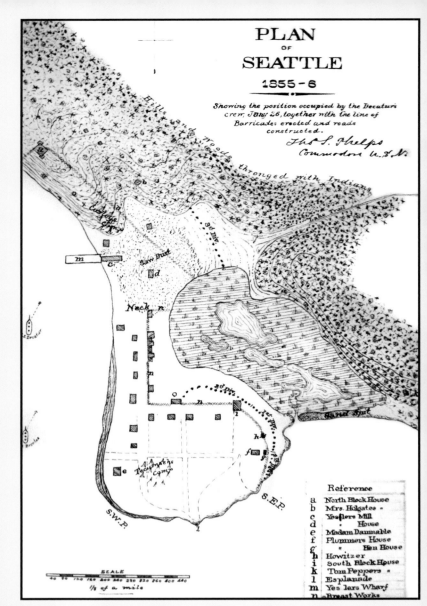

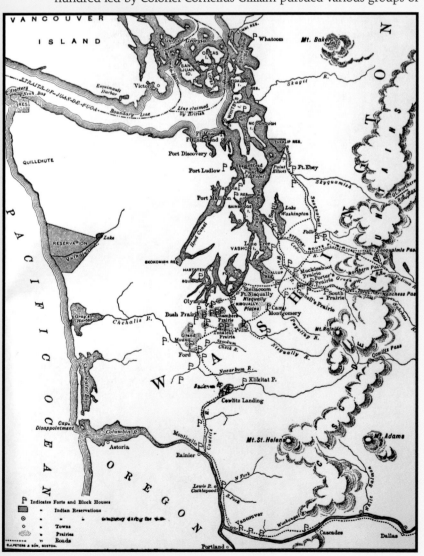

Cayuse and Paloos, bringing superior firepower with them but fighting an adversary that used guerilla tactics against them. By May 1848 Lee's force numbered over four hundred men. Despite this large show of force, those responsible for the Whitman massacre were not found; five surrendered to the authorities two years later, and—despite the likelihood that all were not the actual murderers—were promptly tried by a military commission and executed. The Cayuse War, as this period of unrest became known, continued with varying intensity until about 1855, destroying all trust between Indian and settler.

The most well-known Indian conflict in Washington and Oregon is the attack on the new settlement of Seattle, but this was really a minor foray during two other "wars"—the Puget Sound War and the Yakima War—periods of conflict following on from the Cayuse War.

In 1854 the first governor of the newly created Washington Territory, Isaac Stevens—who was also charged with Indian affairs in the territory—negotiated the Treaty of Medicine Creek with the Nisqually tribe, by which they agreed to move to reservations, leaving, of course, the rest of the country free for EuroAmerican settlers. The land on the reservations, however, proved unacceptable to the Nisqually, who principally fished for food, and they began an uprising led by their chief, Leschi.

In May and June 1855, Stevens, now with superintendent of Indian Affairs for Oregon Territory, Joel Palmer, negotiated two further treaties, these with the Indians east of the Cascades. Again the tribes

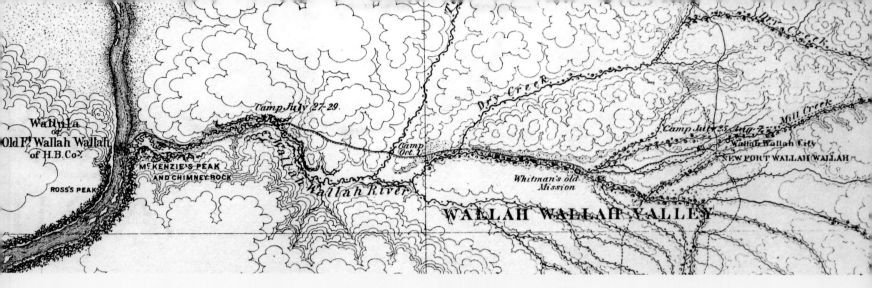

were coerced into moving onto reservations. The Walla Walla, Umatilla, and Cayuse were to move to a reservation in northeastern Oregon, and the Yakima (since 1994, the original form, Yakama, has been
used) were to move to another reservation, south of the present-day
city of Yakima. But gold was found on the Yakima reservation, and a
conflict immediately erupted as the Yakima tried to keep gold seekers
off their land. A more general uprising resulted, with all Indian groups
uniting behind Yakima chief Kamiakin (or Kamaaiakun). Small groups
of settlers, government agents, and military forces were attacked by
Yakima, and the region became perilous for travelers.

In January 1856 it seems that some Yakima and other interior
tribes decided to cross the Cascades and bring the war to the new settlements along Puget Sound. Here they joined with other hostile Natives, led by Leschi, to prosecute the attack. But they had picked the
wrong time. Moored offshore was the U.S. naval ship *Decatur,* which
had been sent to Puget Sound to protect the settlers against depredations from Haida and Tlingit, from much farther north. *Decatur*'s men
and guns, including howitzers, allowed the successful defense of the
little town against a prolonged attack over many hours. Because many
tribes took part, they had to use their common language, Chinook,
to communicate, but the settlers also understood Chinook and were
thus able to anticipate what was going to happen. Estimates of the
attacking force have ranged from 150 to 2,000. The Indians retreated, but not before sending word that they would soon return with

MAP 141 (*far left, top*).
This well-known plan of Seattle shows the situation during the Indian attack on 26 January
1856. The plan was drawn by Thomas Phelps,
an officer on the USS *Decatur*. The configuration of the coastline has changed, but the map
shows approximately the area of today's Pioneer Square District. See also MAP 153, *page 71*.

MAP 142 (*far left, bottom*).
A later map (1909) shows the location of Indian reservations, forts, and blockhouses in the
Puget Sound region about 1856, during the period known as the Puget Sound War.

MAP 143 (*above*).
The location of the new *Fort Wallah Wallah*—the site of today's city of Walla Walla—on
a map drawn in 1858, the year it was built, as
part of the campaign to fight Indians in eastern
Washington and Oregon. The *Old Fᵗ· Wallah
Wallah*, the Hudson's Bay Company trading
post, is shown at left.

MAP 144 (*left*).
A military map of the attack on Colonel Edward
Steptoe's forces at Ingossomen Creek (now Pine
Creek), in the Palouse Hills of Eastern Washington, in May 1858. Steptoe, with about 150 soldiers, found himself facing over 600 Spokane,
a situation from which he wisely retreated. The
formation of Steptoe's camp is shown at top
right, atop a butte later named after him.

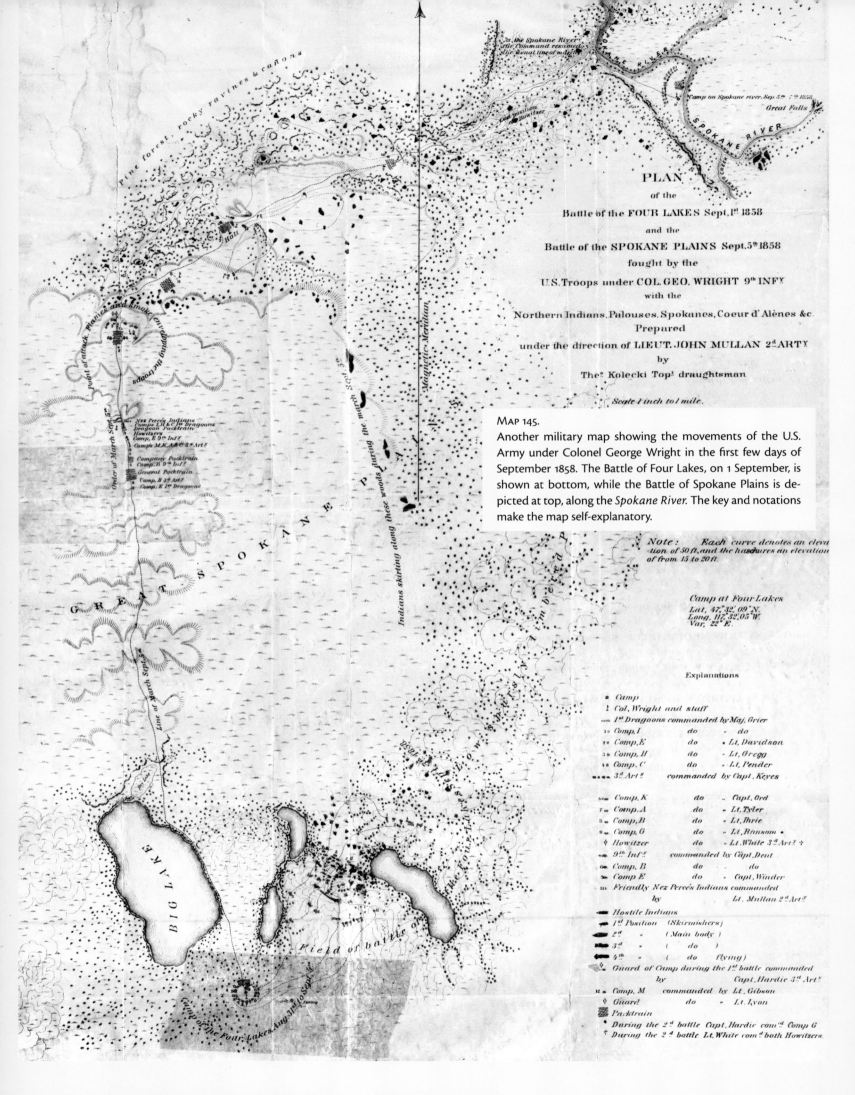

MAP 145.
Another military map showing the movements of the U.S. Army under Colonel George Wright in the first few days of September 1858. The Battle of Four Lakes, on 1 September, is shown at bottom, while the Battle of Spokane Plains is depicted at top, along the *Spokane River*. The key and notations make the map self-explanatory.

twenty thousand warriors. Subsequently, the Puget Sound area was blanketed with blockhouses and forts (MAP 142, *page 64*) and in Seattle blockhouses were built and the forest cleared to remove any hiding places. A number of U.S. naval vessels arrived for support. Governor Stevens persuaded some settler-friendly Indians to take on the role of bounty hunters and pursue members of the offending tribes. At any rate, there was no further concerted attack on settlements in the Puget Sound area.

About this time tensions reached a peak in southern Oregon, especially in the Rogue River Valley, where settlers and miners had appropriated Indian lands and resources. Settlers there even went as far as attacking Indian villages, and in October 1855 the U.S. Army had to position men between the two to prevent bloodshed. By 1856, after the Indians had turned on the Army, they were defeated and forced onto reservations.

The northern conflict continued, however, in Eastern Washington the following year with what was really an extension of the Yakima War—chief Kamiakin having again persuaded some other tribes to join his effort to repel the usurpers of their land—though it is often referred to as the Coeur d'Alene War. Fort Walla Walla (MAP 143, *page 65*) was built in 1858 to act as a base for operations. In May that year 164 dragoons and infantrymen under Colonel Edward Steptoe ventured north from the fort intending to present a show of force to the eastern Indian tribes. In the Palouse Hills he suddenly found himself facing between six hundred and a thousand Spokane, Paloos, and Coeur d'Alene, from which, badly outnumbered, he was lucky to escape (MAP 144, *page 65*).

Later that year a much larger army of about 690, including men brought in from California—set out northward, commanded by Colonel George Wright. A second group of about 300 was to proceed south from Fort Colvile. Wright's instructions were to "attack all hostile Indians with vigor, make their punishment severe, and persevere until the submission of all is complete." Wright's force was attacked by considerable numbers of warriors of several tribes as it camped at a place called Four Lakes, about twenty miles south of the Spokane River (MAP 145, *left*). But the superior rifles and military tactics of Wright's men won the day. One of the injured was Yakima chief Kamiakin. After camping for three days Wright's army resumed its march north, and the Indians attacked once more but this time were decimated by howitzers and were driven north to the Spokane River. This was the Battle of Spokane Plains shown on MAP 145.

The Indians were all but defeated, and several of their chiefs approached Wright to try to negotiate terms, but after Wright's men also captured about eight hundred of their horses, the Spokane surrendered outright. A few days later the Coeur d'Alene did likewise. Kamiakin, Yakima instigator of the Indian coalition, fled to British Columbia. The following year, the region was declared open for settlement.

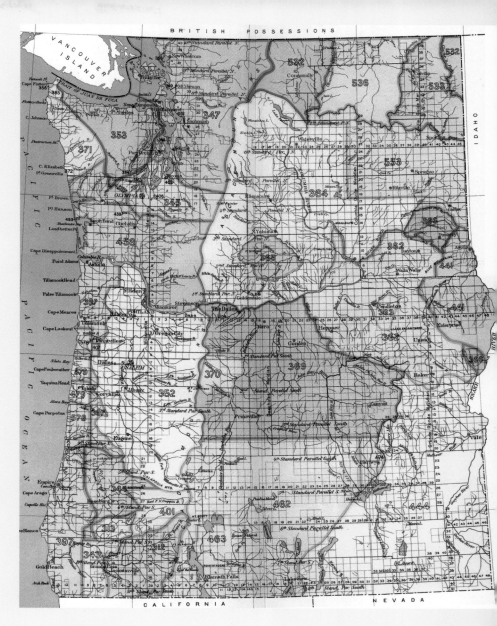

MAP 146 (*above*).
In 1896 Charles Royce of the Bureau of American Ethnology compiled maps of the entire United States showing Indian land cessions. This map is a composite of the Washington and Oregon maps. Some of the numbered areas represent cessions as follows: *364/365*, the Yakima, and *362/363* the Cayuse, Umatilla, and Walla Walla, from a treaty signed at Camp Stevens, in the Walla Walla Valley, on 9 June 1855; *345*, the Treaty of Medicine Creek, 26 December 1854, with the Nisqually and others; *369*, with the Confederated Tribes of Middle Oregon, at Wasco in 1855; *347*, with the Duwamish and Suquamish, at Point Elliot in 1855; *352*, with the Kalapuya in 1855; *462/463*, with the Klamath, Modoc, and Snake in 1864; and *444*, with the Shoshone in 1863.

MAP 147 (*right*).
Indian reservations (shaded orange) in Washington and Oregon in 1870, shown on a census map published in 1874. Hunting grounds (shaded light orange) are also indicated, as are the areas of EuroAmerican settlement (shaded gray-blue).

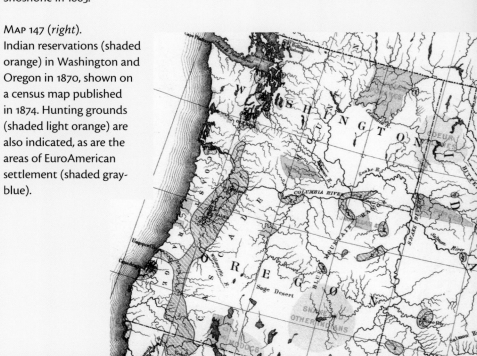

SOUTHERN GOLD

Hot on the heels of the California Gold Rush of 1849 came the discovery of gold in the Rogue River Valley of southern Oregon in 1851, resulting in an immediate influx of gold seekers that created tensions with the Indians of the region, leading to open warfare in what has been called the Rogue River Wars.

Placer gold was first found on Josephine Creek in May 1851, the first of many finds in the region. Two pack-train drivers, James Cluggage and John R. Pool, discovered gold by accident in Rich Gulch during the winter of 1851–52 while on their way to the California gold fields. Cluggage filed a Donation Land Claim, and a tent city quickly formed; he called it Table Rock after the nearby mesa, but it soon became known as Jacksonville, likely from Jackson County, created in January 1852 (and itself named after Andrew Jackson, seventh president of the United States). The gold-mining area attracted many Chinese miners, and Jacksonville soon had the Northwest's first Chinatown, though later Oregon imposed many prohibitions and taxes on the Chinese.

By 1854 most of the easy gold had gone, and many gold seekers moved on to other western promises of instant riches. Yet gold continued to be found and extracted, sometimes using environmentally damaging hydraulic methods. The Briggs Mine, near the California state line, produced two thousand ounces of gold in 1904, the date of MAP 151 (*right*).

MAP 148 (*above*).
The gold *Mines* and *Jacksonville* are shown astride the Oregon and California Trail in the *Siskiyou Mts.* of southern Oregon on this map published in 1853.

MAP 149 (*below*).
The southern part of the *Diagram of a Portion of Oregon Territory*, published in 1852, shows the *Rogues River*, with *Jacksonville* at the center of the *Gold Regions*, with *Mines* dotting the area. Names undoubtedly having a story behind them include *Gallows*, *Grave*, and *Jump off Joe* creeks. *Kenyonville* is Canyonville, at the northern end of the defile and trail leading to the valley of the South Umpqua River.

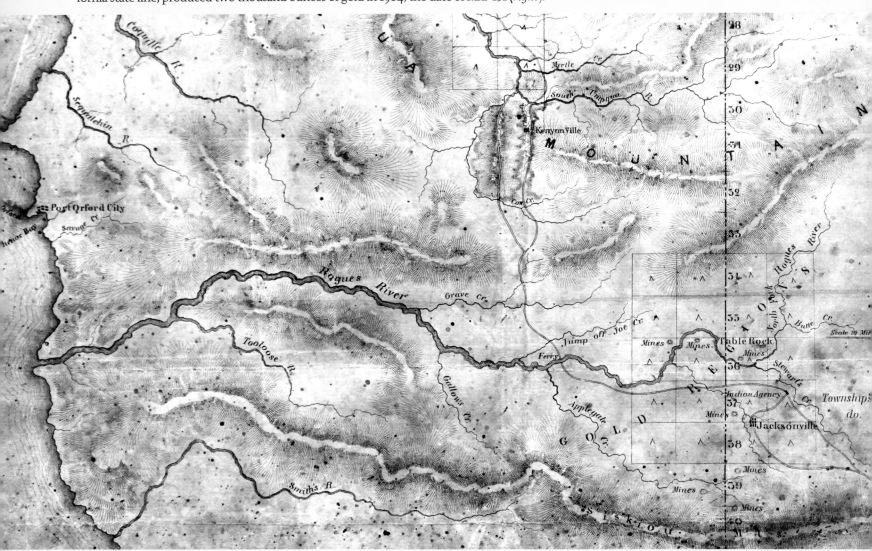

MAP 150 (right).

Gold diggings and a *Water-duct* are noted west and southwest of *Jacksonville* on this 1854 survey map, made to accommodate the many land claims in the region. Of particular interest here is *One-horse Town*, of which there is little record other than this map, at a location that became Kanaka Flats, which has a reputedly wild history. It derived its name from the presence of Hawaiians who came to the region for the gold rush but, ineligible to file land claims, later left. Kanaka Flats Road still exists at this location.

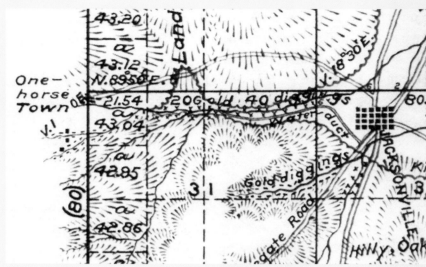

MAP 151 (below).

Gold workings, shown as small circles, litter the mountains of southern Oregon on this map, which dates from 1904, after the railroad—by then the Southern Pacific—had passed through in 1883. The reason for the rise of *Medford* at *Jacksonville*'s expense is immediately obvious; exactly the same fate befell *Kerby*, which lost its position as Josephine County seat to *Grants Pass* in 1886. The squares are the township survey grid.

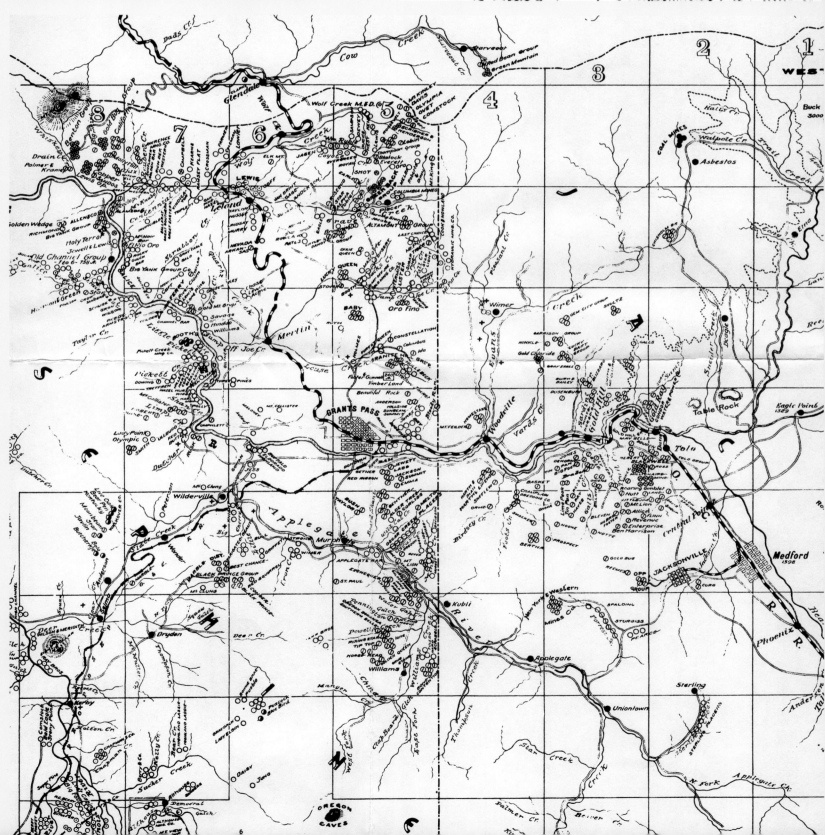

MOVING NORTH

There were few EuroAmerican settlers north of the Columbia when Oregon Territory was created in 1848, and those who did live there were concentrated in the Lower Cowlitz Valley, essentially still a tributary to the trade of the Willamette. The lumber resources of North Oregon and the fine navigable body of water with many natural harbors soon attracted entrepreneurs, and the unpreempted land available around Puget Sound attracted those of agricultural bent. But getting to the sound was not easy and at first involved trusting one's family to Indian canoes up the Cowlitz, which discouraged many a potential settler.

Five families founded New Market, or Tumwater, on the Deschutes River, in November 1845 to take advantage of water power from falls on the river. They were assisted by Hudson's Bay Company traders from Fort Nisqually on orders, contrary to company policy, from John McLoughlin. In October the next year, Edmund Sylvester and Levi Smith claimed land where the Deschutes met Puget Sound, on Budd Inlet, where lumber could be shipped—though in fact the tidal range there made that choice a poor one, as ships might ground on an ebb tide and half the townsite could be flooded at high tide!

Nevertheless, Sylvester's town (Smith was drowned in 1848) was destined for higher things. At a celebration dedicating the town—which was to be called Smithfield—in 1850, Isaac Ebey, a new settler on Whidbey Island, read some poetry: "Olympia's gods might view with grace / Nor scorn so fair a dwelling place," and Olympia it was. The settlement went on to become the customs house for North Oregon in 1851, and when Washington Territory was created in 1853 the first governor, Isaac Stevens, named Olympia as the capital. It had four hundred inhabitants (Map 155, *overleaf*).

Nearby, Steilacoom was established in 1849 following the location there of an Army garrison intended to control the Indians. The settlement around the fort was energized in 1852 when merchant and shipowner Lafayette Balch built a sawmill and turned the place into a major exporter to his sales yard in San Francisco (Map 154, *overleaf*).

Map 152 (*right* and *inset right*).
The top part of the *Diagram of a Portion of Oregon Territory*, published in 1852 by the surveyor general of Oregon. This was the last year that the area shown would be part of Oregon Territory; the following year Washington Territory was created, reflecting the demands of an increasing population north of the Columbia. *Cowlitz Farms*, or Cowlitz Landing, was the location of the first territorial convention, in August 1851; its owner, Edward Warbass, had renamed the place Warbassport, but the name did not appear on any official map. *F^t. Nisqually, Steilacoom,* and the principal population center, *Olympia,* are shown, but neither Tacoma nor Seattle yet appear. The northward extension of this map (inset, at smaller scale) shows no settlements; the first settlement at *Bellingham Bay* was in 1852, the year this map was published. Note the position of the international boundary (see Map 115, *page 51*).

Seattle

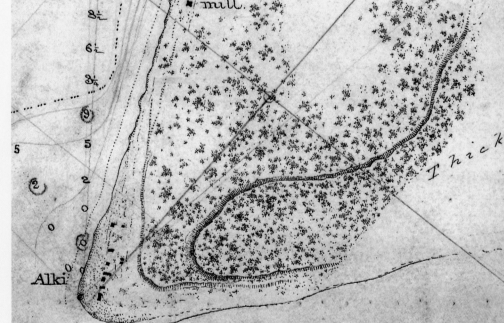

mill

Alki

Thick

MAP 153 (*below, right*), with details (*above* and *right*).
The tiny village of *Seattle* sits on a point on the east side of
Duwamish Bay—named Elliott Bay by the U.S. Exploring Expedition in 1841 and later to reassume this name—on this original 1854 field map from the U.S. Coast Survey (see page 74).
Across the bay is *Alki* (at the very bottom of the small-scale
whole map), the first site claimed by John Low and Lee Terry
in 1851 and continued by Terry after the others left for Seattle.
Note the *mill.* The estuary of the Duwamish River is at right on
the small-scale map. The Coast Survey noted that "the town
of Seattle is on a small point at the NE. part of the bay . . . it
consists of a few houses and stores, and a small saw-mill. It has
but little trade."

Before long, settlement spread farther north. David Denny and John Low, whose families had traveled the
Oregon Trail, scouted north to Puget Sound in 1851, meeting Lee Terry on the way. The latter two staked land to
claim on Alki Point at the entrance to Elliott or Duwamish
Bay (MAP 153, *above*) in September 1851, and Denny sent
word to his older brother Arthur for the rest
of the families to join them. They arrived on
a schooner on 13 November. No cabins were
finished on their arrival, prompting much
discouragement, especially given the time
of year. Cabins were soon built, however,
and in what one writer termed "a burst of
grandiloquent optimism," Denny named
the settlement New York. However, the location proved far too exposed for ships loading
lumber, and in February 1852 Denny, Carson
Boren, and William Bell selected flat land
inside the bay to relocate their community,
naming it Duwamps. Denny, so the story
goes, borrowed his wife's clothesline and

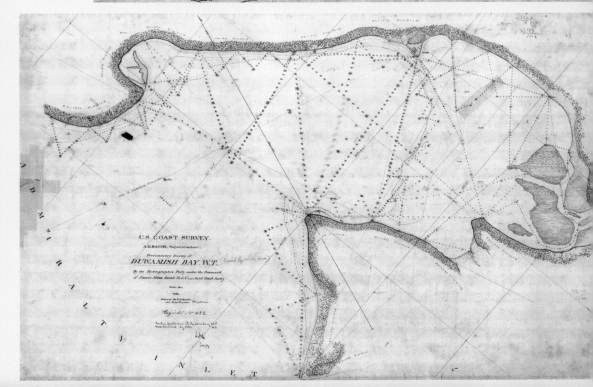

U.S. COAST SURVEY.
A.D. BACHE, Superintendent.

Preliminary Survey of
DUWAMISH BAY. W.T.

By the Hydrographic Party under the Command
of James Alden Lieut. U.S.N. Asst. Coast Survey

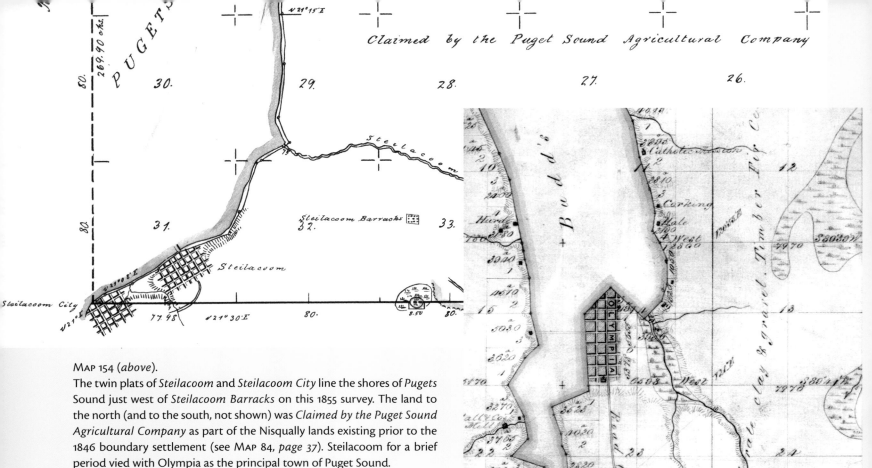

MAP 154 (*above*).
The twin plats of *Steilacoom* and *Steilacoom City* line the shores of *Pugets* Sound just west of *Steilacoom Barracks* on this 1855 survey. The land to the north (and to the south, not shown) was *Claimed by the Puget Sound Agricultural Company* as part of the Nisqually lands existing prior to the 1846 boundary settlement (see MAP 84, *page 37*). Steilacoom for a brief period vied with Olympia as the principal town of Puget Sound.

MAP 155 (*right*).
Olympia and *Budd's* Inlet, shown on an 1854 survey map, with several roads, including one from *Olympia to Nisqually.* Olympia had been named the territorial capital the year before this survey.

tied a weight on the end to carry out soundings of the bay to ensure ships would be able to dock at his chosen site. Families began relocating from Alki Point that April. In October, Henry Yesler built the first steam-powered sawmill on the sound, but early Seattle was largely eclipsed by other settlements on the sound, especially Steilacoom and Olympia.

A number of other settlements on Puget Sound had their beginnings in the 1850s. The Donation Land Claim Act was expected to create farms, but in many cases on the Sound it instead begat cities—or attempts to create cities. Many settlers platted cities that never went beyond the planning stage, activity that was repeated during the real estate boom following the coming of the transcontinental railroad (see page 82).

What would eventually become one of the largest cities on the Sound, Tacoma, was first settled in 1852 when Nicolas Delin, a Swedish immigrant, built a sawmill near the head of Commencement Bay, but the small settlement that grew up around it was abandoned by 1856 because of fear of Indian attack.

Port Townsend was founded in 1852 precisely at the point where its promoters, Francis Pettygrove, from Portland, and Loren Hastings, from San Francisco, calculated that the wind pattern would create a good place for inbound ships to stop. Their idea was to create a processing and loading facility for the export of lumber.

Across Admiralty Inlet, Henry Roeder, a German ship captain, and his associate Russell Peabody had the same idea later the same year. Their selection of location, however, proved problematic. Their

choice, Whatcom Creek in Bellingham Bay, had an irregular flow, and the large tidal range left their settlement, Whatcom, half a mile from deep water. Settlement in Bellingham Bay was saved by the discovery of coal. Edmund Fitzhugh claimed land south of Whatcom and platted a city he called Sehome (MAP 158, *far right*) on land he claimed after he found coal in the roots of an upturned tree. The location turned out to be navigable for ships, and soon Fitzhugh's Bellingham Bay Coal Company was exporting coal south to Portland and San Francisco.

Although Indian attacks had a chilling effect on EuroAmerican settlement on Puget Sound, it was not to be stopped. Free land, cheap land, apparently boundless lumber, and a body of water so preferable to that of the Columbia ensured continued population growth. By 1853 North Oregon counted 3,965 non-Indian inhabitants, and by 1860 Washington Territory had grown to 11,594, almost all west of the Cascades.

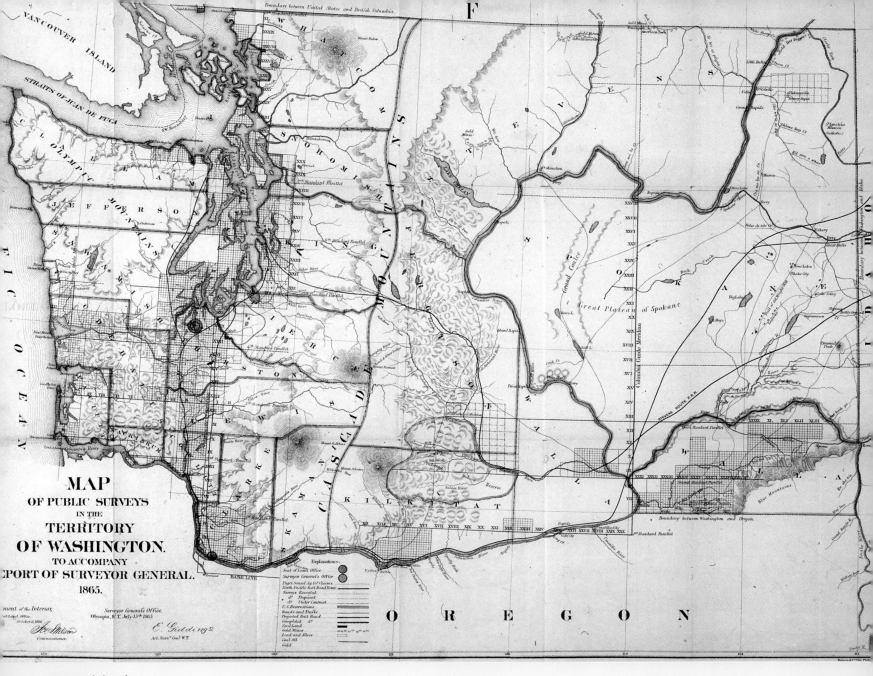

Map 156 (above).

Washington Territory in 1865, on a map showing the progress of the public survey system as of that year. *Monticello*, the location of the second territorial convention in 1852, is shown at the mouth of the *Cowlitz River*, in the location of today's Longview. The map shows a proposed railroad from the *Columbia River* at *Vancouver* to *Steilacoom* on *Puget Sound*, and another along the banks of the Columbia, then branching into two south of Olympia, with lines to *Discovery Bay*, on the *Straits of Juan de Fuca*, and to *Seattle*. A proposed transcontinental line terminates in Seat-

tle; this is the *Stevens Route P.R.R.*, the line recommended by Isaac Stevens from the Pacific Railroad Surveys of 1853 (see page 80). The only railroads in operation at this time were the portage lines along the Columbia (see page 76). Note the *Yakima Indian Reserve*. Settlements at *Whatcom*, *Sehome*, *Port Townsend*, and *Port Madison* (on *Bainbridge I[slan]d*) are shown on Puget Sound and *Admiralty Inlet*. In the interior, *Walla Walla* is shown, and a *Great Falls* and *Ferry* on the *Spokane River* at the location that would become Spokane.

Map 157 (below).

Fort Bellingham, built in 1856–58 by U.S. Army troops under George Pickett (of later Civil War fame) to counter the threat of Indian attacks. *Inset* is a photo, taken somewhat later, of one of the fort's blockhouses.

Map 158 (below).

Edmund Fitzhugh and partners' 1858 plat of *Sehome* in Bellingham Bay; the area is now part of Bellingham (see page 138).

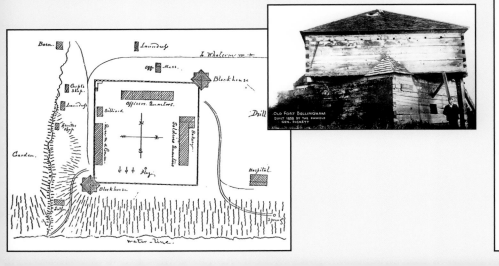

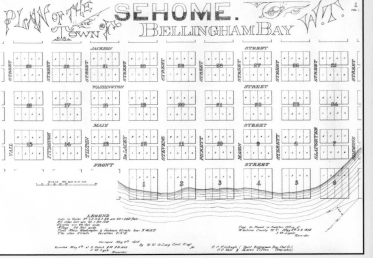

SURVEYING THE COAST

The United States Coast Survey was established in 1807 and charged with surveying the nation's coast to make it safe for shipping. When the United States acquired Oregon in 1846 and California two years later, the Coast Survey was suddenly faced with thousands more miles of coast requiring survey. The California Gold Rush in 1849 made the California coast a priority, but not long after, the survey was continued northward to encompass the entire West Coast, including Oregon, at least in a preliminary way, by 1853 (MAP 163, *below, center right*).

It is not surprising that the first priority north of California was the mouth of the Columbia River, with its treacherous sandbars across its entrance that continued to claim ship after ship. The mouth of the Columbia was surveyed in detail and maps published in 1851 (MAP 159, *right*). The estuary area had to be continually resurveyed because the sandbars kept shifting. Puget Sound, the new, more reliable harbor for the Northwest, soon followed (see the map of Duwamish Bay, 1854, MAP 153, *page 71*).

As is well illustrated by MAP 160, *below,* hydrographic mapping often involved much laborious sailing back and forth, taking manual soundings at close regular intervals. Triangulation was used to maintain the survey accuracy over long and not directly measurable distances from an initial carefully measured baseline (MAP 161, *below, right*).

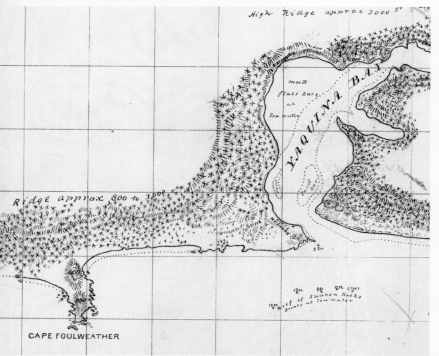

View of the Entrance of Columbia River—Cape Hancock or Disappointment E. by N. (Compass) 12 Statute Miles.

The Coast Survey became the U.S. Coast and Geodetic Survey in 1878 after it was given the added responsibility of mapping inland areas. The coasts slowly became safer for mariners, and the charts—even modern digital ones—used today are direct descendants of these early charts, which have built on each other cumulatively over time.

MAP 159 (*left, top*).
One of the top priorities for the Coast Survey was the treacherous mouth of the Columbia River. This is the earliest of its surveys of the area and shows in some detail the channels that ships needed to use to get into to the river. *Left, center* is a view of the entrance viewed from the sea.

MAP 160 (*far left, bottom*).
Part of the 1869 original field survey of *Yaquina Bay* and *Cape Foulweather*, on the Oregon Coast. The latter was Captain James Cook's first point of contact with North America in 1778, on his third voyage of discovery (see page 21).

MAP 161 (*left, bottom*).
Another pen-and-ink original field map illustrates the considerable amount of work that making a hydrographic survey map sometimes involved. The lines of numbers record depths as the survey ship sailed back and forth. This survey was of *Young's Bay*, just west of Astoria, at the Columbia's mouth, and was carried out in 1889. Note (in red ink) the *Line of Proposed R.R. Bridge.*

MAP 162 (*right, top*).
Survey methods involved triangulation on a large scale. Prominent landmarks such as mountain peaks and headlands were usually used for this. From the 1886 USCGS annual progress report comes this map of Puget Sound and vicinity showing the complex triangulation network.

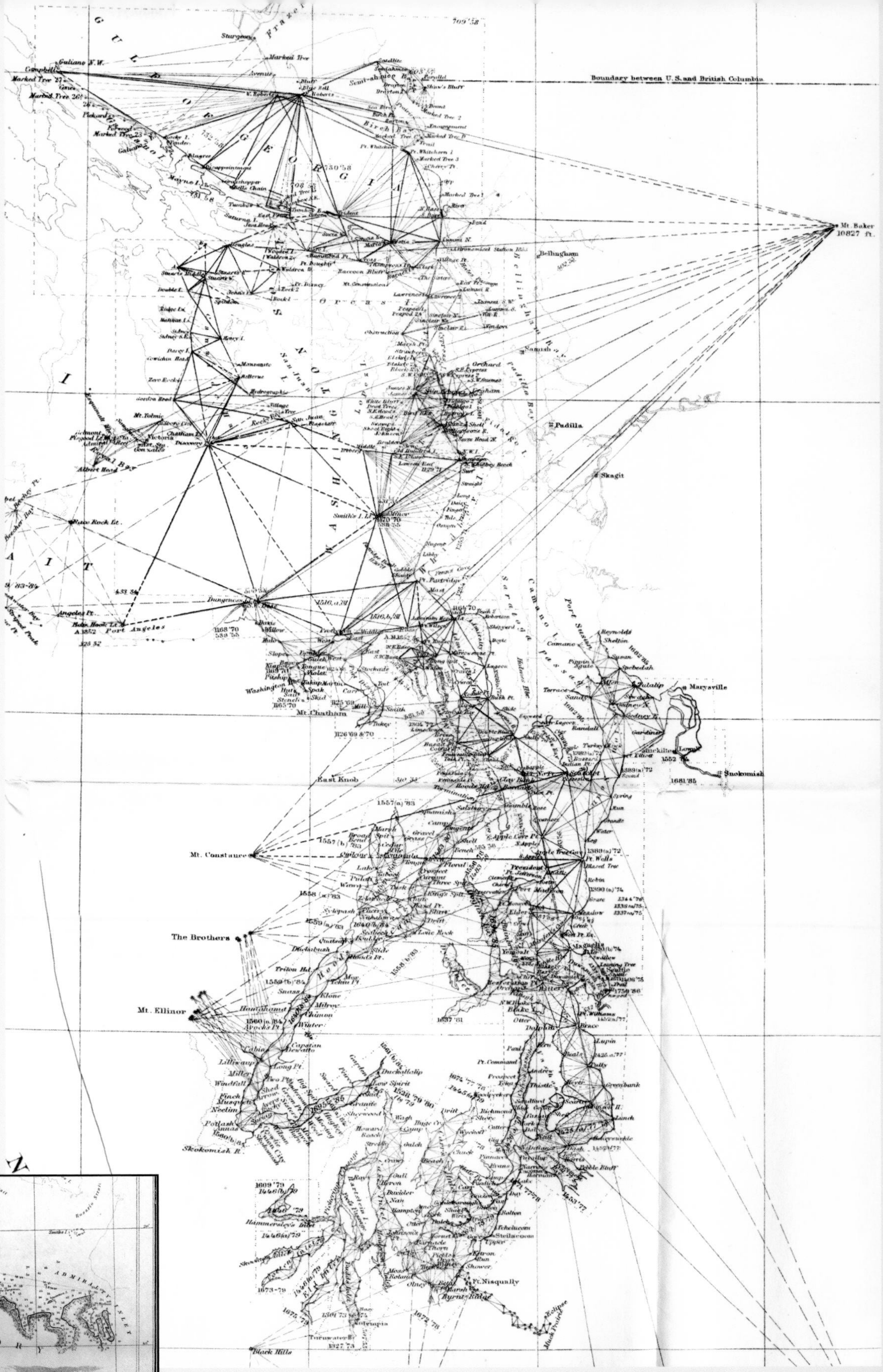

MAP 163 (*below*).
In 1853 the Coast Survey completed its initial survey of the entire West Coast and published it in several sheets, complete with coastal views to aid navigation. This is a detail from the northernmost sheet and shows *Gray's Harbor, Cape Flattery*, and the *Straits of Juan de Fuca.*

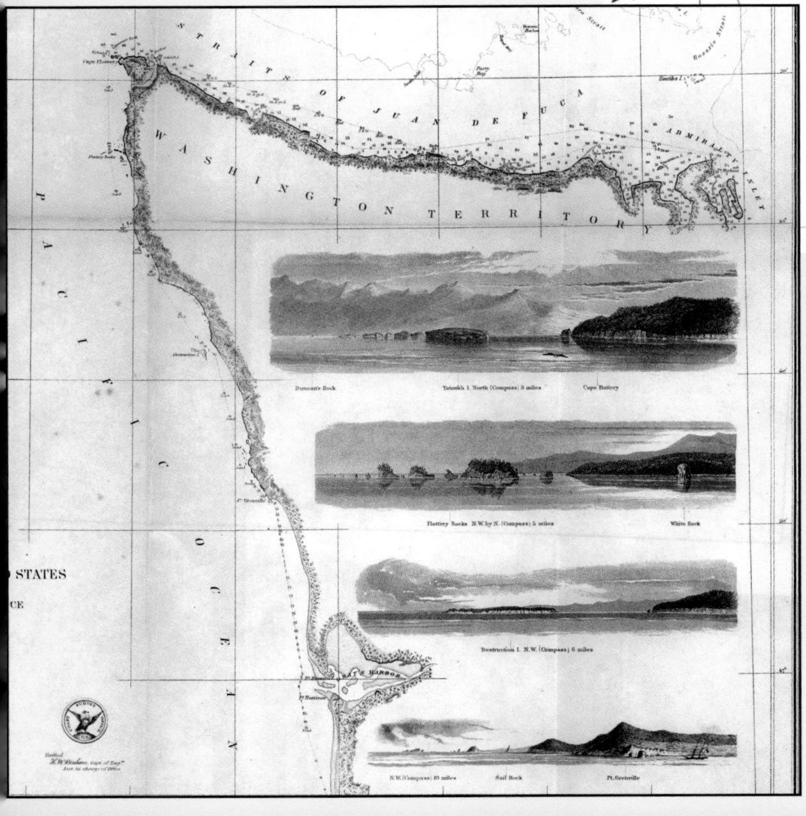

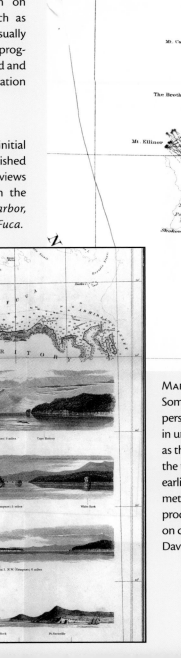

MAP 164 (*right*).
Sometimes Survey personnel were involved in unusual projects such as this plaster model of the topography, and the earliest offshore bathymetry, of the West Coast, produced in 1887 based on data from George Davidson.

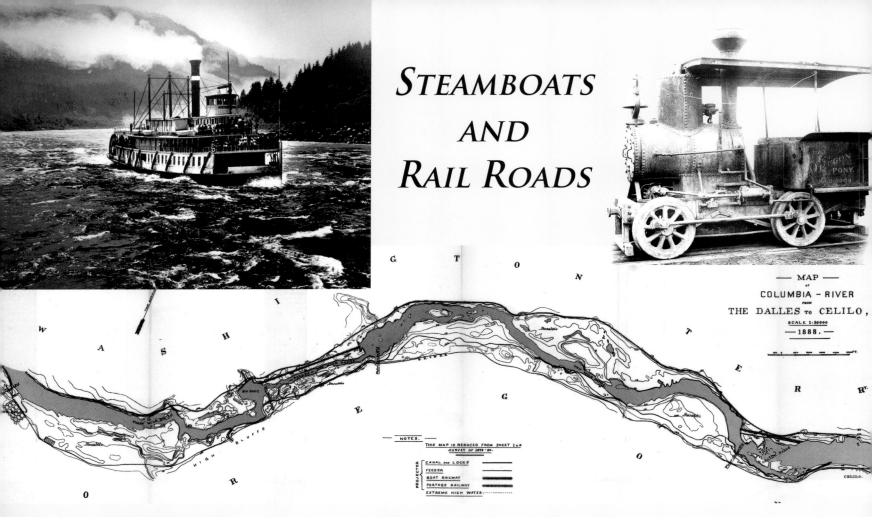

STEAMBOATS
AND
RAIL ROADS

— MAP —
OF
COLUMBIA - RIVER
FROM
THE DALLES TO CELILO,
SCALE 1:30000
— 1888. —

In a new country where roads were at best muddy trails and at worst, nonexistent, it was only natural that commerce would take to the rivers. Steam power made it possible to make way counter to the river's current, and, later, even for an experienced captain to traverse some rapids under the right conditions.

The first steamboat in the Northwest was the Hudson's Bay Company's little *Beaver*, which appeared at Astoria in 1836, though

it was soon found to have too deep a draft to ply the river. The *Columbia,* built in Astoria, was the first steamboat to regularly ply the waters of its namesake river, providing a service from Oregon City to the lower Cascades from 1850 to 1853. Beginning in 1855, side-wheelers gave way to more efficient stern-wheelers.

In 1860, the Oregon Steam Navigation Company was formed and organized a fleet of steamboats on regular schedules, creating an

MAP 165 (*above*).
The multitudinous rocks and rapids of the middle Columbia are well shown on this map, drawn in 1879–80 but published in 1888. The map appeared in a report by the U.S. Army Corps of Engineers analyzing the feasibility of various canals, portage railroads, and even a boat railway, both at The *Dalles or Five Mile Rapids* and *Celilo Falls.* The boat railway would have winched boats on rails up an inclined plane past the rapids, thus eliminating the need for the time-consuming transshipment of freight; the proposed track locations are shown by the thicker crosshatched lines (see key). But a boat railway was expensive to build, with a projected cost of $1.4 million; these projects were shelved in favor of the Celilo Canal, completed in 1914.

MAP 166 (*right, center*) and MAP 167 (*below, center*).
These two maps are details of part of the map of Washington Territory in 1865, MAP 156, *page 73.* They show the location of the two portage railroads operating around the rapids and falls of the Columbia River at that time. MAP 166 is the *Cascades Rail Road,* on which the *Oregon Pony* (*above, top right*) operated between 1862 and 1863. MAP 167 is the portage *Rail Road* built on the Oregon side between The *Dalles* and *Celilo.* Thirteen miles long, it was completed in 1863 and is now the oldest part of the Union Pacific system, pre-dating even the first transcontinental line. Note that the lines shown on the north bank of the Columbia on this map are only projected lines at this time. The photograph at *right* shows the Oregon Steam Navigation's locomotive *J.C. Ainsworth,* shipped to Oregon in 1863, with a train on the portage line from The Dalles to Celilo in 1867. MAP 168 (*far right*) is an 1880 map detail of the location of the photograph.

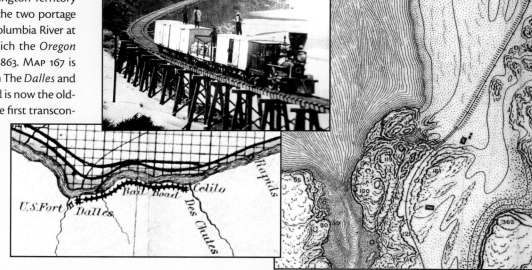

Left.
An ad for the Oregon Steam Navigation Company's services from an 1877 Portland directory. The illustration depicts both steamer and railroad.

Far left, top. The stern-wheeler *Bailey Gatzert* approaches the Cascade Locks about 1901.

MAP 169 (*right*).
A *Portage R.R.* is shown on the north bank of the Columbia at the Cascades on this 1887 map. This was the portage line of the Cascades Railroad, which operated with steam power from 1863 to 1896. *Inset* is a photograph of the line, with steamboat in the background, in 1867. *Cascades Locks,* which would replace the railroad in 1896, was being planned at this time. On the south bank, the Oregon Portage Railway has been incorporated into the Oregon Railway & Navigation Company's line, built in 1879 and at first an extension of the portage railroads along the south bank of the Columbia (see page 90).

effective monopoly by undercutting individually operated boats and then buying them up. Boats and by the mid-1860s steamboats had reached as far inland as Lewiston, Idaho, and allowed miners speedy passage to the new gold mines of Montana and Idaho. Steamboats accessed the middle Columbia (above Cascade Rapids) and the upper river (above Celilo Falls), and the upper Willamette (above Willamette Falls) with difficulty; boats either had to be winched up rapids or were disassembled and rebuilt above the obstructions. Passengers and freight traversed the rapids first using portage roads and then portage railroads. These railroads, illustrated in the maps on this page, at first used horse-drawn wagons and then, beginning in 1862, steam locomotives, the first being the diminutive *Oregon Pony.*

Canals and locks slowly replaced portage railways. The Oregon State Portage Railway between The Dalles and Celilo was completed in 1905 as an interim measure until the Celilo Canal was completed in 1914. The future, however, belonged not to waterways but to railroads; by 1908 there were continuous rail lines on both banks of the Columbia (see page 168).

Above. The sternwheeler *Dalles City* unloads freight on the middle Columbia about 1910.

MAP 170 (*below*).
A U.S. Army Corps of Engineers map showing the planned Cascade Locks. The map appeared in a government report in 1884. The canal and locks were built between 1893 and 1896. Note the north bank *Narrow Gauge Portage Railway* (the Cascades Railroad, also shown on MAP 169) and the south bank *O.R.&N.C^os. Railway.*
Inset, left. Steamboats in the Cascade Locks, as depicted on a postcard from about 1910.

MAP 171 (*below*).
The Oregon State Portage Railway was built between 1903 and 1905 and operated until replaced by the Celilo Canal in 1914. This map shows the line running into *Celilo,* with its *cannery* and *cold storage.* In many places, as here, the state road paralleled the *O.R.&N.* The State Portage Railway lasted until 1915, when the 8.6-mile-long Dalles–Celilo Canal and Locks were completed.

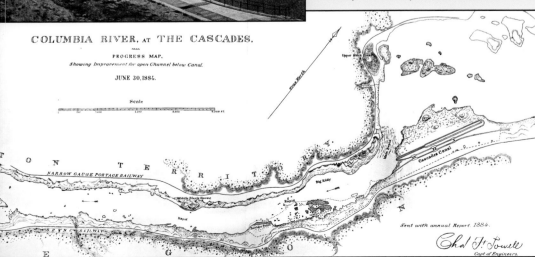

Columbia River Steamers in Cascade Locks. 90

COLUMBIA RIVER, AT THE CASCADES.
PROGRESS MAP.
Showing Improvement for open Channel below Canal.
JUNE 30, 1884.

Sent with annual Report, 1884.

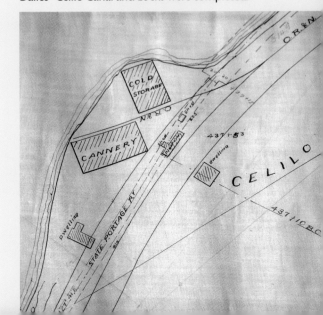

MILITARY WAGON ROADS

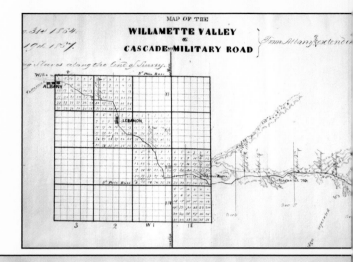

The first roads were but trails, paths created by the passage of travelers themselves. Some of the more used trails were made into roads, improved enough to allow for the passage of horse-drawn wagons, but not likely without travail, especially in the winter rains. The Barlow Road, the connection of the Oregon Trail to the Willamette Valley (see page 53) was one of the first such long-distance wagon roads.

In the latter case the territorial government had paid roadmakers to construct the road. In local cases groups of settlers might get together to improve a trail to wagon road standard for their common good. Between 1865 and 1869 Congress gave land grants to the State of Oregon to finance companies that would build military wagon roads, theoretically to facilitate the movement of the U.S. Army to deal with Indian raids. In reality the military use of the roads was minimal, but they

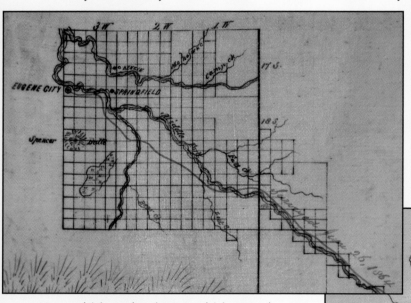

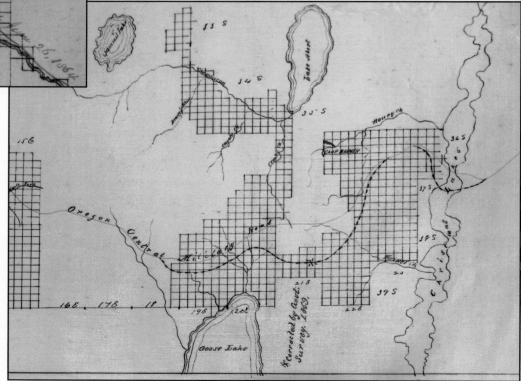

MAP 172 (*right, top*) and MAP 173 (*right, center*).
Two 1867 maps, both with the same title, showing the *Willamette Valley & Cascade Mt. Military Road*. MAP 172 shows the road's route between *Albany* and *Lebanon* and eastward; land to be claimed in the Willamette Valley is shown. MAP 173 shows the road's eastern end at the confluence of the *Malheur River* and the *Snake River*.

MAP 174 (*above* and *right*).
Two parts of the same map, drawn in 1869, showing the route of the *Oregon Central Military Road* in the vicinity of *Eugene City* and *Springfield* (*above*) and in south-central Oregon (*right*); the U.S. Army's *Camp Warner* is shown. Both parts of the map show land surveys ready to facilitate land grant claims.
Above, top. Old wagon on display at the Columbia Gorge Discovery Center at The Dalles.

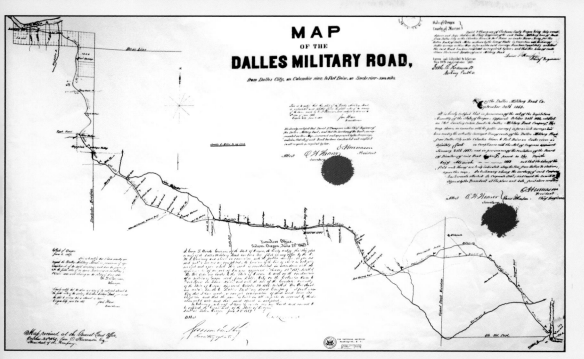

MAP 175 (*left*).
The signed and sealed map showing the survey of the Dalles Military Road from *The Dalles*, at top left, to the Snake River near the *Malheur R.*, at bottom right. The document includes certification by the president of the Dalles Military Road Company attesting that the location of the road is correctly shown.

MAP 176 (*below*).
The *Coos Bay Wagon Road* is shown east of *Coos City* on the Oregon Coast on this somewhat damaged segment of an 1873 map. The wagon road ran east to Roseburg. A *Coal Mine* and associated *Railroad* are just visible at top left.

did encourage settlement. However, the process was often a sham. Road construction companies sometimes received large amounts of land in return for relatively little work.

The first military road was the Oregon Central Military Wagon Road, between Eugene and the Snake River (MAP 174, *left, bottom*). The road was approved in 1865 and certified complete in 1870 but was actually completed two years later. The road provided access to southern Oregon, where in 1866 the Army built Camp Warner as a base to prevent Indian raids. The road took a 420-mile circuitous route designed to encompass the most valuable land for the land grant. The road's route included Stone Bridge, a causeway across marshy land in southern Oregon, built by the Army in 1867.

Other long-distance roads built in this period to receive land grants included the Willamette Valley and Cascade Mountain Military Wagon Road, completed in 1868, which ran east from Albany across the middle of the state (MAP 172 and MAP 173, *left*); and the Dalles Military Road (MAP 175, *above*), completed in 1869, which ran southwest from the Columbia at The Dalles to Boise via Canyon City and the eastern Oregon gold fields (see page 102). In addition, two shorter land grant roads were built: the Corvallis and Yaquina Bay Wagon Road, from Corvallis west to the coast at Yaquina Bay (the grant is shown on MAP 177, *below*), and the Roseburg and Coos Bay Wagon Road, which similarly ran west from Roseburg to Coos Bay (MAP 176, *above right*).

A few people made a lot of money from military road construction. Later some of the land grants were sold to land developers, then known as colonization companies, which then sold land to a new wave of settlers (see MAP 384, *page 157*).

MAP 177 (*below*).
Part of a large 1887 map showing the land grant of the Oregon & California Railroad (MAP 222, *page 94*) also shows the *C. & Y.B. Military Road Grant* and the *W.V. & C.M. Military Road Grant*, the land grants of the Corvallis and Yaquina Bay; and Willamette Valley and Cascade Mountain military roads, respectively. As with the railroad grants, not all land within the zones was granted—only up to three sections per mile in the three-mile-wide zone.

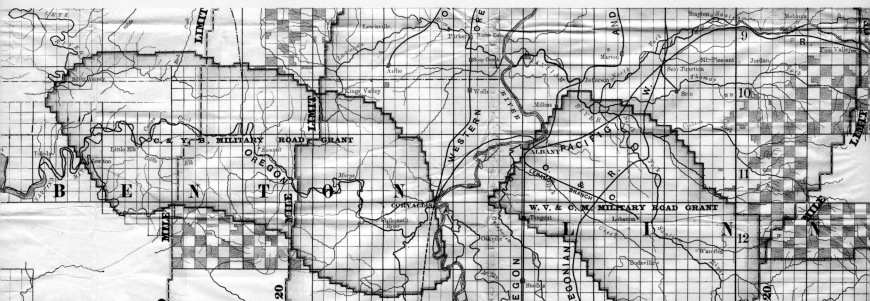

THE RAILROAD COMES TO THE NORTHWEST

The Northwest's first railroad was a simple horse-drawn wooden tramway built in 1846 to connect Abernethy Island with Oregon City, and a number of such lines complemented portage roads on the Columbia, where the first steam-powered trains had run beginning in 1862 (see page 76). The key to the growth and development of the Northwest was, however, a connection to the already extensive network of rail lines east of the Mississippi—the transcontinental.

As early as 1832 the editor of the Ann Arbor *Western Emigrant*, Samuel Dexter, had published a proposal "to unite New York and the Oregon by a rail way," correctly noting that it would "increase the value of the public domain." But it is Asa Whitney, a Connecticut businessman who traded with China, who is generally credited with the transcontinental railroad concept. Following a commercial treaty between China and the United States signed in 1844, Whitney in 1845 proposed a $15 million line to the Pacific funded by a sixty-mile-wide land grant. Principally because the rest of the West was still part of Mexico, Whitney's proposed line ended at Puget Sound. Whitney's forward-looking ideas were sidelined by the sectional bickering leading up to the Civil War, a period during which Congress would never have reached an agreement for a northern line.

In an attempt to avoid sectional interests by finding the technically best route, Congress in 1853 authorized the Pacific Railroad Surveys, a reconnaissance of several route possibilities—northern, southern, and central—principally carried out by U.S. Army topographical engineers.

The northern survey was carried out beginning in June 1853 by the newly appointed governor of the just-created Washington Territory—Isaac Ingalls Stevens, previously with the U.S. Coast Survey—as he made his way west to take up his appointment. Stevens explored west from St. Paul, and Captain George B. McClellan (later a Civil War general)

MAP 178 *(below)*.

The Pacific Railroad Survey map of the Pacific Northwest showing recommended routes to the coast, the first following the north bank of the Columbia River and the second up the valley of the *Yakima River* and across the Cascades at *Snoqualmie Pass*. The map was published in 1859. The former route would much later be used by the Spokane, Portland & Seattle and the latter by the Chicago, Milwaukee, St. Paul & Pacific (the Milwaukee Road), although the Northern Pacific's Cascade Branch would follow the Yakima Valley before traversing Stampede Pass, a few miles south of Snoqualmie Pass. Note that the proposed coastal routes end at *Seattle* and at *Discovery Bay*, on the *Strait of Juan de Fuca*. This map seems to be the first time Mount Adams appeared on a map in its correct location; the name was suggested by Hall Jackson Kelley as one of the summits of his Presidents' Range (the Cascades) but for a different location (see MAP 87, *page 38*).

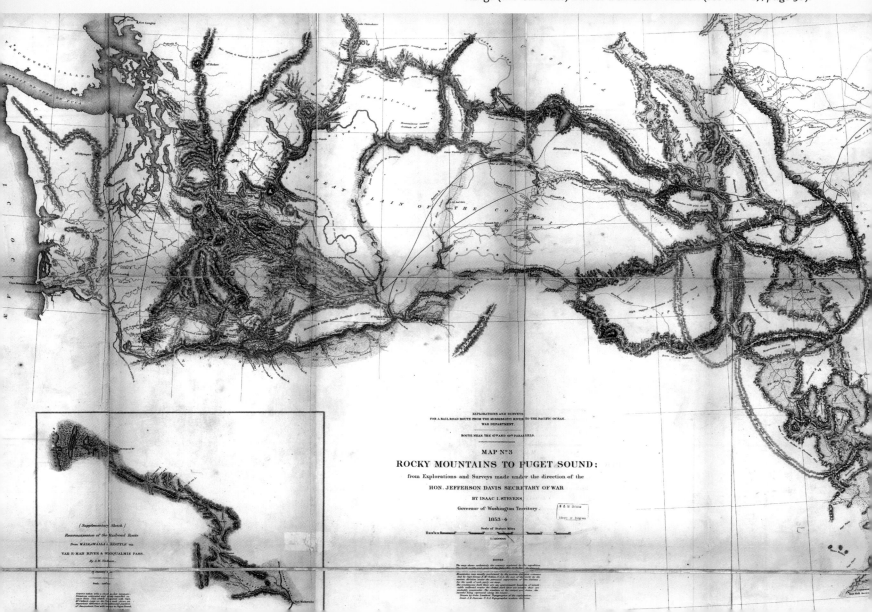

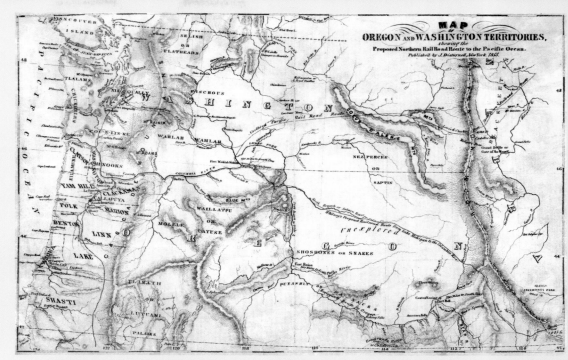

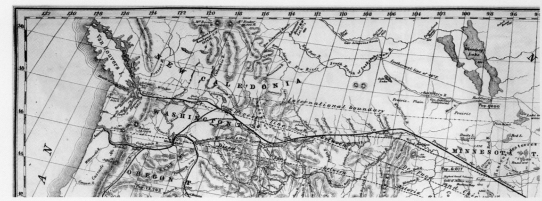

MAP 179 (*above, left*).

Asa Whitney's proposed route for a transcontinental rail line to Puget Sound is shown on this 1849 version of his map; after the U.S. acquisition of California Whitney added a branch to San Francisco Bay.

MAP 180 (*above, right*).

This map published in 1855 by commercial map-maker John Disturnell shows a *Northern Pacific Rail Road Route* approximating the line eventually built by the Northern Pacific but ending at *Seattle*. Also shown is *Whitney's Proposed Railroad Route* via *South Pass* and following the valley of the *Salmon River*, a tributary of the Snake.

MAP 181 (*right, center top*).

Railroad engineer Edwin Ferry Johnson published his ideas for routes to the Pacific in the *American Railroad Journal* in 1853, before the results of the Pacific Railroad Surveys were known, with the proposed route superimposed on dashed great circle lines representing the shortest distances between *Astoria and Chicago* and *De Fuca and Chicago*. Fourteen years later Johnson was appointed chief engineer of the Northern Pacific.

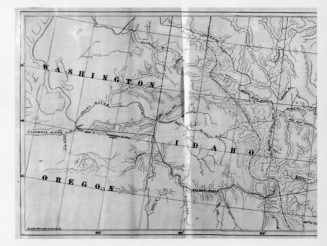

MAP 182 (*right*).

The western portion of Edwin Johnson's planning map of the proposed route of the Northern Pacific to the Columbia River, produced in 1870. The route, shown in red ink, follows the Salmon River before running due west to Wallula; no route is yet indicated beyond that point.

MAP 183 (*right*).

When Jay Cooke began selling Northern Pacific bonds in 1870, this map accompanied the prospectus. The map bears the stamp of a Portland real estate agent and a handwritten notation, "Property of D.S. Baker, Walla Walla." Dorsey Syng Baker, who owned this map, was at the time proposing to build the Walla Walla & Columbia River Railroad (see page 84) to connect Walla Walla to the steamboat service on the Columbia. The proposed routes here differ markedly from those finally built: the main line of the Northern Pacific seems to pass directly through *Walla Walla* on its way to *Portland*; the line to *Puget Sound*, which was from the beginning intended as only a branch, separates from the main line in *Montana*. Note that the map shows neither the Rockies nor the Cascades—why give potential investors any cause for alarm?

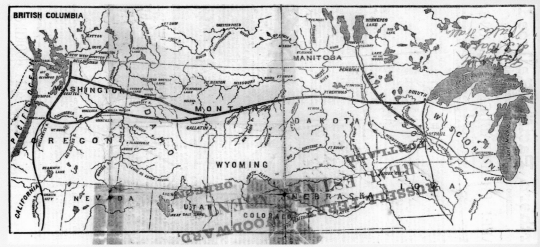

explored from the west, trying to find passes across the Cascade Mountains. Stevens, of course, wanted his northern route to be the one chosen and produced a glowing report that detailed five passes over the Continental Divide. McClellan found nothing, having refused to test the snow depth in Snoqualmie Pass, the one actually suitable for a railroad; the snow was not as deep as it appeared. This pass was later used by the Milwaukee Road. The Washington Territory legislature, suspicious of Stevens's report, commissioned a civilian engineer, Frederick West Lander, to survey another route to Puget Sound. His work, in which he found a feasible route to the Northwest west of South Pass, was included in the final edition of the Pacific Railroad Survey Reports, a massive twelve-volume summary of all the survey results, the last of which was not published until 1861.

THE NORTHERN PACIFIC

President Abraham Lincoln signed the act creating the Northern Pacific Railroad in 1864, giving the company land grants and a right-of-way to build a railroad and telegraph from Lake Superior to "some point on Puget Sound." Little was accomplished until 1869, when Jay Cooke, famous for financing much of the Union effort during the Civil War, became the fiscal agent for the Northern Pacific. Bonds were issued in 1870 (MAP 183, *previous page*), but they had lots of competition and did not sell well. By 1873 the line had reached the Missouri at Bismarck, Dakota. Construction also began at Kalama, on the north bank of the Columbia just west of Portland, in 1870, building north toward Puget Sound.

Later in 1873 the railroad announced that Commencement Bay was to be its terminus, much to the chagrin of Seattle and Olympia. But the rails almost did not make it to Puget Sound at all, for Jay Cooke became insolvent—there was a general financial collapse in 1873, and Cooke had previously lost anticipated German sources of funding because of the Franco-Prussian War—dragging the Northern Pacific toward bankruptcy. Twenty-two miles remained to be completed to reach New Tacoma, the settlement that had sprung up on Commencement Bay in anticipation of the railroad's arrival. Ten more miles were completed before work stopped because the tracklayers had not been paid. The day was saved by one-time steamboat captain John C. Ainsworth of the Oregon Steam Navigation Company, now the railroad's Pacific manager. He provided a personal loan;

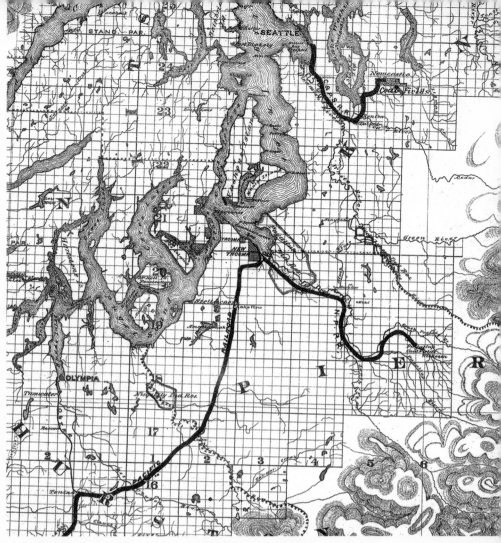

MAP 184 (*above*).
The line of the Northern Pacific into *New Tacoma* is shown on this 1881 map. Also shown is the Northern Pacific extension to the mines at *Puyallup Coal Fields Wilkeson*, and the Seattle & Walla Walla line from Seattle to the *Newcastle Coal Fields*. The latter line is labeled *C.&P.S.R.R.*, the Columbia & Puget Sound Railroad, the name it was given following its sale in 1880.

MAP 185 (*right*).
The southern end of the coastal Northern Pacific line, on another part of the same 1881 map. A gap remains between *Kalama* and the Oregon Railway & Navigation line into *East Portland*; a connection to Goble, across the river from Kalama, would be completed two years later.

the remaining track was laid, and the first train steamed into New Tacoma on 16 December 1873.

Seattle businessmen, not to be outdone by their rival city Tacoma, decided to build their own connection eastward and in 1875 incorporated the Seattle & Walla Walla Railroad (see next page).

No further work on the transcontinental line was possible; by 1875 the railroad was bankrupt, but the threat to the company land grant impelled the directors, led by Frederick Billings, to find a way out of the financial mess. Preferred shares were offered to bondholders, and a new company, also called the Northern Pacific Railroad Company, was created. This new company managed to finance a thirty-one-mile extension from Tacoma to coal mines at Wilkeson (MAP 184, *above,* and MAP 199, *page 88*), enabling it to earn revenue by exporting coal to San Francisco.

Seattle's Own Railroad

Most communities sidetracked by the railroad accepted their fate. But not Seattle. When the Northern Pacific decided on Commencement Bay as its terminus, a group of Seattle businessmen determined to build their own connection with the transcontinental and link their city with the up-and-coming eastern center of Walla Walla to boot. Their line was to be built across Snoqualmie Pass and in so doing demanded the land grant made to the Northern Pacific for its proposed Cascade Branch.

Construction began in May 1874, entirely with volunteer labor, and twelve miles of track were laid, but it was not until 1877 that the line reached the coalfields at Renton, finally reaching the coal mines of Newcastle (Map 201, *page 88*). But the line was undercapitalized and never got anywhere near its intended destination of Walla Walla. The railroad was sold to the Oregon Improvement Company, a subsidiary of the Oregon Railway & Navigation Company, in 1880 and renamed the Columbia & Puget Sound Railroad. Nevertheless, the creation of the railroad demonstrated to all that Seattle was a force to be reckoned with, and the Northern Pacific would extend its line to Seattle in 1884 and complete its Cascade Branch—securing its land grant—in 1887.

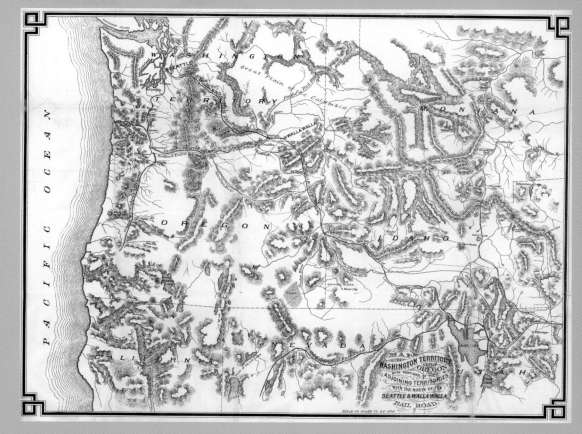

Map 186 (*above*).
The planned route of the *Seattle & Walla Walla Rail Road* is shown on this 1874 map. The railroad connects to the transcontinental Union Pacific line via a proposed *Portland, Dalles & Salt Lake Rail Road* beyond *Walla Walla*.

Map 187 (*below*).
The planned entrance into Seattle of the *Seattle & Walla Walla Rail Road* is displayed on this 1875 plat map of the city.

Map 188 (*below*).
The proposed route of the *Seattle and Walla Walla Railroad* in the Yakima Valley of southeast Washington is depicted on this 1878 map.

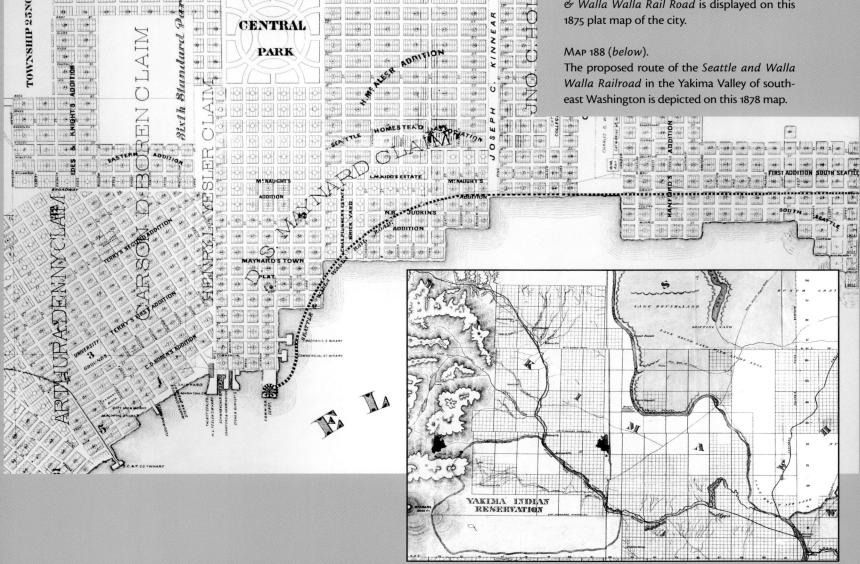

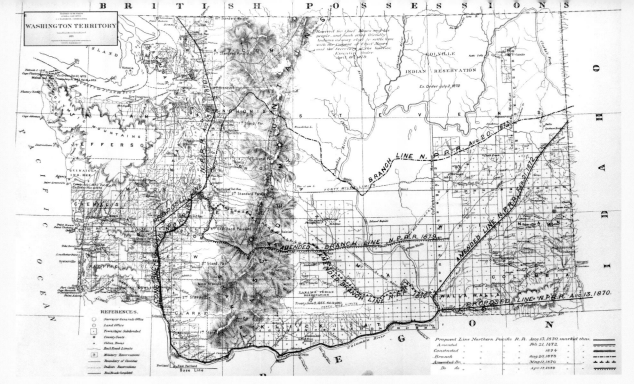

The Northern Pacific by 1879 had once again resumed building west from Dakota Territory under the leadership of Billings, and that year the railroad completed its first hundred miles west of the Missouri River. By the following year the line had reached Wyoming. But then the company's plans were derailed.

A German businessman and financier, Henry Villard, had acquired control of the Oregon Railway & Navigation Company (OR&N) in 1879. That company, which incorporated the Oregon Steam Navigation Company (see page 76) and the Walla Walla & Columbia River Railroad (see below), intended to connect the portage lines of the Columbia and build east from Portland. Villard realized that the Northern Pacific's direct line to Tacoma would bypass the OR&N sys-

tem and endanger Portland's preeminence in the Northwest. Thus, in 1881 he engineered a hostile takeover of the much larger Northern Pacific using a now-famous "blind pool," where the investors did not know what their money would be used for but trusted Villard to make good. Both the OR&N and the Northern Pacific came under the control of Villard's holding company, the Oregon & Transcontinental, and Villard could ensure that the transcontinental connected with the OR&N. The Northern Pacific's intended line along the north bank of the Columbia was eliminated.

The Northern Pacific resumed construction, building both westward and eastward, and on 8 September 1883 the lines met at Gold Creek, in western Montana. Although the path to the Pacific

Walla Walla's Railroad

Walla Walla's own railroad got off to an earlier start than Seattle's because of the dogged persistence of Dorsey Syng Baker. The farmers of the booming Walla Walla area wanted a railroad to transport their grain to the Columbia and had proposed building a railway to Wallula as early as 1861. The project, however, engendered much bickering over costs and proved beyond their means. Baker, a doctor-turned-businessman and owner of a Walla Walla financial company, determined to push the project through. He had the forty-six-mile route surveyed in 1871 and spent the next four years driving his concept to fruition. Built as a narrow-gauge line with iron-faced wooden rails

to save money, Baker finally had to force the residents of Walla Walla to subscribe to his road and hand over three acres of land for a terminal. The first train steamed proudly into Walla Walla on 23 October 1875. A branch to the south was begun in 1879. The Oregon Railway & Navigation Company purchased the line in 1881 and converted it to standard gauge.

MAP 190 (below).
The line of the Walla Walla & Columbia River Railroad from *Wallula*, on the Columbia, at left, along the valley of the *Walla Walla River*, is marked in red on this 1878 map. A planned branch southward leaves the main line at *Whitman*, and a proposed extension is shown north from *Walla Walla*.

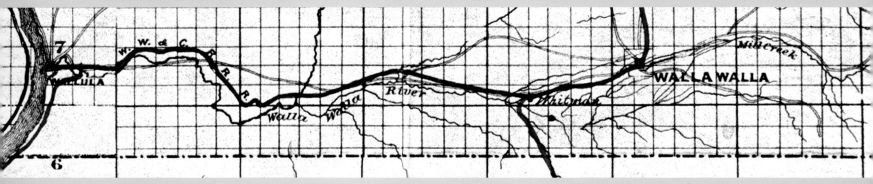

Map 191 (*right*).
The Northern Pacific's Cascade Branch is shown under construction on this map published by the railroad in 1882. The transcontinental line is complete on the section shown, though a gap still exists farther east. Also shown as under construction is the line from *Portland* to *Rainier*, though as built this line would run to Goble, just south of Rainier and opposite *Kalama*. At *Wallula J[unction]* the transcontinental begins to use the tracks of the Oregon Railway & Navigation Company, shown in blue. An Oregon & Transcontinental (a Villard company) branch from *Tacoma* to *Seattle* is under construction and would be completed in 1884. OR&N lines extend south down the Willamette Valley and southeast to *Baker City*, with an extension to the *Snake River* being indicated; the latter line would soon meet with the Union Pacific at Huntington (see page 91).

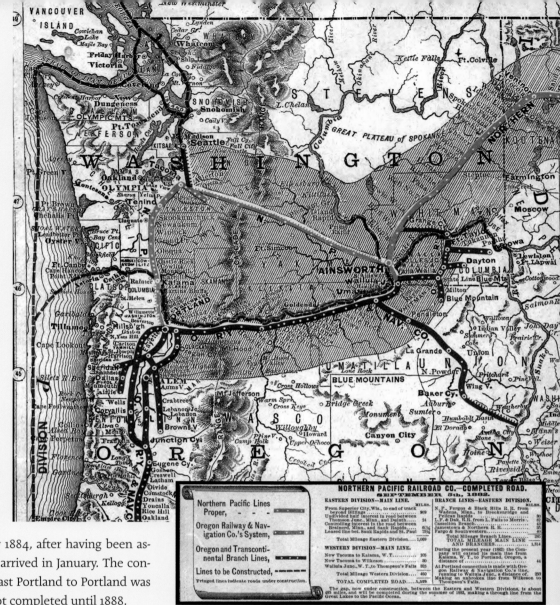

had been created, some details remained before a truly continuous transcontinental would exist. A bridge over the Snake River at Ainsworth (Map 192, *right, center*) was completed in April 1884. A line bridging the gap between Portland and Goble, across the Columbia from Kalama (Map 185, *page 82*), was completed in September 1883, but a train ferry was not in place until July 1884, after having been assembled in Portland from components that had arrived in January. The construction of a bridge over the Willamette from East Portland to Portland was delayed owing to navigation concerns and was not completed until 1888.

By that time the final link to Puget Sound was finally complete. The Northern Pacific's Cascade Branch had two completions: the first using a temporary switchback over the summit and the second when the tunnel was completed. Work on the Cascade Branch began in earnest in early 1884, and that summer the decision was made to use Stampede Pass rather than the alternatives, the Naches Pass or the Snoqualmie. The railroad arrived near the existing town of Yakima in December 1884, but the railroad decided not to stop there, instead platting a new town called North Yakima where it owned much of the land (see page 144). Most of the buildings in the original town ended up being removed to the new one.

Labor was becoming hard to find, and the contractors made an agreement with Chinese agents to bring in 1,500 Chinese men to work between Ellensburg and the summit. In April 1886 the railroad decided to build a switchback over Stampede Pass so that trains might operate—and bring in much-needed revenue—while a tunnel was being bored. The eight-mile-long switchback required gradients of over 5 percent, and two massive specially built locomotives were constructed to operate it, one at each end of only a few passenger cars at a time. On 1 June 1887, with celebrations at the summit, the switchback was complete and with it the first completion of the Cascade Branch. Contractor Nelson Bennett had meanwhile been boring the two-mile-long Stampede Tunnel under the summit of Stampede Pass. The tunnel was driven with considerable accuracy and speed. Work began in January 1886, and new steam-driven air compressors were used to power pneumatic drills. The tunnel, bored from both ends, was broken through on 3 May 1888 and opened to traffic on 27 May. The Cascade Branch was now truly completed. Bennett dismantled his air compressors and shipped them off to a tunnel site on the Siskiyou Mountains section of the Oregon & California Railroad (see page 92).

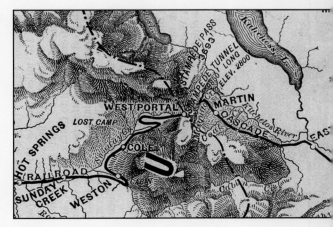

Map 192 (*right, center*). This 1881 map shows the yet-to-be-bridged crossing of the Northern Pacific at *Ainsworth*, just south of present-day Pasco, and the junction with the OR&N at *Wallula*.

Map 193 (*right*). Detail from an 1888 Northern Pacific map shows the *Stampede Tunnel 9850' Long Elev. 2800'*. The dashed line is the line of the summit of the Cascades. Stampede Pass was found by engineer Virgil Bogue in 1881. Bogue would later draw up a comprehensive plan for Seattle (see page 123).

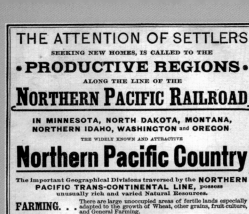

The Northern Pacific Land Grant

Railroad land grants, first introduced in 1851, were seen at first as a win-win way of accelerating development of western lands. Government lands remaining after grants could be sold for double the value of the land given away, since access increased the land value, and settlement would return taxes to the government for previously untaxed land.

The Northern Pacific (including the Cascade Branch) acquired a land grant of 41 million acres, about 31 percent of the total 131 million acres of land granted to railroads in the United States (not including another 43 million acres granted but not taken up). Nearly 9 million acres of the land granted to the Northern Pacific was in Washington.

There were rules governing what land the railroad could claim; generally these were alternate sections within a six-mile-wide belt, but land reserved for government sale or use and land already claimed or owned by individuals was exempt. In many cases this meant that the land close to the rail line was not available, and to compensate for this the railroad was allowed to select sections—but not all the land—from within a much wider belt, depending on the circumstances and geography. The Northern Pacific's selection belt extended in some cases to fifty miles from the rail line (MAP 195, *below*), and it was these wide alternate selection belts that caused many people to misinterpret the land grant, a misconception abetted by the government's own exaggerated maps (MAP 194, *left, center*). MAP 198, *far right, bottom*, shows one such case, published in a broadsheet in 1892. The Northern Pacific grant in fact covered just under 20 percent of Washington's land, far less than the 50 percent this map suggests.

MAP 194 (*left, center*).
The land grants (for both railroads and military or wagon roads) were depicted far too broadly on many maps, especially those at a small scale. This federal government map published in 1878 seems to have begun the widely believed myth that railroads had been granted huge swaths of land; most of Washington appears to have been given away.

Left, top. One of the advertising panels on the back of MAP 196.

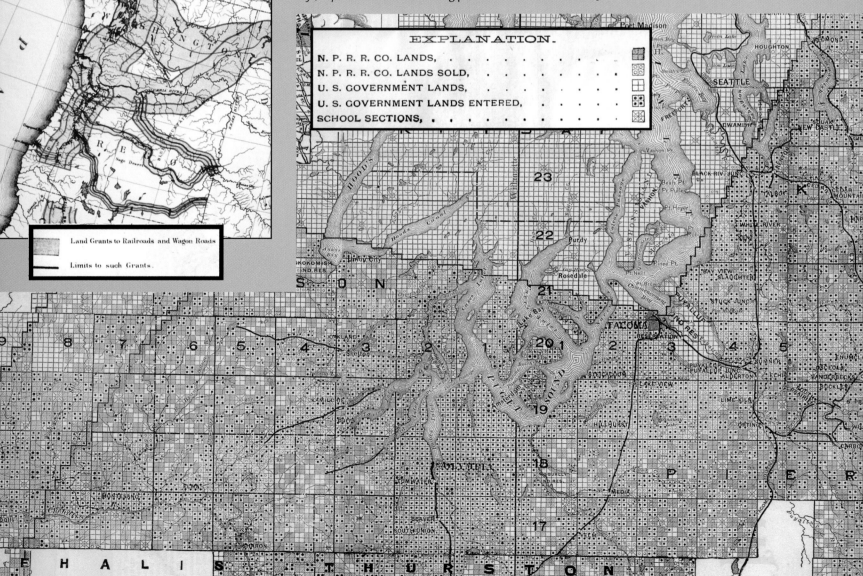

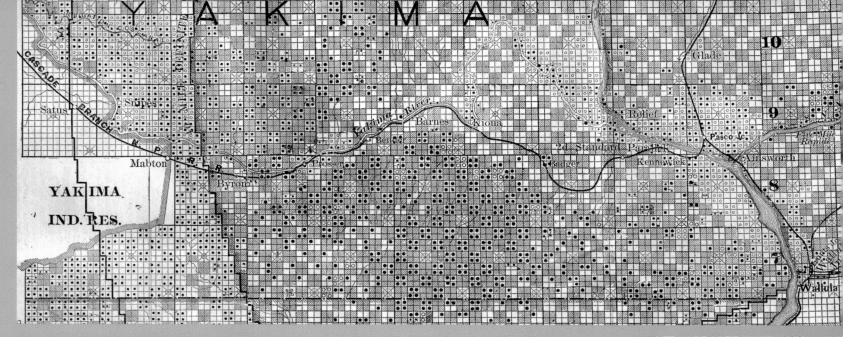

MAP 195 (*left, bottom*) and MAP 196 (*above*).
These maps were used to market the Northern Pacific's land grant. MAP 195 dates from 1888 and shows land for sale in the southern Puget Sound area. The jagged lines at left are the forty-and-fifty-mile limits for selection of alternate sections. These same limits are shown on MAP 196, published in 1891; here they pertain to the original route without the Cascade Branch. White areas on MAP 195 are unsurveyed, often mountainous land. On MAP 196 the white unsurveyed area is the *Yakima Ind[ian] Res[ervation]*. The key (inset on MAP 195), applies to MAP 195, MAP 196, and MAP 197 and is important, as it allows distinction between land granted to the Northern Pacific (both since sold and unsold at the date of the map) and government land (both "entered," that is, claimed, or unclaimed or sold); and land reserved for schools.

Left. Terms of sale for Northern Pacific lands, from the back of the 1891 map.

MAP 198 (*below*).
In 1883, just as the Northern Pacific was completing its first transcontinental link with Puget Sound, many people were upset at what they thought was a massive giveaway of land. This broadsheet, which completely misrepresents the situation, was published by the Seattle Chamber of Commerce. This map's creator did not realize—or chose to ignore—that the black "cloud" represented only the limits of the zone within which the railroad could select sections that were unclaimed and only where sections nearer the rail line were unavailable.

MAP 197 (*right*).
This section of the 1891 Northern Pacific land grant map also shows the line and land grant of the Spokane Falls & Northern Railway, built by railroad entrepreneur Daniel Corbin in 1888–93 but partially operating north of Spokane Falls by 1889. The line, initially friendly to the Northern Pacific, was leased to that railroad in 1898 but the following year was sold to rival Great Northern.

NORTHERN PACIFIC RAILROAD
LANDS FOR SALE.

The Northern Pacific Railroad Company has a large quantity of highly productive agricultural and grazing lands for sale at LOW RATES and on EASY TERMS. These lands are located along the line in the States traversed by the Northern Pacific Railroad, as follows: Over 1,250,000 acres in Minnesota; 6,800,000 acres in North Dakota; 17,450,000 acres in Montana; 1,750,000 acres in Northern Idaho; 9,375,000 acres in Washington and Oregon, aggregating over

36,600,000 ACRES.

These Lands are for sale at the LOWEST PRICES ever offered by any Railroad Company, ranging chiefly

From $2 to $6 per Acre

For the Best Wheat Lands, for the Best Farming Lands, for the Best Grazing Lands, in the World.

TERMS OF SALE.

The price of Agricultural Lands in **Washington and Oregon** ranges chiefly from $2.60 to $6 per acre. If purchased **on five years' time,** one fifth cash. At end of first year the interest only on the unpaid amount. One fifth of principal and interest due at end of each of next four years. Interest at 7 per cent. per annum.

On 10 years' time.—Actual settlers can purchase not to exceed 320 acres of agricultural land **on 10 years' time,** at 7 per cent. interest, one tenth cash at time of purchase, and balance in nine equal annual payments, beginning at the end of the second year. At the end of the first year, only the interest is required to be paid. Purchasers on the 10 years' credit plan are required to settle on the land purchased, and to cultivate and improve the same.

GRAZING LANDS are sold **at from $1.25 to $2.50** per acre, according to location and quality, and on from one to ten years' time.

TIMBER LANDS.—Cash, or by special agreement.

For terms of sale of lands in **Minnesota, North Dakota, and Montana,** see page 5.

NOTICE.—The Land Department of the Northern Pacific Railroad Company employs no agent or others along its line who are authorized to receive or receipt for any moneys for the Company or to bind the Company by any agreements or acts whatsoever.

All applications for the purchase of Northern Pacific Railroad Lands in Washington, Idaho, and Oregon, and all payments thereon, must be made to **PAUL SCHULZE,** General Land Agent, at Tacoma, Wash.

The following named real-estate dealers have been furnished plats and maps, and they will be pleased to give intending settlers information and assistance in selecting Railroad and Government Lands:

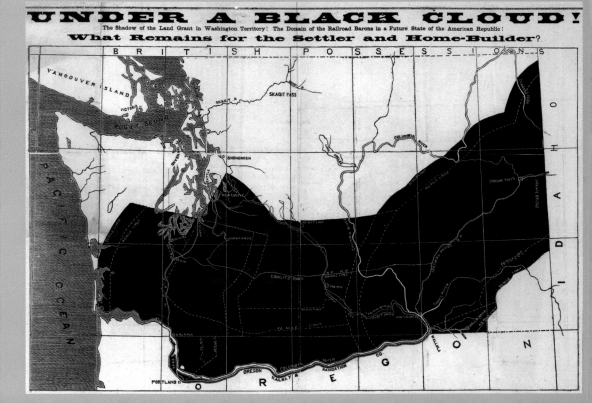

UNDER A BLACK CLOUD!
The Shadow of the Land Grant in Washington Territory! The Domain of the Railroad Barons in a Future State of the American Republic!
What Remains for the Settler and Home-Builder?

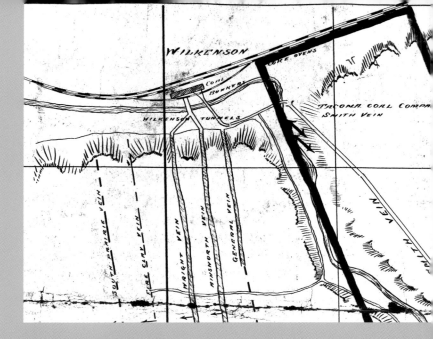

Coal for Railroad and City

Coal was the lifeblood of both railroads and cities. The discovery of coal in the vicinity of both Seattle and Tacoma provided an economic boost to the local economies and allowed the Northern Pacific to earn much-needed revenue before its connection to the transcontinental line. The Seattle & Walla Walla, created with the intention of linking Seattle to the eastern part of the territory, finally found salvation as a little line hauling coal from the mines at Renton and Newcastle, reaching the latter in 1877. It did so well that the line was purchased by Henry Villard's Oregon Improvement Company in 1880, becoming the Columbia & Puget Sound. The Northern Pacific, near bankruptcy and with a line from Kalama to Tacoma otherwise isolated, completed a spur to the mine of Samuel Wilkeson, also in 1877. Coal was hauled to tidewater at Seattle or Tacoma and shipped to San Francisco or used locally.

When the Northern Pacific was surveying its Cascade Branch, coal was found in the upper reaches of the Yakima Valley. For a railroad running on steam it was a fortuitous find on land within its land grant; the railroad was able not only to use the coal to power its operations but also to export the coal using its new line. The cost of coal mined at Roslyn proved to be less than that from the mines near Seattle and Tacoma. By 1890, the Roslyn mines were producing over 55,000 tons of coal each month, two-thirds of which was used to fuel Northern Pacific locomotives as far east as Billings, Montana. Union Pacific searched for coal in eastern Oregon but, finding nothing workable, was also buying coal from Roslyn by 1890.

MAP 200 (*below*), MAP 201 (*bottom right*), and MAP 202 (*far right*).
The coal mines and railroads for the Puget Sound area shown on a 1913 geological map. Note that some of the railroads are later additions. MAP 200 shows the mines of the *Renton* area, at the southern end of Lake Washington; MAP 201 shows the *Newcastle Mines*, southeast of the lake; and MAP 202 shows the *Wilkeson* and *Carbonado* mines east of Tacoma.

MAP 199 (*above*).
The mines of the Tacoma Coal Company at *Wilkenson* (Wilkeson), on the flanks of Mount Rainier, shown on an 1875 map converted from a blueprint.

MAP 203 (two details, *below*).
This 1892 map showed the coal mine west of Seattle, but very inaccurately. The *Coos Bay Coal Mine*, with self-explanatory label, was not served by the railroad.

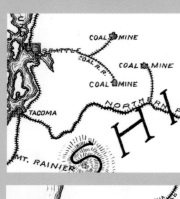

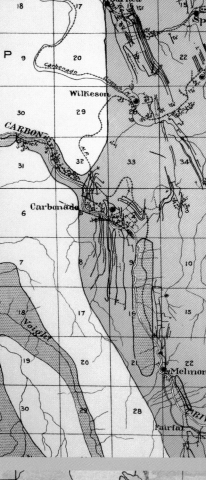

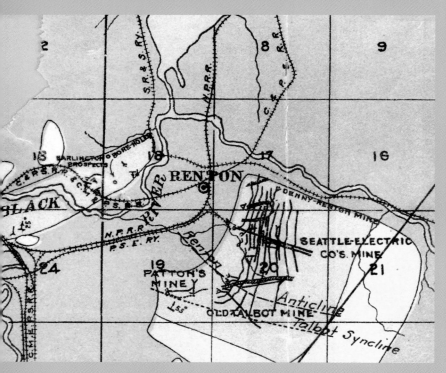

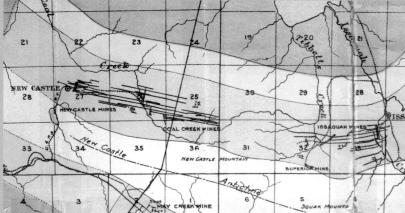

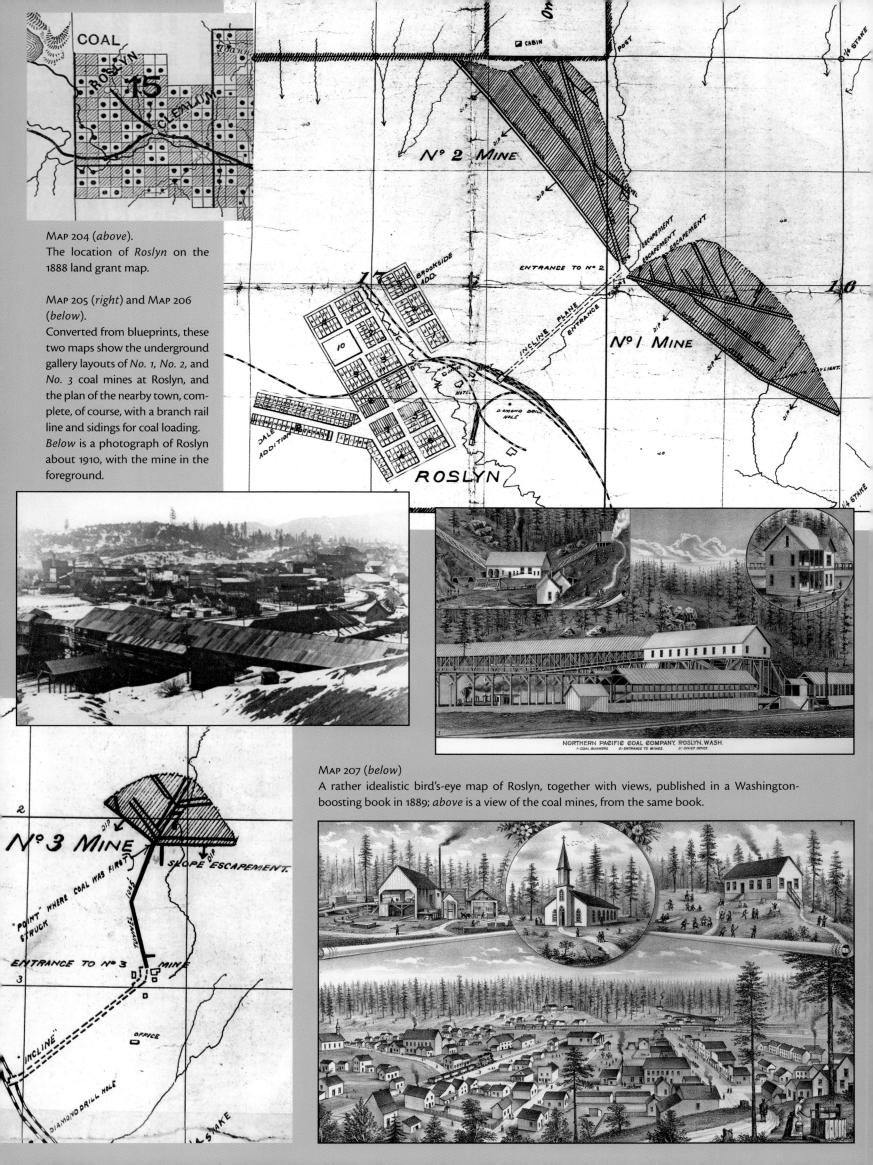

MAP 204 (*above*).
The location of *Roslyn* on the 1888 land grant map.

MAP 205 (*right*) and MAP 206 (*below*).
Converted from blueprints, these two maps show the underground gallery layouts of *No. 1, No. 2,* and *No. 3* coal mines at Roslyn, and the plan of the nearby town, complete, of course, with a branch rail line and sidings for coal loading.
Below is a photograph of Roslyn about 1910, with the mine in the foreground.

MAP 207 (*below*)
A rather idealistic bird's-eye map of Roslyn, together with views, published in a Washington-boosting book in 1889; *above* is a view of the coal mines, from the same book.

NORTHERN PACIFIC COAL COMPANY, ROSLYN, WASH.
1: COAL BUNKERS   2: ENTRANCE TO MINES.   3: CHIEF OFFICE.

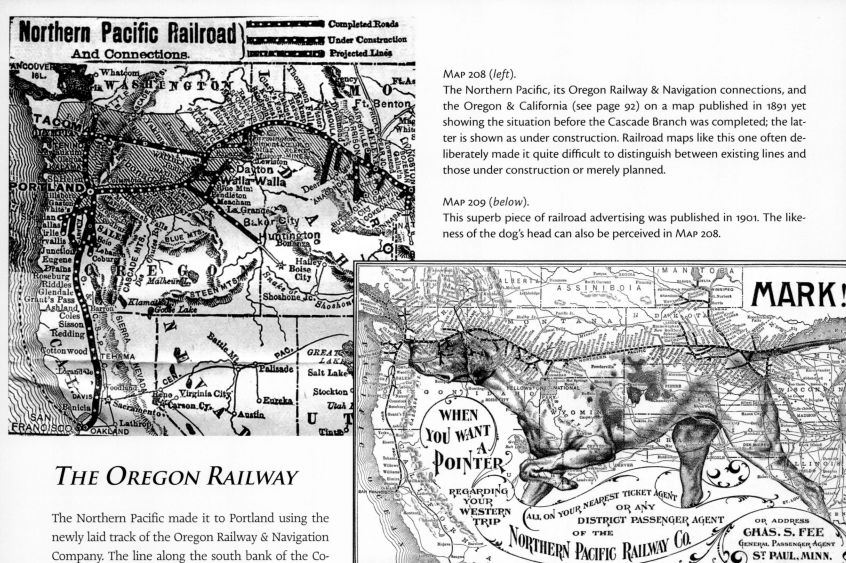

Map 208 (*left*).
The Northern Pacific, its Oregon Railway & Navigation connections, and the Oregon & California (see page 92) on a map published in 1891 yet showing the situation before the Cascade Branch was completed; the latter is shown as under construction. Railroad maps like this one often deliberately made it quite difficult to distinguish between existing lines and those under construction or merely planned.

Map 209 (*below*).
This superb piece of railroad advertising was published in 1901. The likeness of the dog's head can also be perceived in Map 208.

The Oregon Railway

The Northern Pacific made it to Portland using the newly laid track of the Oregon Railway & Navigation Company. The line along the south bank of the Columbia was difficult to construct where the underlying lava beds jutted out into the river, and in a number of places multiple tunnels had to be driven. Rail service between East Portland and what railroad promotional literature called "the tributary empire lying to the east of the Cascade Mountains" began on 20 November 1882; less than a year later it was part of the transcontinental link.

But when the OR&N had first been created, Henry Villard had intended that it would eventually connect with the Union Pacific, via a branch north from that company's main line (the first transcontinental) and had reached what is usually referred to as an "understanding" to that effect. Because Union Pacific had to hand over its California traffic to the Central Pacific at Ogden, Utah, it craved a more secure Pacific outlet—and one it might ultimately make its own—in the Northwest. Villard's opportunistic takeover of the Northern Pacific, however, put him at odds with Union Pacific's plans. Now in possession of his own transcontinental, he no longer needed Union Pacific either. But Union Pacific was threatened by a potential diversion of its traffic to a new transcontinental line built by the Southern Pacific (controlled by the same "Big Four" who had built the Central Pacific; see page 92). Thus Union Pacific moved to build northwestward from its transcontinental line at Granger, Wyoming, incorporating a subsidiary, the Oregon Short Line Railway, in April 1881. Despite Villard's new ambitions, Union Pacific managed to come to an agreement to connect with the OR&N at Baker City.

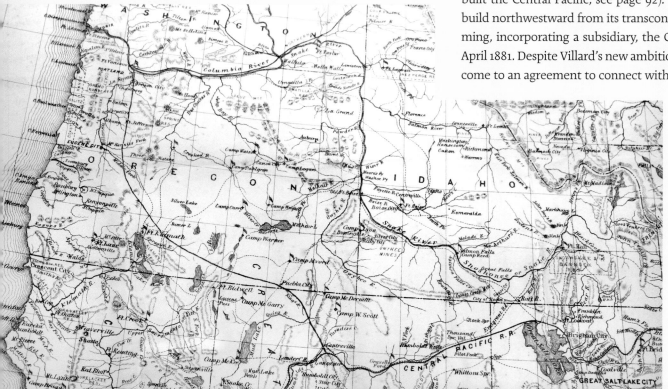

Map 210 (*left*).
One of the first proposals for a transcontinental line to Portland is shown on this map published in 1868. Here the line left the *Central Pacific R.R.* in the valley of the *Humboldt R.* in California and reached Portland via the valley of the *Willamette R.*

The OR&N gave priority in manpower and money to the transcontinental line. Then, short of money, it had difficulty raising funds to pay for the line to Baker City; the line was not completed to that point until September 1884. By that time, however, the OR&N had determined to minimize the encroachment of Union Pacific into its turf and had graded well beyond Baker City, taking control of the critical narrow canyon of the Burnt River, forcing the Oregon Short Line to connect at the Snake River. The OR&N deliberately stopped at Huntington, just before the river, and thus avoided having to pay for the cost of the bridge. The last spike was driven on 25 November 1884, finally connecting the OR&N with the Oregon Short Line and giving Union Pacific access to the Northwest, which it has held on to ever since.

MAP 211 (*right, top*).
The proposed OR&N line from the Columbia to *Baker City* is shown on this 1881 map. As it happened, the railroad did not stop at Baker City but reached the Snake River at Huntington.

MAP 212 (*right, center*).
The completed OR&N line from *Umatilla,* on the Columbia, to meet Union Pacific's Oregon Short Line at *Huntington.* The line was originally planned to meet the Union Pacific at *Baker C[it]y.* The map is dated 1891.

MAP 213 (*below, left*).
The OR&N moved to service the rich farmland of the Palouse country in the 1880s and acquired some land grants in the process. This map, published in 1881, shows existing and planned OR&N lines in eastern Washington and Oregon; the Northern Pacific main line is also shown. The OR&N's Palouse land grant is shown between *Colfax* and *Steptoe.* Standard-gauge track had reached *Walla Walla* in 1881, and from that point the railroad built both south to *Pendleton* and north to *Dayton* and Colfax—with continuations eastward across the Idaho boundary to *Lewiston* and *Moscow*—and farther north to Winona and ultimately Spokane. Some of this expansion was achieved by the purchase of other railroad franchises. Union Pacific leased the OR&N in 1887 and purchased it two years later.

MAP 214 (*below*).
The OR&N line south from *Walla Walla* to *Pendleton,* completed in 1887, is shown on this 1891 map. Also shown are lines of the locally promoted Oregon & Washington Territory Railroad (*O&WT*). This line was completed between Wallula and Centerville in 1888 and to Pendleton the following year.

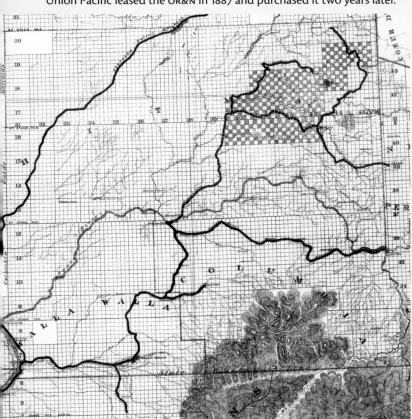

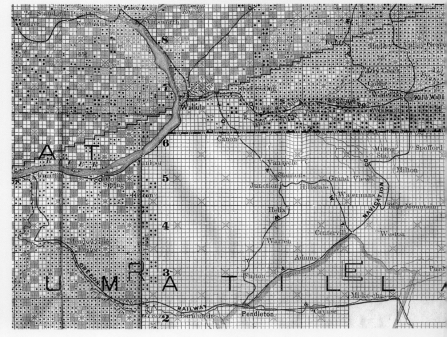

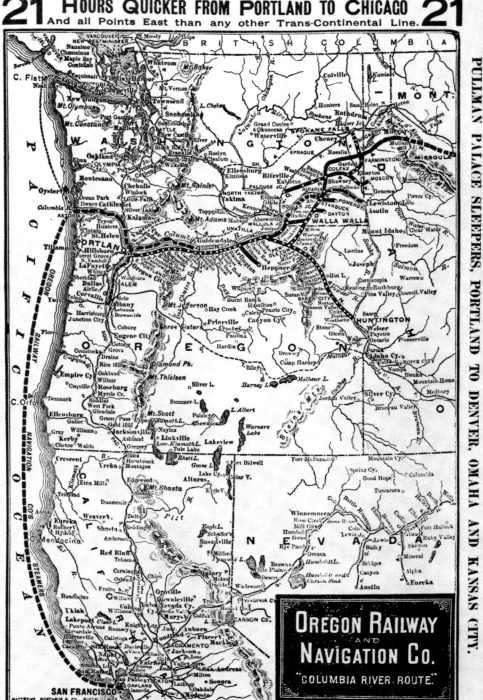

FREE FAMILY SLEEPING CARS ON EXPRESS TRAINS TO MISSOURI RIVER WITHOUT CHANGE.

PULLMAN PALACE SLEEPERS, PORTLAND TO DENVER, OMAHA AND KANSAS CITY.

OREGON RAILWAY
AND
NAVIGATION CO.
"COLUMBIA RIVER ROUTE."

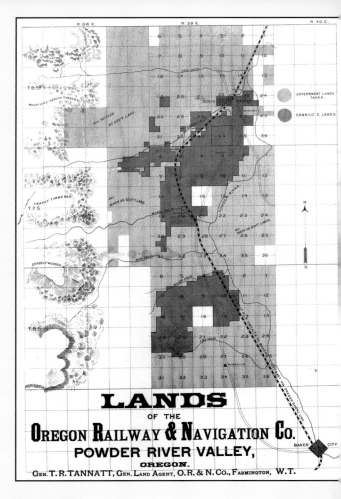

LANDS
OF THE
OREGON RAILWAY & NAVIGATION CO.
POWDER RIVER VALLEY,
OREGON.
GEN. T. R. TANNATT, GEN. LAND AGENT, O.R. & N. CO., FARMINGTON, W.T.

MAP 215 (*left*).
The OR&N system about 1891. Now owned by Union Pacific, the side banners advertise Pullman Palace Sleepers from *Portland* through to Denver, Omaha, and Kansas City and tout a speed advantage over "any other Trans-Continental Line"—competitor Northern Pacific. Also shown is the route of the company's coastal steamships from The *Dalles* and Portland to San Francisco.

MAP 216 (*above*).
The land grant of the OR&N in the Powder River valley of eastern Oregon, shown on an 1888 map produced by the railroad for sales purposes. *Baker City*, at the center of the gold-mining district of eastern Oregon, is at bottom.

CONNECTING WITH CALIFORNIA

Two companies, both confusingly named the Oregon Central, began building south from Portland in 1868. Both were after land grants authorized by Congress in 1866. A west-side company (that is, west of the Willamette River), the Oregon Central *Railroad,* completed a line from Portland to St. Joseph in 1871. An east-side company, the Oregon Central *Rail Road,* owned by stagecoach and transportation tycoon Ben Holladay, completed a line south from East Portland, reaching New Era, just south of Oregon City, on 30 December 1969, completing its first twenty miles in time to claim its land grants (MAP 217, *right, top*).

In 1870 the east-side Oregon Central joined with the Oregon & California Railroad, another Holladay company begun that year with the intention of linking with Californian lines to the south. In California, the so-called Big Four of the Central Pacific—Collis Huntington, Leland Stanford, Mark Hopkins, and Charles Crocker—had in 1867 acquired the California & Oregon Railroad, which was building north up the Sacramento Valley. The following year the Big Four acquired the as-yet-unbuilt Southern Pacific, and the California & Oregon became part of their Central Pacific–Southern Pacific system. By 1879 their line had reached Redding.

Holladay's Oregon & California built south to Salem in 1870, reached Eugene in October the following year, and arrived at Roseburg by December 1872. But then the railroad ran out of money, and it was not until 1883 that construction began once more; Grants Pass saw its first train in December 1883, followed by Ashland in May 1884. The California line was reached in December 1887, after the construction of a series of hairpin bends and tunnels to carry the line over the Siskiyou Mountains. Southern Pacific acquired control of the Oregon & California in stages between 1884 and 1887, so that by the time the

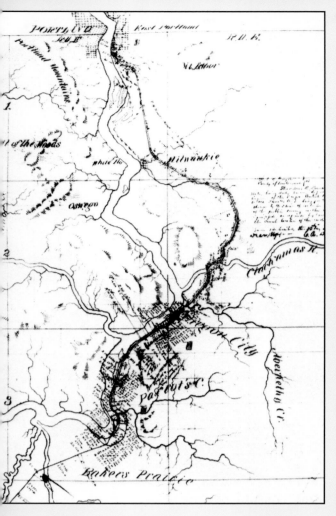

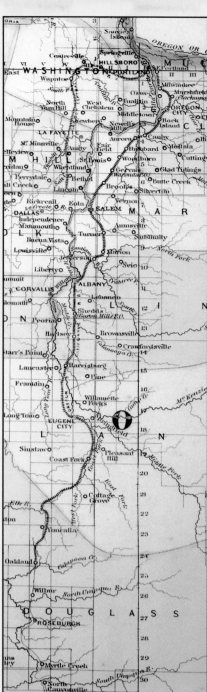

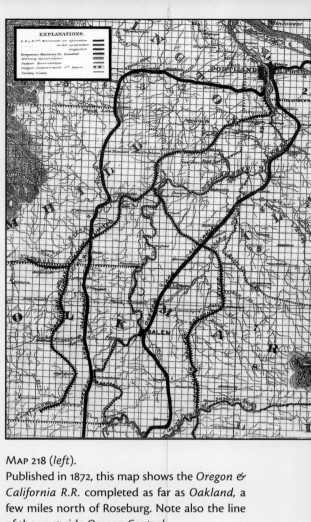

Map 217 (above).
This unfortunately poorly reproduced map of the Willamette Valley between *Portland* and *Oregon City* was attached to the certificate sent to the governor of Oregon by the president of the Oregon Central Railroad on 30 December 1869. It shows the line and certifies the completion of the first twenty miles of the Oregon Central between *East Portland* and *Parrot's C[reek]* (now New Era) just in time to be eligible for a land grant on either side of the track (see Map 222, *overleaf*).

Map 218 (*left*).
Published in 1872, this map shows the *Oregon & California R.R.* completed as far as *Oakland*, a few miles north of Roseburg. Note also the line of the west-side *Oregon Central*.

Map 219 (*above*).
The Oregon & California took over the west-side Oregon Central in 1880, and this 1881 map shows its lines south of Portland under that name. The crosshatched lines are those of the Oregonian Railroad, which built a small network of lines south of Portland beginning in 1880. The company was sold to the Oregon & California in 1893 and merged with that system.

Map 220 (*far left*).
The line between *Roseburg* and *Redding* is shown as under construction (with short lines along one side) on this 1879 railroad map.

Map 221 (*left*).
By 1884, the date of this map, the line is shown as complete in California but still only built to Roseburg; it was in fact completed as far south as Ashland by the middle of the year.

line connected with the one being built north from California, the whole of the San Francisco–Portland route was in Southern Pacific hands.

The Oregon & California gained a large land grant, a total of 3.7 million acres, in the usual checkerboard pattern within a zone up to sixty miles wide, twenty sections for each mile of track laid (Map 222, *overleaf*). The company was required to sell the land to "actual settlers" for no more than $2.50 per acre. But "actual settlers" interested in the mainly forested and steep slopes were hard to come by, and the railroad resorted to selling land for its timber value, in contravention of the legislation. Further, in 1903, with a change of policy brought in by new owner and "robber baron" Edward H. Harriman, the railroad stopped selling land altogether,

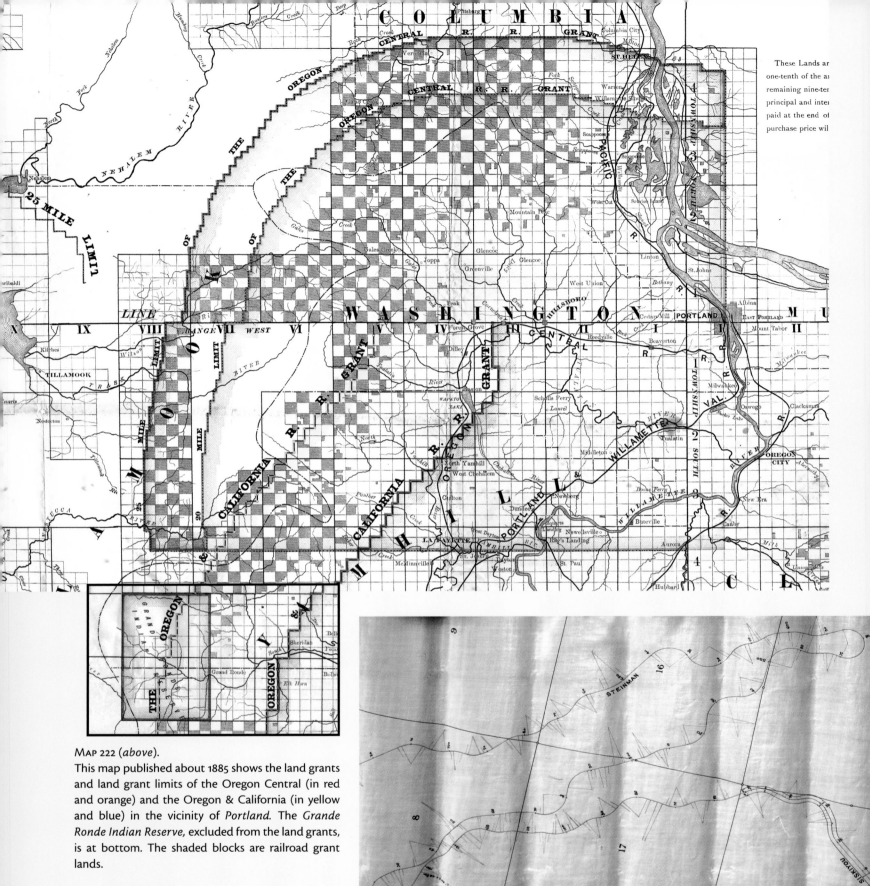

These Lands ar[...]
one-tenth of the a[...]
remaining nine-te[...]
principal and inter[...]
paid at the end of[...]
purchase price wil[...]

MAP 222 (*above*).
This map published about 1885 shows the land grants and land grant limits of the Oregon Central (in red and orange) and the Oregon & California (in yellow and blue) in the vicinity of *Portland*. The *Grande Ronde Indian Reserve*, excluded from the land grants, is at bottom. The shaded blocks are railroad grant lands.

MAP 223 (*right*).
The line over the Siskiyou Mountains was the last, and the most difficult, of the sections built to complete the line from Portland to San Francisco. This map, drawn in 1889, illustrates the extreme nature of the pathway required to cross the mountains. Detail of the *Siskiyou Summit Tunnel*, bottom right on the main map, is shown in the *inset*. The line was effectively replaced in 1926 when a new route via the Natron Cutoff, a line south via Klamath Falls, was completed, with new track being laid to allow the Southern Pacific to run its premier train, the *Cascade*, between San Francisco and Portland.

preferring instead to keep it for its own use. It was a major error; after a decade of protracted legal battles, Congress in 1916 confiscated the remaining land grant, some 2.9 million acres; it was the largest one-time loss of granted lands by any railroad in the United States.

ORPHAN ROAD

As the Northern Pacific's transcontinental moved ever closer to its connection with Tacoma, the citizens of Seattle chafed at the coming victory for their commercial rival. In October 1881 Henry Villard and his Oregon Improvement Company promised to build a line connecting Seattle with Tacoma and the transcontinental. But money was tight, and the wheels of progress turned slowly. The Northern Pacific built north from its branch line to Wilkeson, just seven miles to Stuck Junction. From there the Puget Sound Shore Railroad, another Villard creation, this time a subsidiary of his Oregon & Transcontinental holding company (and whose obligations were later assumed by his Oregon Improvement Company), constructed fourteen miles to Black River Junction, and from that point used the same rail bed as the Columbia & Puget Sound (which was narrow gauge) for the ten miles into Seattle; the line was completed by June 1884. The route is clearly shown on MAP 224, *right*.

But the double ownership of the line led to both confusion and apathy; no trains ran, and the press dubbed the silent extension the "Orphan Road." On 17 June 1884 the first standard-gauge train steamed into Seattle, a special train bearing Northern Pacific vice-president Thomas Oakes. But sporadic, half-hearted service and deliberately inconvenient train schedules continued until late 1889, when the line was purchased by Northern Pacific subsidiary Northern Pacific & Puget Sound Shore Railroad and, after ratification of the sale in January 1890, integrated into the Northern Pacific system, finally ending Seattle's branch line status.

SEATTLE, LAKE SHORE & EASTERN

Until that time, Seattle businessmen hated the disadvantage bestowed by the inefficient Orphan Road and worried that the Northern Pacific's Cascade Branch would not be completed. Not for the first time, they determined to do something about it.

The idea this time was to allow the resources of the Cascade Mountains, especially coal, to be brought to Seattle for use or export and to connect the city with its interior hinterland. This being a large enterprise, eastern capital had to be courted, but financial backing was found, and the railroad, named the Seattle, Lake Shore & Eastern (SLS&E) began construction in Seattle in March 1887. The plan was to build through Snoqualmie Pass to Ellensburg, cross the Columbia at Priest Rapids, where

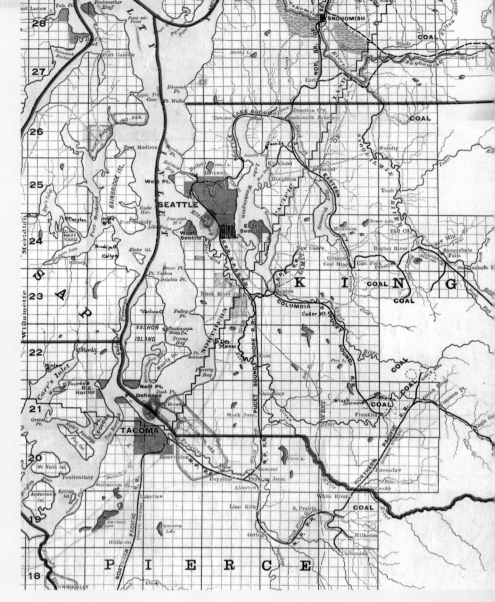

MAP 224 (*above*).
This map, published in 1890, shows the location of Seattle's "Orphan Road." The southern part runs from *Puyallup Junc.* (Meeker Junction) to *Stuck Junc.* and is marked *N.P.R.R.* The middle section, from Stuck Junction to *Black River Junc.*, is marked *Puget Sound Shore R.R.*, while the northern section into *Seattle* is the combined narrow and standard tracks of the *C.&P.S.R.R. & P.S.S.R.R.* (Columbia & Puget Sound [see page 83] and Puget Sound Shore railroads). The map also shows the Northern Pacific's *Cascade Division*, the transcontinental link to Puget Sound completed in 1887, connecting with the existing lines to the coal mines of *Wilkeson* and *Carbonado*; the *Columbia & Puget Sound R.R.* line to the coal mines at *Franklin*; and the *Seattle, Lake Shore & Eastern Ry.* line from Seattle to *Snoqualmie* and farther up the valley, built in 1887–88. The *Gilman Coal Mine* is shown, at the location of today's Issaquah.

the railroad would connect with steamboats, and then continue east to Spokane Falls. Within six months, forty-two miles of line had been built east around the northern end of Lake Washington, connecting the Gilman Coal Mine with the city (MAP 224, *above*). Sallal Prairie (near North Bend) was reached the following year.

Construction was also started in the interior. After a fund-raising drive in Spokane Falls, work began in May 1888, building west. Later that year the Northern Pacific, sensing a challenge to its monopoly, began building northwest from its main line at Cheney. Grading teams from the two companies nearly came to blows at the Grand Coulee, with both fighting for a crossing place to the disadvantage of the other. SLS&E track reached Wheatdale, five miles east of Davenport, by December, but the line was only completed to Davenport because the town's citizens subsidized and assisted

MAP 225 (*right*).

The Northern Pacific in Eastern Washington in 1891. By this date the Northern Pacific had purchased the Seattle, Lake Shore & Eastern, and the line from Spokane to Davenport had been incorporated into its own system. The Northern Pacific's line to Coulee City was completed in 1890.

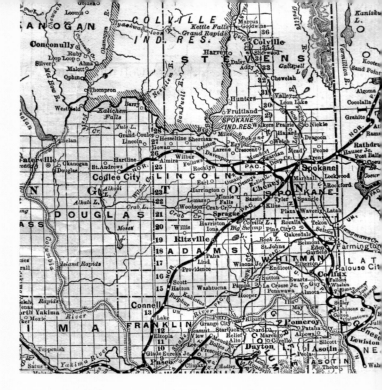

in building the last five miles, not wanting to be hostage to the monopolistic whims of the Northern Pacific, which completed its line to the Grand Coulee in 1890, where it platted Coulee City (MAP 225, *right*).

The reason the SLS&E stopped work in the interior was partly due to that perennial issue in early railroad building, running out of money, and partly due to a change in priorities, for the railroad had been offered an opportunity to link with the newly completed transcontinental to the north, the Canadian Pacific.

In March 1888 the SLS&E acquired the Seattle & West Coast Railway, which was building a line from Snohomish Junction (Woodinville) to Snohomish; this line was completed in July that year. In May 1889 the SLS&E began building north from Snohomish, concentrating its effort on this potentially more lucrative line and at the same time stopping work in the interior. The line to the Canadian border was completed in April 1891.

However, three months later the Northern Pacific purchased a controlling interest in the SLS&E. This move did not do it any good, for two years later, in the general financial debacle of 1893, the Northern Pacific was unable to meet the bond obligations, and in June the SLS&E and in October the Northern Pacific both slid into bankruptcy.

After reorganization and a name change to the Seattle & International Railway the railroad was sold in 1898 to an itself reorganized Northern Pacific, preventing it from falling into the hands of the Canadian Pacific.

MAP 226 (*below*).

From a Seattle, Lake Shore & Eastern–sponsored book published in 1889 comes this map of the intended network of the railroad. The main line between Seattle and Spokane is now through Cady Pass rather than the originally intended Snoqualmie Pass, a route change that led a syndicate of Seattle investors to purchase much of the site of the new intended river crossing, today's Wenatchee. There is also a coastal line north to meet the Canadian Pacific main line in the Lower Fraser Valley. The line to the Canadian border was completed in 1891 and became the principal achievement of the SLS&E.

MAP 227 (*left*).
The Seattle, Lake Shore & Eastern Railway on an 1889 map emphasizing the railroad's connection to the Canadian Pacific main line at *Mission*.

MAP 228 (*right*).
This 1889 map shows the SLS&E line to Canada projected north of *Snohomish*.

The Northern Pacific continued to build in Washington. Acting principally in response to a threat from a smaller system, the Oregon & Washington Territory Railroad, which completed a line from Centralia to Montesano, on the Chehalis River, in 1891, the Northern Pacific in 1892 completed a line to Grays Harbor, where the company carried out a maneuver similar to the one it did in Yakima: not wanting to be restricted by servicing the existing main city, Aberdeen, where it owned no land, the railroad instead terminated its line on the south side of Grays Harbor at a new city where it owned the land—a place it grandly named Ocosta-by-the-Sea, later just Ocosta. Soon after the announcement, in May 1890 some three hundred lots were sold for a total of $100,000. Yet almost immediately the railroad decided it had erred and offered to build a spur to Aberdeen if the city would contribute to the cost. The city could not raise the money but later built the line and gave it to the Northern Pacific; the first train steamed into Aberdeen in April 1895.

MAP 229 (*below*).
The reorganized Northern Pacific's system in the Northwest in 1900. In addition to the main line via what it called the Cascade Extension to Puget Sound and the even older Portland-to-Tacoma line, there is now the ex–Seattle, Lake Shore & Eastern, ex–Seattle & International line, north to *Sumas*, where a short connection was made to the Canadian Pacific main transcontinental line at *Mission*. Also shown are branches to the *Grays H[arbo]r* cities of *Ocosta*, *Hoquiam*, and *Aberdeen*; and another to *South Bend*, on Willapa Bay (see next page). The proposed line north from *Coulee City* was never built, and the route from *Lind* to *Ellensburg* in central Washington was the one belatedly utilized by the Chicago, Milwaukee & St. Paul (see page 166). The quality of the red overprinting on this map leaves something to be desired, being offset to the right and bottom. Such were the quality controls of the printing process a century or more ago.

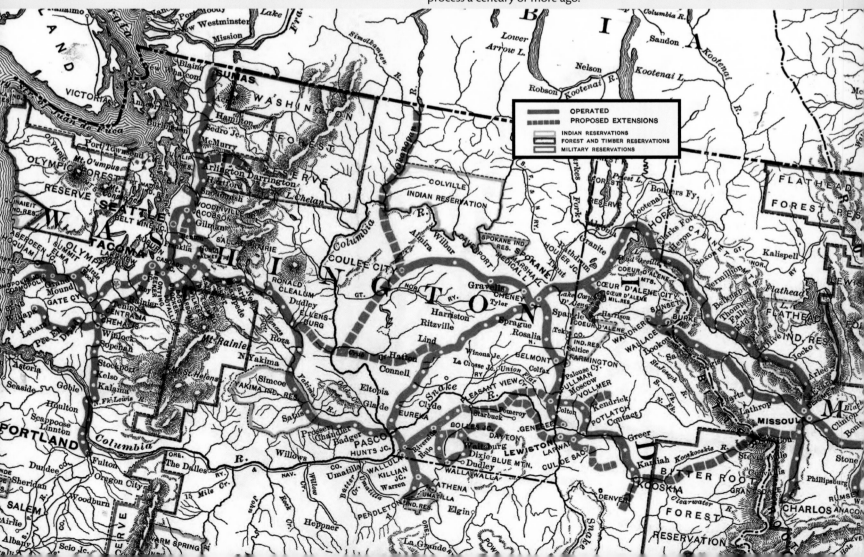

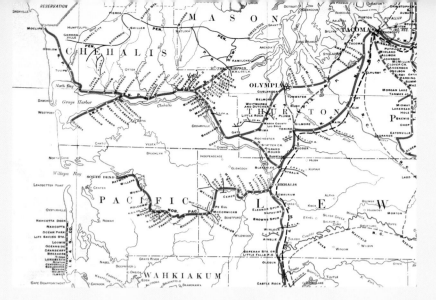

MAP 230 (*left*).
This 1892 map depicts the SLS&E line to Canada as a *Canadian Pacific R.R. (Extension)*. This was premature; although the Canadian Pacific was indeed interested in purchasing this line, it was ultimately acquired by the Northern Pacific in 1898 to prevent this incursion.

MAP 231 (*above*).
The Northern Pacific lines to *Grays Har[bor]* and *Willapa Bay* depicted on an 1891 railroad map.

MAP 232 (*above, right*).
A 1908 railroad map reveals the extension of the Northern Pacific line from Grays Harbor to the little resort town of Moclips. By 1898 the Northern Pacific had acquired the lines from Centralia to Montesano built by the Oregon & Washington Territory Railroad. Also shown on this map is the line of the *I.R.&N.Co.*, the Ilwaco Railway & Navigation Company (see MAP 233, *below*).

MAP 233 (*below*).
The Ilwaco Railway & Navigation Company (incorporated as the Ilwaco, Shoalwater Bay & Grays Harbor Railroad in 1883) operated for over forty years on the Long Beach peninsula. This isolated narrow-gauge road, completed in 1889, connected with steamboats from Astoria at Ilwaco and delivered passengers and freight to other steamboats at Nahcotta, on Willapa Bay; it was, therefore, a true portage railroad. This map shows detail of *Nahcotta* dock in 1912, with, *inset*, an 1892 ad and an 1893 photo.

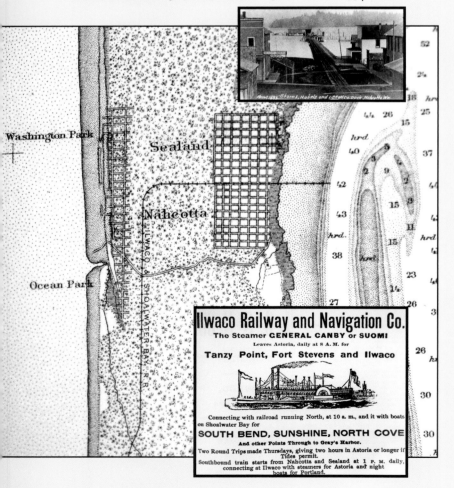

A similar line to the coast was built just to the south, to South Bend, on Willapa Bay. Here the railroad was the beneficiary of a land donation of half the land of the South Bend Land Company, which, of course, stood to more than double the value of the rest of its land with the coming of the railroad. The railroad was completed to South Bend in 1892.

BUILDING AN EMPIRE

In 1889, James Jerome Hill, principal of the St. Paul, Minneapolis & Manitoba Railway, which had already built a line as far west as western Montana, decided that the road should extend all the way to "some suitable point" on Puget Sound. For this purpose, he used another railroad charter he owned, that of the Minneapolis & St. Cloud, changing its name to something altogether more grand and appropriate: the Great Northern Railway.

Hill's plan was to create a high-grade transcontinental line that would be efficient to operate and connect this on Puget Sound with a north–south line that would funnel traffic to the transcontinental. For the Puget Sound line he would buy controlling interests in existing or planned lines wherever possible.

Even while the transcontinental line was being located, others were building tributary railroads on the coast. And in the spring of 1890 Hill picked Seattle as his "suitable point" for the western terminus, although frequent threats to go elsewhere if he did not get what he wanted kept the issue in doubt for some time. A million dollars were spent surveying a new line from Portland to Seattle in cooperation with Union Pacific subsidiaries the Oregon Short Line and the Utah & Northern before the project was shelved.

To the north Hill bought the Fairhaven & Southern in 1889. This was a railroad begun by Nelson Bennett—the contractor who had built the Stampede Tunnel for the Northern Pacific—between Bellingham Bay and coalfields a few miles east. Hill extended it to the Canadian border and also obtained from Bennett a charter for a connecting line in British Columbia, the New Westminster & Southern, to carry rails to the Fraser River opposite New Westminster, where a connection could be made with the Canadian Pacific (MAP 235, *right*). Not surprisingly, the Canadian Pacific was not particularly thrilled by this move.

To connect with the Fairhaven & Southern, Hill obtained the charter of the Seattle & Montana Railway, which built seventy-eight miles of track north from Seattle. The last spike of the northern section was driven at Blaine on 14 February 1891. Hill also obtained control of the Seattle & Northern, which was building a line from the originally Northern Pacific–sponsored port of Ship Harbor (the location of today's ferry terminal at Anacortes; see page 140) to connect with the Seattle, Lake Shore & Eastern and coalfields of the Skagit Valley. This west–east line was complete from Anacortes to Hamilton by February 1891 but would not reach Rockport, farther up the Skagit, until 1901. In November 1891, an official train ran from Seattle to the Fraser River, where celebrations were held to mark the coastal line's completion.

In Seattle, sixty acres of land were purchased on the northern edge of the city, between Smith's Cove and Salmon Bay, for rail yards, repair shops, and the like. The Great Northern also received a gift of land south of the business district, and city council granted the company half of Railroad Avenue, 60 feet of a 120-foot-wide strip, to connect the two, in an effort to ensure that Seattle was unchallenged as the terminus for the transcontinental.

Map 234 (right).
Part of the rail yards on the northern edge of Seattle, from an 1897 plan. The *S&M Ry* (Seattle & Montana) *Round House* is shown.

Map 235 (below).
This 1890 map shows the proposed (crosshatched) line (unnamed) of the Great Northern subsidiary, the Fairhaven & Southern, completed from *Sehome*, on *Bellingham Bay*, to the coal mines and then north into British Columbia, where it met the New Westminster & Southern. The line is not yet shown running south to meet the line of the Seattle & Montana. The Seattle, Lake Shore & Eastern's line (which would become Northern Pacific line in 1898) runs north to Sumas. The Seattle & Northern runs from *Ship Harbor* (Anacortes) as far as the Skagit Valley coalfields.

Map 236 (below, right).
The completed coastal system of the Great Northern, shown on a map from about 1908. The transcontinental line joins at *Everett*; the line from *Anacortes* is shown reaching *Rockport*, though in fact it did not make it that far up the Skagit Valley until 1901. The dashed line connecting from the east in Canada at *Sumas* is a line from the Kootenay region of British Columbia to the coast that Hill intended to build but did not complete, finding it cheaper to ship the resources of southeastern British Columbia out using existing lines in Washington.

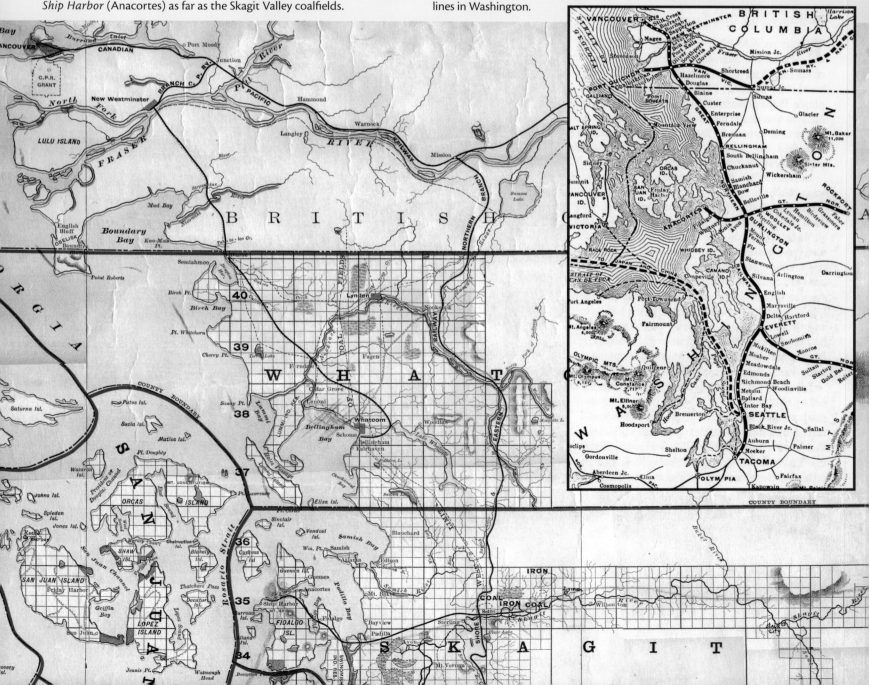

MAP 237 (*left*).
A map prepared to inform potential German immigrants about the lands along the route of the Great Northern Railway in 1892 (but likely drawn about 1890) shows the line still but projected in the Northwest. The route runs north of the route finally taken, north of *Spokane Falls* rather than through it, as Hill later negotiated, and north of *Wanatchee*. It seems clear that the exact route was not known to the creator of this map at the time it was drawn.

Hill assigned Elbridge H. Beckler, his chief engineer in Montana, to locate the route for the Pacific Extension. One of his locating engineers, John F. Stevens, in December 1889 found Marias Pass, a way across the Continental Divide that would not require tunneling. Stevens, with his assistant, Charles Haskell, also found the railroad's path across the Cascades, which Haskell named Stevens Pass. Unfortunately this crossing required a two-and-a-half-mile-long tunnel, which took seven years to complete; in the meantime the Great Northern had to use a complicated set of switchbacks (MAP 238 and MAP 239, *below*).

The original plan called for a northerly route that would have missed Spokane altogether, but in February 1892 Hill visited the city and persuaded the citizens to contribute a right-of-way, and so the line was routed through Spokane, though initially using the tracks of the SLS&E and the OR&N. The Great Northern arrived at Hillyard, now a suburb of Spokane, in June 1892.

The Columbia River was crossed at Wenatchee and a ferry brought in to serve until a bridge could be built; the bridge was completed in May 1893. The construction crews from east and west met at Madison, later renamed Scenic, on 6 January 1893, and Seattle's transcontinental was complete. The first train from St. Paul steamed into Seattle in June, via the line from Everett of subsidiary Seattle & Montana.

Hill not only built a railroad empire but also an economic one—industry and agriculture, after all, would fill his freight trains. Immigration programs swung into high gear. By 1897 Hill had control of the Northern Pacific and, with it, effective control of a major part of the Northwest's rail traffic. In 1899, the Northern Pacific sold 900,000 acres of its Washington and Oregon land grant as timberland to up-and-coming forest giant Weyerhaeuser Timber Company; Frederick Weyerhaeuser just happened to be Hill's neighbor in St. Paul. The transaction was of dubious legality—grant land was supposed to be sold to settlers—but it held, with profound future consequences for the economic development of the Northwest. Hill also attracted apple grower Julius Beebe to Washington; the latter's crop would fill many eastbound Great Northern trains. In 1929, the Great Northern named its premier transcontinental train to Seattle *The Empire Builder* to honor Hill.

MAP 238 (*left*) and MAP 239 (*below*). Switchbacks hardly ever found their way onto railroad literature because they were not something a railroad wanted its patrons to know about. MAP 238 is an 1899 map and views from an independently produced postcard. MAP 239 was published by the Great Northern, but only in 1900, as the 8-mile-long traverse required by the switchbacks was about to be replaced by the 2.6-mile-long Cascade Tunnel. The array of switchbacks, especially on the steeper west slope of the *Summit of Cascade Range*, is quite staggering. Trains had to be split into seven-car sections, each requiring three locomotives (*inset* photo). Construction of the tunnel did not begin until 1897. Smoke in the long tunnel then became a major threat to train crews, and in 1909 the tunnel was electrified. In 1929 it was replaced with a much longer new tunnel (see page 170).

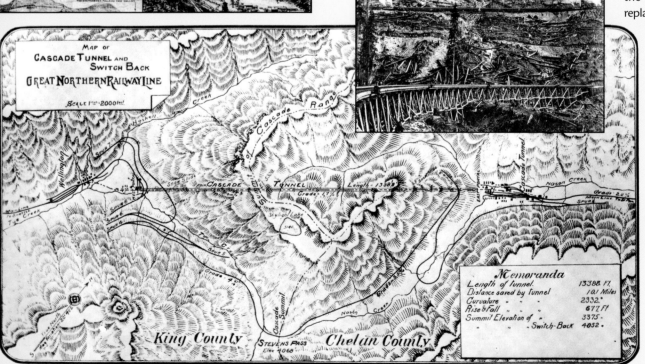

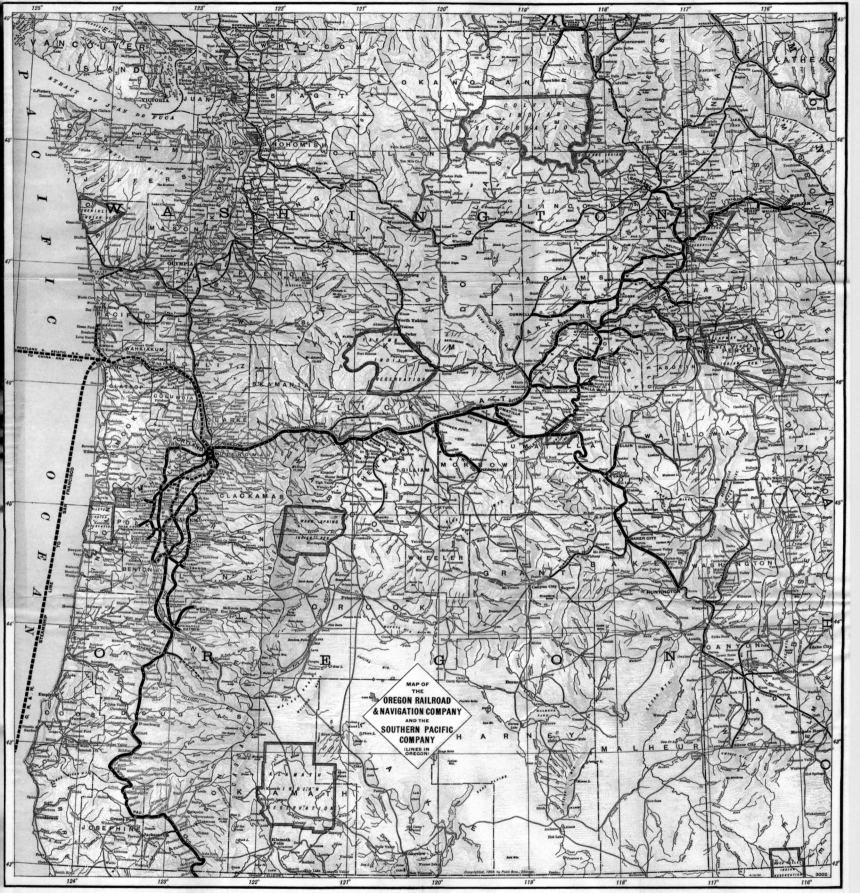

MAP OF
THE
OREGON RAILROAD
& NAVIGATION COMPANY
AND THE
SOUTHERN PACIFIC
COMPANY
(LINES IN
OREGON)

FOR GRAINS, GRASSES, FRUIT, LIVE STOCK AND MINERALS THE TERRITORY EMBRACED IN THIS MAP IS UNEQUALLED.

MAP 240.

The railroad map of Washington and Oregon in 1904. The Southern Pacific line runs south to California; Union Pacific, which now owns the Oregon Railway & Navigation system, runs along the Columbia and then southeast to connect with the Union Pacific main line to Omaha; and the Northern Pacific and Great Northern lines run east from Eastern Washington. Coastal lines link Portland with Vancouver, B.C.; lines serve the mining districts of northeast Washington and the rich agricultural lands of the Palouse. Yet more railroads were on the horizon; as in the rest of the United States, railroads in the Northwest would ultimately overbuild, and many lines would fall into disuse with the rise of competitive forms of transportation.

GOLD AND SILVER

A long series of scrambles for precious metals provided valuable impetus for economic growth in the Pacific Northwest, even when the find was elsewhere. The California Gold Rush of 1849 created demand for food from the Willamette Valley; gold rushes of the 1860s in Idaho and Montana led to the creation of the Oregon Steam Navigation Company and the beginnings of large-scale steamboat navigation on the Columbia; and the Klondike Gold Rush of 1897 was a boost for Seattle.

Major gold rushes and longer-term gold- and other metal-mining activity occurred in southern Oregon in the 1850s (see page 68); in eastern Oregon from 1862 to the 1890s; in northeast Washington between about 1890 and 1920; and in northwest Washington in the 1890s. When the gold ran out, communities died, creating ghost towns. But gold continues to be found and mined in the Pacific Northwest.

In 1862 would-be Confederate soldiers making their way south to California found placer gold in eastern Oregon. They named the location Sumter, after Fort Sumter, the federal fort in Charleston Harbor that was attacked by the Confederates, beginning the Civil War the previous year. The name was later somehow changed to Sumpter.

News of the gold discovery spread rapidly, and hundreds of gold seekers converged on the region. Decades later hard rock deposits were found and set off another round of intense gold mining. By 1900 Sumpter had more than three thousand inhabitants, and both mining and placer dredging were in full swing. By that time the

MAP 241 (*below, main map*).
This stunning map with shaded topography gives an excellent idea of the geomorphology of eastern Oregon. The map was produced in 1885 and shows the extensive gold fields of the region (in yellow) together with a large number of placer mines (red dots) and hard rock (quartz) mines (black circles). Also shown are the wagon roads and trails (in red) and railroads (in black). The Oregon Railway & Navigation's line from the Columbia through *Baker City* to *Huntington* and its connection with the Union Pacific's Oregon Short Line had been completed in 1884, but the narrow-gauge *S.V.R.R.* (Sumpter Valley Railroad) from Baker City to *Sumpter* was not completed until 1891. Although the map title says *Oregon Rail Road and Navigation Co.* the company used the term "railway" until it was reorganized in 1896.

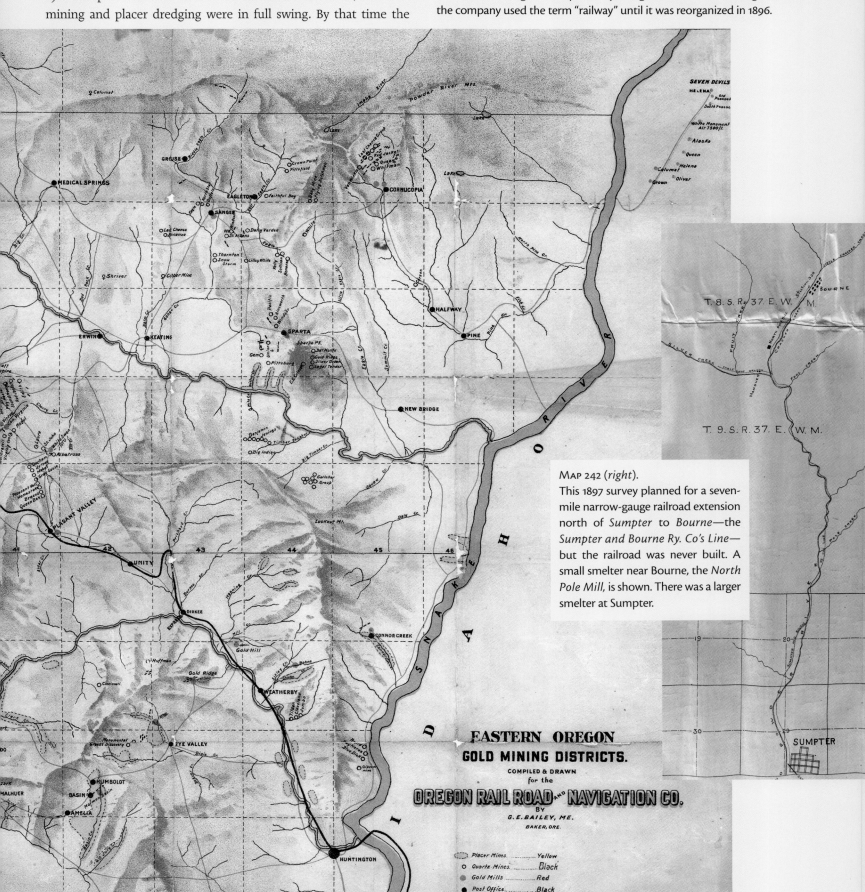

MAP 242 (*right*).
This 1897 survey planned for a seven-mile narrow-gauge railroad extension north of *Sumpter* to *Bourne*—the *Sumpter and Bourne Ry. Co's Line*—but the railroad was never built. A small smelter near Bourne, the *North Pole Mill*, is shown. There was a larger smelter at Sumpter.

EASTERN OREGON
GOLD MINING DISTRICTS.
COMPILED & DRAWN
for the
OREGON RAIL ROAD AND NAVIGATION CO.
BY
G. E. BAILEY, ME.
BAKER, ORE.

Placer Mines	Yellow
Quartz Mines	Black
Gold Mills	Red
Post Office	Black
Roads	Red

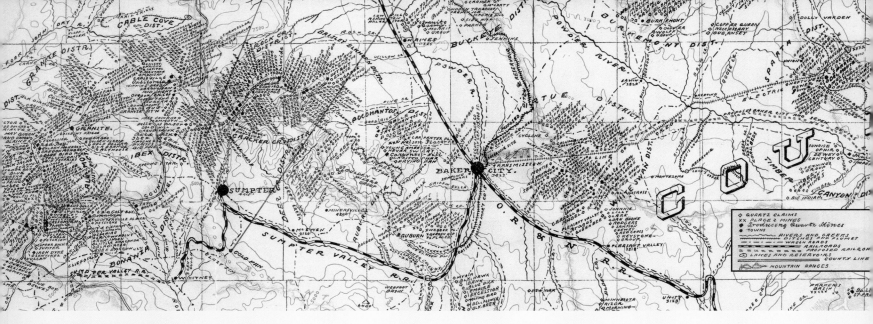

MAP 243 (*above*).

A later map of the gold fields in the *Baker City* and Sumpter Valley area of eastern Oregon, this one dating from 1904, shows even more mines than MAP 241. But by this time many have been worked out. Placer mines are shown by an *X*, and hard rock (quartz) mines are shown by a diamond, filled in if producing; there are more not producing than are. The *Sumpter Valley R.R.* has been completed to *Sumpter* and a branch built to *Whitney*, with a projected line, never built, beyond that.

MAP 244 (*below, left*).

Hundreds of locations in the northwest produced maps like this: a plot of mineral claims concentrated on the areas where gold or other minerals were found or were thought likely to be found. This map is the area around *Republic*, in northeastern Washington, with the Great Republic claim shown at the northwest corner of the *City Limits of Republic*, a townsite plat named after the mine. This map is from a 1910 report.

MAP 245 (*below, right*).

A similar map, this one dated 1888, of silver-mining claims in the area of *Salmon City*, (at left) later Conconully, in north-central Washington. Silver was found in mineralized ledges in this area in 1886. Conconully stills exists, but nothing is left of *Ruby City* (at right on the map) except a few foundations; at one time it had the reputation of being the wildest city in Washington Territory. The map also shows a planned town, *Arlington* (at top of map), but this never existed on the ground.

region was serviced by the railroad, which had reached Baker City in 1884 (see page 91). A narrow-gauge railroad to Sumpter from Baker City, up the South Fork of the Powder River, was completed in 1891, though this was for logging purposes rather than to service the gold mines. In 1916 a fire destroyed most of the town of Sumpter.

Gold was found at a number of locations in northeastern Washington and in adjacent British Columbia in the 1890s. Gold was found near the Canadian border in March 1896 at the Great Republic claim by prospectors Thomas Ryan and Philip Creasor (MAP 244, *below, left*). Creasor later platted a town he called Eureka Gulch, but the name of the town was changed to Republic at incorporation in 1900. The gold rush in this area lasted from 1896 to about 1900 and saw the building of two competing railroads, both completed (in 1901 and 1902) just as the gold rush was over. The region did produce a considerable amount of gold, but clearly not all the gold was found at the time, however, because more than a hundred years later, in 2008, a new gold mine was opened.

In northwest Washington, placer gold had been found in small quantities for decades when, in 1885, more significant amounts of gold were found on the South Fork of the Nooksack River, setting off a minor gold rush, but the major find was of hard rock gold on the North Fork of that river in 1897 (MAP 246 and MAP 247, *right*). The Mount Baker Gold Rush involved perhaps as many as two thousand gold seekers scrambling to stake their claims and emptying out the nearest settlement, Sumas, on the Canadian border. New Whatcom city council combined with the board of trade to produce a prospectus of the gold district, *Whatcom County, the Klondike of Washington,* which they claimed to be free of overstatements but which, of course, was not. And as usual, it was the merchants outfitting the gold seekers that made more money than most of the seekers themselves. Settlements appeared in abundance on maps, if not on the ground: Gold Hill, Trail City, Union City, and Gold City (MAP 249, *far right*)—and there were others. A new town, Shuksan, that sprang up around a supplier's store, was soon reported to have 1,500 inhabitants. The original mine was sold to a Portland syndicate and fortunes were made and lost, but like all gold rushes, it soon came to an end. Nevertheless, a few mines continued operating into the 1930s (MAP 248, *right*). Only a few struck it rich. Of over three thousand claims staked in Whatcom County only two are known to have paid off.

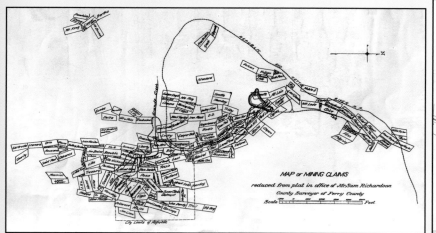

MAP OF MINING CLAIMS
reduced from plat in office of Mr Sam Richardson
County Surveyor of Ferry County

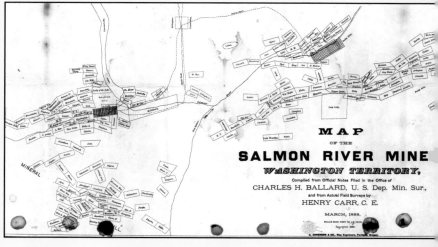

MAP
OF THE
SALMON RIVER MINE
WASHINGTON TERRITORY,
Compiled from Official Notes Filed in the Office of
CHARLES H. BALLARD, U. S. Dep. Min. Sur.,
and from Actual Field Surveys by
HENRY CARR, C. E.
MARCH, 1888.

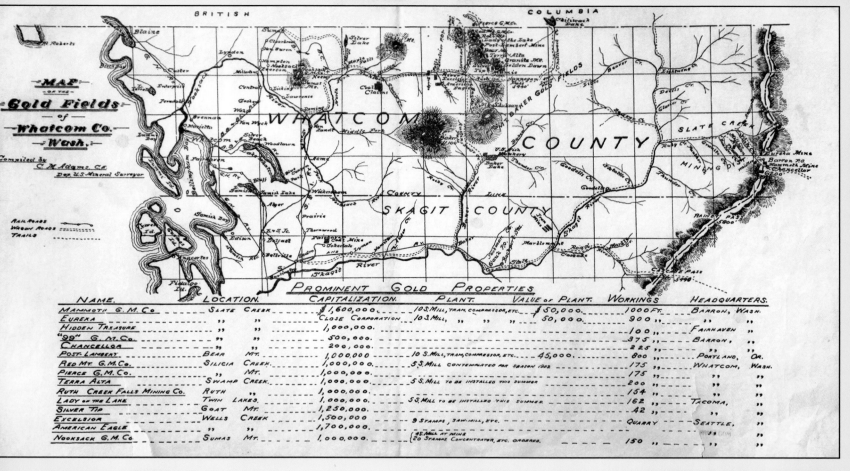

MAP of Gold Fields of Whatcom Co. Wash.

Compiled by C. H. Adams, C.E.
Dep. U.S. Mineral Surveyor

RAILROADS
WAGON ROADS
TRAILS

PROMINENT GOLD PROPERTIES

NAME.	LOCATION.		CAPITALIZATION.	PLANT.	VALUE OF PLANT.	WORKINGS	HEADQUARTERS.
MAMMOTH G.M.Co.	SLATE	CREEK	$1,600,000.	10 S. MILL, TRAM, COMPRESSOR, ETC.	$50,000.	1000 FT.	BARRON, WASH.
EUREKA	"	"	CLOSE CORPORATION	10 S. MILL, "	50,000.	900 "	"
HIDDEN TREASURE	"	"	1,000,000.			100 "	FAIRHAVEN
"99" G.M.Co.	"	"	500,000.			375 "	BARRON, "
CHANCELLOR	"	"	200,000.			225 "	" OR.
POST-LAMBERT	BEAR	MT.	1,000,000.	10 S. MILL, TRAM, COMPRESSOR, ETC.	45,000.	800 "	PORTLAND, OR.
RED MT. G.M.Co.	SILICIA	CREEK	1,000,000.	5 S. MILL CONTEMPLATED FOR SEASON 1902		175 "	WHATCOM, WASH.
PIERCE	"	MT.	1,000,000.			175 "	" "
TERRA ALTA	SWAMP	CREEK	1,000,000.	5 S. MILL TO BE INSTALLED THIS SUMMER		200 "	" "
RUTH CREEK FALLS MINING Co.	RUTH		1,000,000.			154 "	" "
LADY OF THE LAKE	TWIN	LAKES.	1,000,000.	5 S. MILL TO BE INSTALLED THIS SUMMER		162 "	TACOMA, "
SILVER TIP	GOAT	MT.	1,250,000.			42 "	" "
EXCELSIOR	WELLS	CREEK	1,500,000.	9 STAMPS, SAW-MILL, ETC.		QUARRY	SEATTLE. "
AMERICAN EAGLE	"	"	1,700,000.	5 S. MILL AT MINE		"	"
NOOKSACK G.M.Co.	SUMAS	MT.	1,000,000.	20 STAMPS CONCENTRATOR, ETC. ORDERED.		150 "	" "

MAP 246 (*above*) and MAP 247 (*right*).

The Mount Baker gold fields of Whatcom County, in northwest Washington, on maps drawn in 1902. MAP 246 shows *Fairhaven* and *Whatcom,* two of the three towns that would join in 1903 to form Bellingham. The *B.B. & B.C. R.R.* (Bellingham Bay & British Columbia Railroad) was completed between *Sumas* and *Bellingham Bay* in 1891. The proposed railroad extension was not built east of Glacier (not shown on MAP 247 but located seven miles east of *Maple Falls,* where *Glacier Cr.* joins the Nooksack). MAP 247 is a detailed view of the gold-mining area. The *Post-Lambert Group* was the site of the original gold find on the North Fork of the Nooksack in 1897, by prospectors Jack Post and Russ Lambert.

MAP 248 (*below, left*).

A later plan of one of the mining subareas, *Swift Creek,* between *Mt. Baker* and *Mt. Shuksan.* This map appears to have been drawn in the 1930s.

MAP 249 (*below, right*).

A map produced by commercial groups in the Bellingham Bay cities. In addition to the mining claims, it shows the towns platted in the area—*Gold Hill, Trail City, Wilson's Townsite, Union City,* and *Gold City*—none of which exist today.

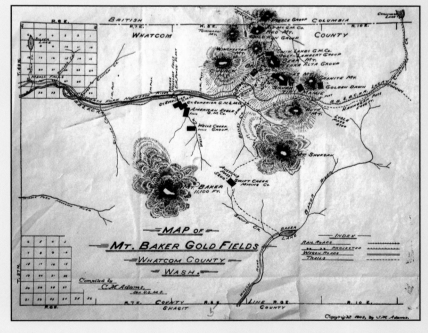

MAP of Mt. Baker Gold Fields Whatcom County Wash.

Compiled by C. H. Adams, Dep. U.S.M.E.

INDEX
RAIL ROADS
WAGON ROADS
TRAILS

Copyright 1902 by C. H. Adams.

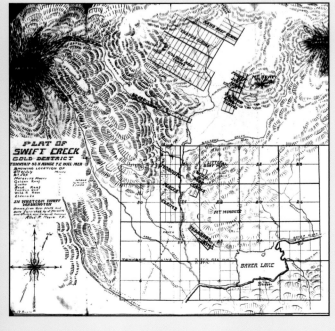

PLAT OF SWIFT CREEK GOLD DISTRICT

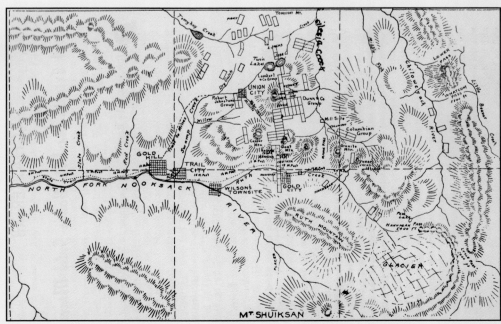

PORTLAND PREEMINENT

Still regional rivals, Portland and Seattle have tried to best each other for more than a hundred years. But, until about 1910, Portland was the more populous. It was the coming of the direct transcontinental railroad to Seattle and Puget Sound that allowed that city to become larger than Portland, confined as it was by the vagaries of the Columbia River.

Portland's settlement had begun in 1843, with the land claim of William Overton and Asa Lovejoy (see page 60), and was incorporated as a city in 1851, soon emerging as the principal settlement of the Willamette Valley because of its position as the highest practical point of navigation for oceangoing ships. The city became the focal point of a network of river steamboats in the 1860s (see page 76), and the first transcontinental railroad reached the city in 1883 (see page 84).

When in 1891 Portland, East Portland, and Albina (the latter two on the east bank of the Willamette) voted to amalgamate, a city of seventy thousand was created, and, as the *New York Times* hastened to point out, the *forty-first* city in size in the nation. Further mergers, with Linnton and St. Johns, took place in 1915, such that Portland then burgeoned to a city of over a quarter million inhabitants. Many external economic factors, of course, drove this growth, but two important internal improvements—bridges and streetcars—facilitated it.

MAPS 250–253 (four maps loosely fitted together, *below*).
The Donation Land Claim of David and Lucinda Hill grew into the little settlement of *Hillsborough*, now Hillsboro and part of the western suburbs of metro Portland. The Hills and several others arrived in 1841, and the settlement, at first named East Tualatin Plains and then Columbia, was renamed Hillsborough in David Hill's honor in 1850. These maps all date from 1852 and are the corners of four separate maps, being the surveys done at that time by the General Land Office surveyors (see page 57). Hillsborough sat astride the Portland Base Line at the corner of four townships. Similar survey maps for the area that is now downtown Portland are shown on page 61.

MAP 254 (*left, bottom*).
Survey maps from 1854 and 1855 show the area now covered by the eastern suburbs of Portland and the city of Gresham but also show that at that time there was very little settlement here. The left (west) side of this map continues from the right (east) side of MAP 134, page 61.

MAP 255 (*below, top*).
This 1860 survey shows the plat of the town of *Vancouver*, Washington. The area west of Fort Vancouver was claimed by American Henry Williamson in 1845 and named Vancouver City. Also shown is the U.S. *Military Reserve*, established in 1849.

MAP 256 (*below, bottom*).
A swath of what is today the built-up area of Vancouver was *Claimed by the Hudson Bay Company under the Treaty of 1846* on this 1856 survey map. The claims of the company were not settled until 1863, when it was awarded $200,000 for all its claimed lands south of the international boundary.

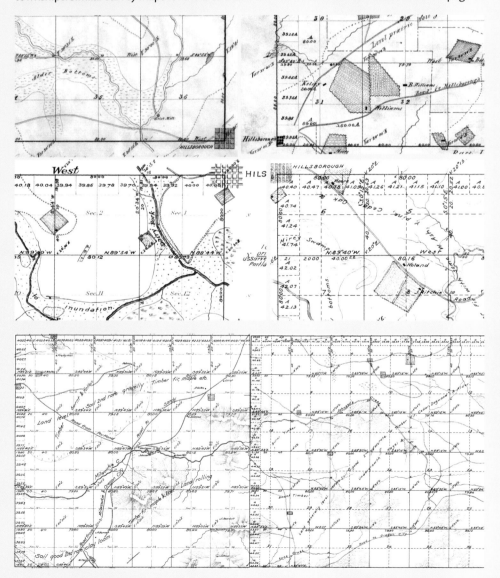

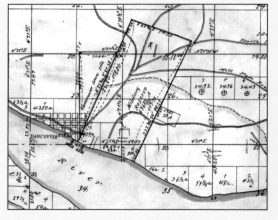

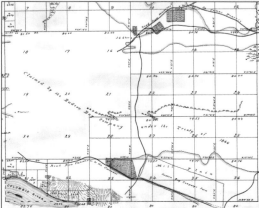

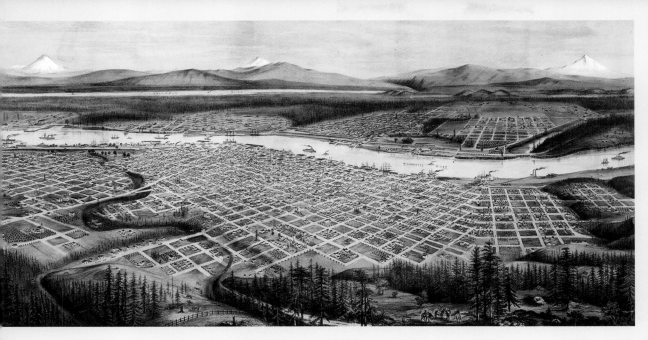

MAP 257 (*left*).
A bird's-eye map of Portland in 1881, two years before the arrival of the transcontinental railroad. The view is from Portland Heights, looking east to East Portland and Albina.

MAP 258 (*below*).
A more utilitarian bird's-eye map of Portland in 1888, again looking east toward the Willamette River and East Portland.

In common with just about every other wooden-building, woodstove-heated western city of note, Portland suffered from the contagion of fire. In 1873 a great fire consumed twenty blocks before it was contained, with fire companies coming from Vancouver by steamboat and from Salem by train to assist. After debating at length what could be done to prevent such calamities in the future, city council made a decision—to purchase a louder fire bell!

Portland has so many bridges that it is sometimes called the City of Bridges. Up to 1887, only ferries crossed the Willamette. The first bridge, the Morrison Street Bridge, was completed in 1887. It was a private bridge and as such levied tolls—20 cents for a wagon with two horses. A railroad bridge, but with an upper deck for road traffic, was built by the Oregon Railway & Navigation Company in 1888. This Steel Bridge was the first built of steel (rather than the wrought iron normally used at this time) on the West Coast. An-

other private bridge, the Madison Street Bridge, was completed in 1891. With the merger of Portland with East Portland and Albina in 1891, citizens clamored for a toll-free river crossing. A committee of citizens purchased the Madison Bridge late that year and removed the tolls. A further toll-free bridge, the swing-span Burnside Bridge, was completed by the committee in 1894.

To the north, the Columbia presented a greater barrier. Until 1917 the river was crossed by a ferry (MAP 261, *overleaf*), but that year a major vertical-lift truss bridge was opened to great fanfare, finally connecting Vancouver to Portland in a fashion that would eventually permit the two to grow into a single urban area. The bridge carries Interstate 5 across the river today, with the help of a nearly identical duplicate span completed in 1958.

Before the advent of the automobile, a streetcar system was absolutely vital for the growth of suburbs in any aspiring city. Developers

MAP 259 (*right*).
This finely detailed map of Portland is part of a Coast and Geodetic Survey map of the Columbia and Willamette published in 1909. Added in pen and pencil are annotations to 1913. The Steel Bridge was replaced in 1912 with a new structure in a slightly different location; these are shown as the *New Bridge* and the *Bridge Removed*. At top, the bridge at *Saint John* is a railroad span built by the Northern Pacific in 1908; the St. Johns road bridge, Portland's only suspension bridge, was not completed until 1931. Going south, the bridges shown are the Burnside Bridge, opened in 1894; the second Morrison Bridge, completed in 1905; and the Hawthorne Bridge, which replaced the Madison Bridge in 1910. On the south side of *Guilds Lake* is the site (unnamed) of the 1905 Lewis and Clark Centennial Exposition (see MAP 264, *overleaf*). Note the new channel *Dredged to 8´ 300´ wide 1911* on the west side of *Swan Island*, the only channel used today, the one to the east of Swan Island having been reduced to a closed basin by landfill in the 1920s.

MAP 261 (*below*).
Dated 1908, this map shows the ferry in place before the Interstate Bridge was completed in 1917. It also does not show the railroad bridge shown on MAP 260.

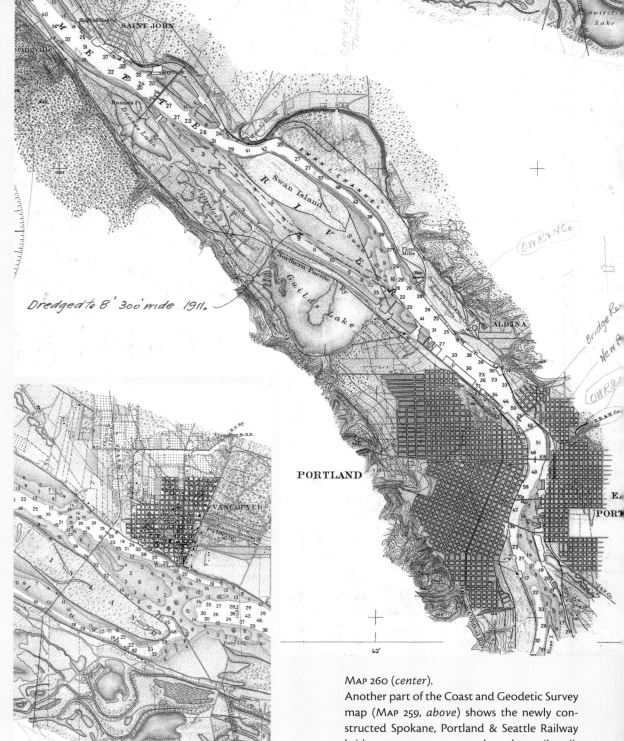

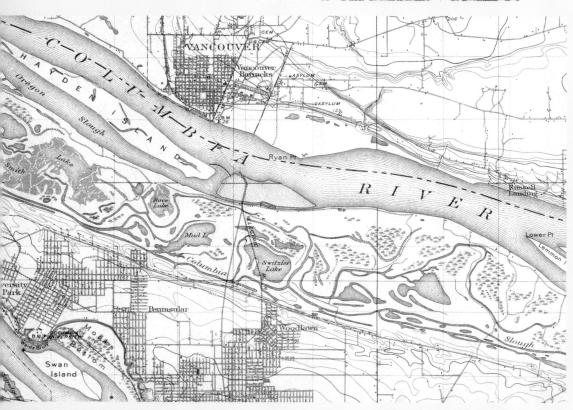

MAP 260 (*center*).
Another part of the Coast and Geodetic Survey map (MAP 259, *above*) shows the newly constructed Spokane, Portland & Seattle Railway bridge, necessary to complete that railroad's line along the north bank of the Columbia into Portland. The bridge was opened in November 1908. Note the ferry landings.

MAP 262 (*above*).
One of the finest of the bird's-eye-view maps of any city in the West, this map was published in 1890. It was an advertising item, a valuable graphic aid in the days before television and color newspapers. Businesses and residents purchased space around the perimeter of the map to depict their business or fine residence, and the map was sold by subscription. The map displays the classic "super-busy" look, with the river full of ships, railroads full of trains, and many smoking chimneys—a sign of prosperity. The Steel Bridge is at left, the Morrison and Madison bridges at right.

MAP 263 (*below*).
This enlarged detail of the bird's-eye map above shows the Portland Cable Railway Company's cable car incline up Chapman Street to Portland Heights; the photo (*inset, left*) dates from about 1900. The incline of over 20 percent was too steep for any other type of transit. It was 1,040 feet long and 247 feet high and was in use from 1890 to 1904.

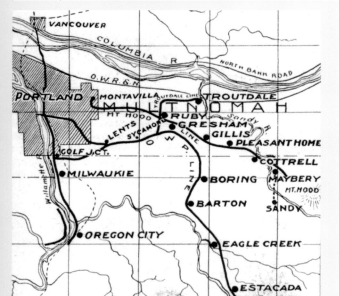

Map 264 (*above*).
In 1905 Portland hosted an extravagant fair intended to boost the economic fortunes of the city's businesses; the centennial of Lewis and Clark's visit in 1805 (see page 27) seemed like a perfect theme. This is a bird's-eye-view map of the elaborate buildings constructed for what was officially the Lewis and Clark Centennial and American Pacific Exposition and Oriental Fair, a title designed, it seems, to cover all bases. The exposition was located on the south side of Guild's Lake. The Willamette is shown beyond the lake.

Map 265 (*left*).
The interurban lines of the Portland Railway, Light & Power (PRL&P) are shown on this 1910 map. Narrow-gauge lines are depicted, appropriately enough, by thinner lines.

Map 266 (*left, bottom*).
The PRL&P network two years later additionally shows the Mount Hood Railway line, completed in 1911 and electrified in 1914. Today this right-of-way is utilized by the eastside MAX line.

would pay handsomely for streetcar service, since it ensured the viability of their project. The Portland Street Railway began horse-drawn service in 1872, and in 1887 steam-powered "dummies" were introduced. Made to appear like passenger cars, they were thought less likely to frighten horses.

In 1889, the first electric streetcar was introduced, running across the Steel Bridge into Albina. A cable car to Portland Heights opened in 1890 (MAP 263, *previous page*). From the 1890s on, a network of competing lines, both steam and electric, facilitated suburban development in a way not previously possible. By 1895 the East Side Railway was operating a service from Portland to Oregon City, considered to be the first true interurban in the United States. In 1906, after a series of mergers and takeovers in the two years before, a number of streetcar and interurban lines were consolidated into Portland Railway, Light & Power, which operated both a city streetcar system (as shown on MAP 267, *right*) and interurban lines (MAP 265, *above, center,* and MAP 266, *left*) as well as the electrical supply system. The company operated 375 cars over nearly two hundred miles of track.

By 1936 buses had begun to replace streetcars on many routes, and the last streetcar ran in 1950, the result of what proved to be a shortsighted decision, given the modern rebirth of rail transit.

The Gothenburg Plan

Portland in its early days was, as many cities were at that time, a less than perfect place. In 1889 the *Oregonian* called Portland "the most filthy city in the Northern states," a reference to its unsanitary sewer system. Prostitution and illegal saloons serving round-the-clock liquor created social problems.

Many attempts at reform were made, and one of these, recognizing the near-impossibility of outright temperance, promoted the idea of government control and effective monopoly of the liquor business; since saloons and prostitution were inextricably linked, this would have dealt with two problems at a single stroke.

The Gothenburg System originated in the 1890s in, as its name would suggest, Sweden. The system called for the creation of a company that was owned by private investors or the city government. In Portland, the company would have received a monopoly license to dispense liquor and control the hours that saloons were open. All employees would have been salaried, and all profits would accrue to the city at a prescribed rate of interest, thus eliminating all motive for promoting liquor sales. The number of saloons would have been dramatically reduced right away, with no increase allowed for ten years. Shareholders would be concerned only with the stability of the system in order to preserve their dividends. From an investment point of view, this made a lot of sense, with high returns and relatively low risk.

The attempt at reform failed, but in 1905 reformers elected Harry Lane mayor of Portland, and he did carry out a program of raiding illegal saloons and shutting them down. However, he was not generally supported by his council, whose members were more easily influenced by special interests.

And a footnote: today Portland boasts thirty-eight breweries—more than any other city in the world.

WOULDN'T THIS BE AN IMPROVEMENT?

If you are in favor of
- Fewer saloons in Portland
- Higher license and more revenue for the city
- Stricter regulations
- Putting the sale of liquor in strong responsible hands
- Eliminating all saloons from the resident districts forever
- Midnight closing
- Taking the saloon completely out of politics

THEN VOTE FOR THE GOTHENBERG PLAN

MAP 267 (*below*).
The Gothenburg Association's map of Portland in 1905, showing in red the areas to which saloons would be confined under its Gothenburg Plan. The base map used also shows the extensive streetcar network in place by this time and shows Portland's growth as influenced by this network. Developers often promoted their lots by contracting with a streetcar company to provide service, but all too often the company went out of business before economically feasible traffic levels could be achieved. It took streetcar consolidation in 1906 to partly address this issue. *Above* and *below, left* are text panels from the Gothenburg broadsheet.

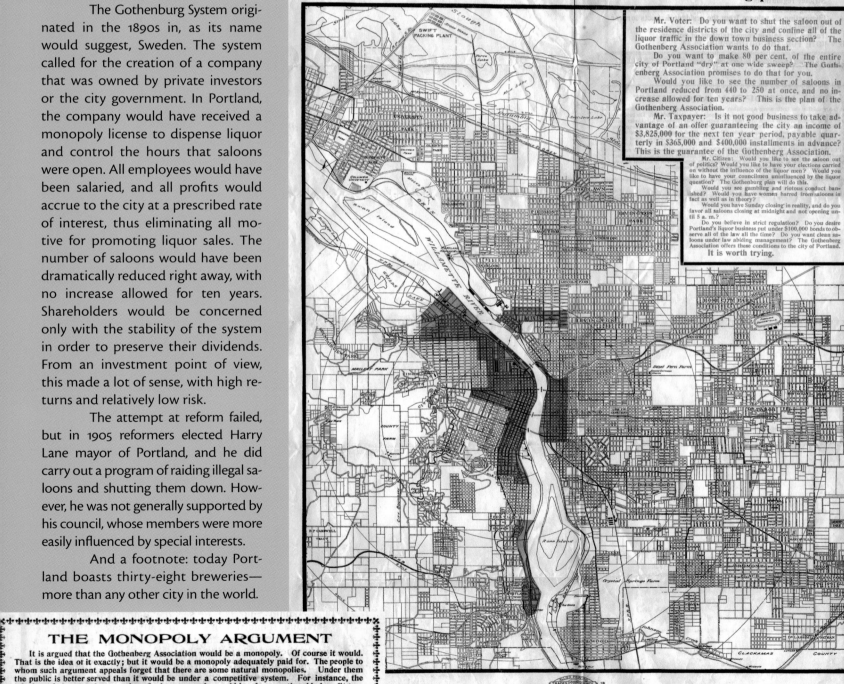

Map of Portland, Oregon. The red ink area shows the business districts to which saloons will be limited under the Gothenberg plan.

Mr. Voter: Do you want to shut the saloon out of the residence districts of the city and confine all of the liquor traffic in the down town business section? The Gothenberg Association wants to do that.

Do you want to make 80 per cent. of the entire city of Portland "dry" at one wide sweep? The Gothenberg Association promises to do that for you.

Would you like to see the number of saloons in Portland reduced from 440 to 250 at once, and no increase allowed for ten years? This is the plan of the Gothenberg Association.

Mr. Taxpayer: Is it not good business to take advantage of an offer guaranteeing the city an income of $3,825,000 for the next ten year period, payable quarterly in $365,000 and $400,000 installments in advance? This is the guarantee of the Gothenberg Association.

Mr. Citizen: Would you like to see the saloon out of politics? Would you like to have your elections carried on without the influence of the liquor men? Would you like to have your councilmen uninfluenced by the liquor question? The Gothenburg plan will do this.

Would you see gambling and riotous conduct banished? Would you have women barred from saloons in fact as well as in theory?

Would you have Sunday closing in reality, and do you favor all saloons closing at midnight and not opening until 5 a. m.?

Do you believe in strict regulation? Do you desire Portland's liquor business put under $100,000 bonds to observe all of the law all the time? Do you want clean saloons under law abiding management? The Gothenberg Association offers these conditions to the city of Portland.

It is worth trying.

THE MONOPOLY ARGUMENT

It is argued that the Gothenberg Association would be a monopoly. Of course it would. That is the idea of it exactly; but it would be a monopoly adequately paid for. The people to whom such argument appeals forget that there are some natural monopolies. Under them the public is better served than it would be under a competitive system. For instance, the telephone is a natural monopoly. Such a monopoly would be of unquestionable benefit to a community if it were properly paid for, as the Gothenberg franchise proposes. Most of the evils of the liquor business arise from competition. IT IS NOT A BUSINESS IN WHICH COMPETITION OUGHT TO BE ALLOWED TO EXIST. The liquor business is a natural monopoly. IT OUGHT TO BE A MONOPOLY under the strictest regulations and subject to control by the people at all times. Do not be misled by arguments of interested persons or interested newspapers. ☞ Read the enclosed franchise yourself.

MAP 268 (*above*).
A bird's-eye view promoting one of the dock facilities in the lower Willamette in 1923. This is Municipal Terminal No. 4, a mile or so upstream from the southern tip of Sauvie Island. The Columbia is shown in the background.

MAP 269 (*right*).
In 1931 Portland Harbor was feeling the pain of the Depression along with most of America. This ad from that year promotes the position of Portland as the gateway to the entire Columbia River basin, with the "walls" of the "basin" no doubt deliberately nearly obscuring the competing port of Seattle.

MAP 270 (*below*).
A 1926 map of Portland for visiting motorists, showing the location of the *Auto Camp Grounds* and main roads.

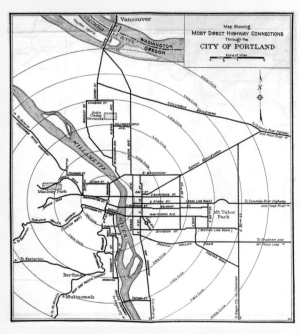

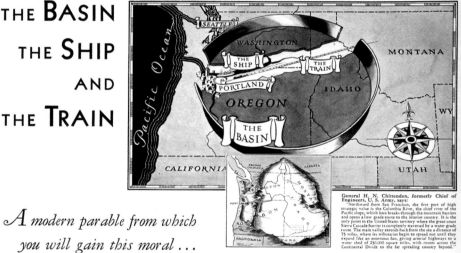

Port	Distance to Ocean	Annual Fog Hours	
SEATTLE	117 MILES	1091	
PORTLAND	110 MILES	541	10-YEAR AVERAGE
SAN FRANCISCO	10 MILES	1089	
LOS ANGELES	24 MILES	841	

FEWER FOG HOURS AT PORTLAND'S PORT OF ENTRY

The Hutchinson Maps

Most of the classic bird's-eye-type maps were produced in the late nineteenth century, but they were such vivid aids to visualization that they were often used to illustrate the impact of a proposed project. They were a favorite of engineering departments for this reason, and none used them better than the Oregon State Highway Commission, which employed a talented graphic artist named Frank Hutchinson.

Between the mid 1930s and the 1950s he produced a stream of superb bird's-eye and other illustrations to depict proposed highways and other projects. The set he produced for a highway project in Portland in 1939, three of which are shown here, also provide us with a historical record of the city at that date. Other Hutchinson drawings are shown on pages 164, 165, and 208.

MAP 271 (*above*), MAP 272 (*right*), and MAP 273 (*below*). Three bird's-eye-view maps of Portland drawn by Frank Hutchinson in 1939. The endeavor was a proposed new bridge across the Willamette and its associated connectors, a project that failed to gain voter support and was not completed until 1973 as the Fremont Bridge, which was built in almost the same position, carrying Interstate 405 across the river. Apart from the proposed bridge and new roads, the view is an accurate representation of Portland in 1939, probably because it was drawn from an aerial photo, a luxury that early bird's-eye map creators did not have. Note the airport, on Swan Island, just north of the bridge. A larger replacement airport on the banks of the Columbia was nearing completion at this time. Harbor Drive, partly shown on MAP 272, is now Governor Tom McCall Waterfront Park.

EXTENSION · ARTERIAL · HIGHWAY · SYSTEM · INTO · AND · THROUGH · PORTLAND | OREGON · STATE · HIGHWAY COMMISSION ~ R.H. BALDOCK, HIGHWAY ENGINEER.

PROPOSED · TRAFFIC · INTERCHANGE · BETWEEN · FRONT · AVE · HARBOR · DRIVE · & · STEEL · BRIDGE

XTENSION · ARTERIAL · HIGHWAY · SYSTEM · INTO · AND · THROUGH · PORTLAND | OREGON · STATE · HIGHWAY COMMISSION ~ R.H. BALDOCK, HIGHWAY ENGINEER.

SEATTLE ASCENDANT

By 1870 the little town of Seattle had grown into the third-largest city in Washington Territory. Its population of 1,100 was exceeded only by those of Walla Walla (1,394) and Olympia (1,203). A decade later Seattle had 3,533 inhabitants; by 1890, with the promise of a direct transcontinental, the Great Northern, it had ballooned to 42,837, and this doubled again by the turn of the century.

From 1891 Seattle grew also by annexations of smaller cities or other settlements that had grown up around it, communities that often voted to join their larger neighbor for its superior water supply and other services. Seattle has a large and complex history of growth by annexation. Seventeen square miles, including Magnolia and Fremont, were annexed in 1891 (Map 278, *page 116*); South Seattle joined the city in 1905; and West Seattle and Ballard in 1907.

Like many early western cities—including Portland (see page 107)— Seattle experienced a catastrophic fire, a common occurrence because of the extensive use of wood for buildings and wood- and coal-burning stoves for heating.

Seattle's great fire began with a pot of glue. On 6 June 1889 a newly arrived Swedish immigrant in a woodworking shop at Front and Madison streets overheated glue on an iron stove, and the glue pot exploded; his attempts to put out the resulting fire with water simply led to its being further

Right, top.
By 1865 Seattle had grown into a small town. Of particular interest in this photograph are the poles spaced along the street; these are to carry telegraph wires soon to link the community with the outside world. The view looks north from Main Street.

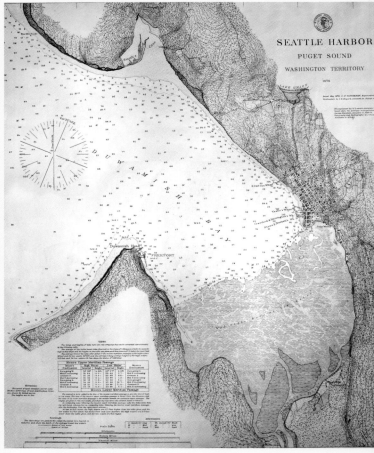

Map 274 (*above, right*) and Map 275 (*below*).
Map 274, an 1879 map of Seattle Harbor published by the U.S. Coast and Geodetic Survey, shows Seattle to still be a compact little settlement. The map's style can be contrasted with that of Map 275, which is a bird's-eye treatment published the year before. The bird's-eye was used for promotional purposes and hence is embellished with much industry, shipping, and trains; the railroad shown is Seattle's own Seattle & Walla Walla Railroad (see page 83).

SEATTLE, WT.

spread around. Soon the building was engulfed in flames, which quickly spread to adjacent buildings and moved down Front Street, claiming the roof of a new brick opera house. Decisive management of firefighting resources might still have saved the day, but the fire chief was out of town and his inexperienced deputy was in charge. As more hoses were connected to hydrants the water pressure dropped precipitously, making firefighting more and more difficult. The mayor organized teams of men to carry goods from threatened stores to the waterfront, but the fire caught up to them and eventually destroyed even the wharves along the shore. Coal bunkers containing perhaps three hundred tons of coal caught fire and burned for days.

MAP 276 (*above*).
Seattle was promoted with a series of bird's-eye maps, many of which are shown on these pages. This was an 1884 version.

MAP 277 (*below*).
Real estate company Llewellyn, Dodge published this bird's-eye map of Seattle in 1889 just weeks before the great fire destroyed many of the buildings shown. After the fire someone drew the boundary of the burned area on the map, thus leaving us a visual record of the fire's extent.

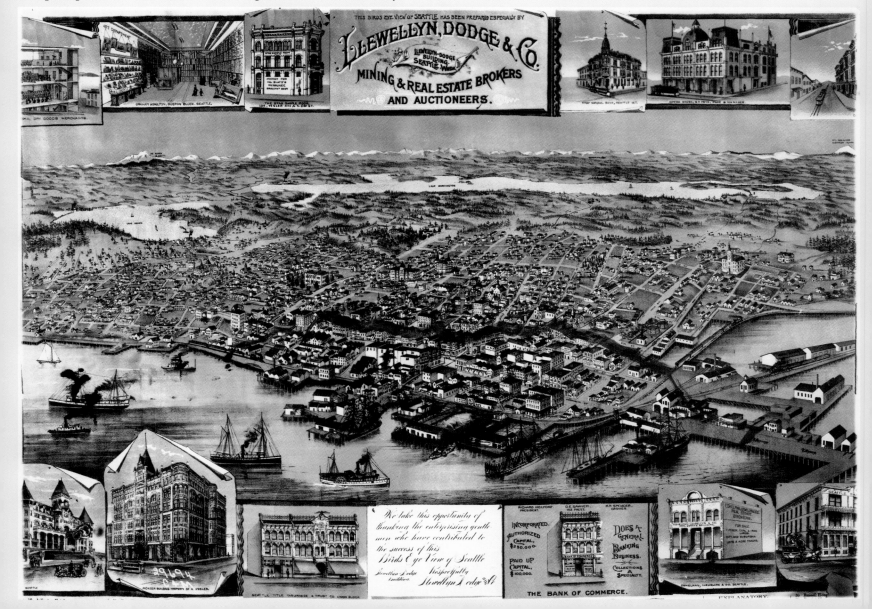

With the commercial area of the city gutted, aid poured in from other West Coast cities as far south as San Francisco. Commercial rival Tacoma, which had dispatched firefighters right away, sent three boats and a train full of food, set up a large tent, and began serving food to Seattle residents made homeless.

Seattle quickly rose from its ashes; within two years the city had rebuilt most of the area lost—but this time the buildings were all made of brick.

Map 278 (*left*).
From the 1891 report of the Seattle city engineer comes this map showing annexations up to that year, when seventeen square miles (here colored dark green) were added to Seattle. This was just the beginning; further additions, including land reclaimed from tidelands, would add close to a hundred square miles to the city's original extent.

Map 279 (*below*).
Over sixteen square miles would be added to Seattle in 1907 with the annexation of West Seattle, shown here on a superb 1899 real estate map. The distance to Seattle, via the *Seattle Terminal Railway*, has been truncated. Note the subdivision of tideland lots north of *Railroad Ave*. At this time a ferry ran between West Seattle and Seattle.

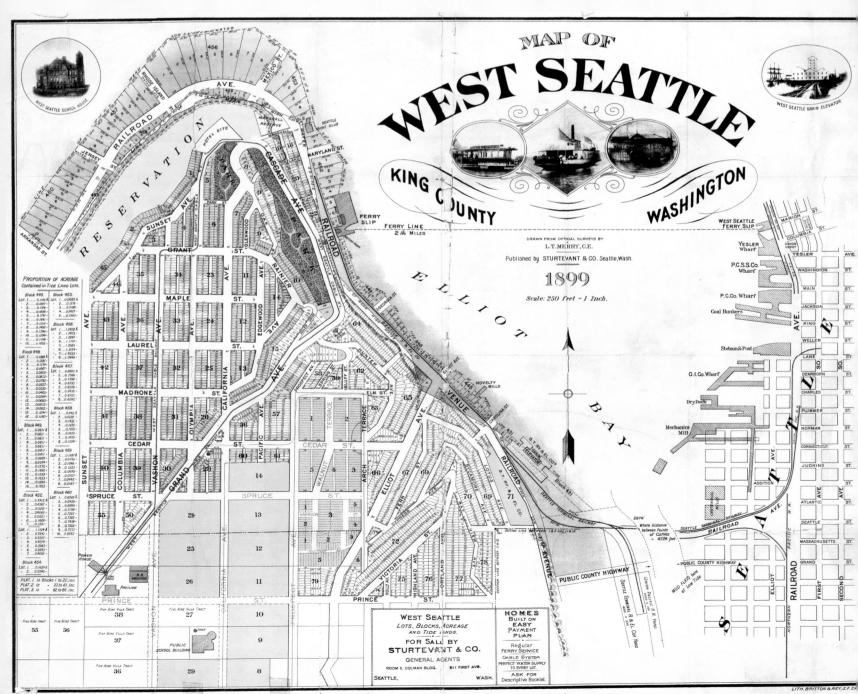

For all cities during this period, the development of a street-car system was of critical importance in facilitating growth. Seattle was no different and relied on the streetcar to integrate annexed areas into the city. A horse-drawn car began operating along Second Avenue in 1884, but it soon became clear that horses would be inadequate to pull streetcars up Seattle's prodigious hills, and for this purpose a cable car, the Lake Washington Cable Railway, began operations in 1888, connecting Pioneer Square with the shores of Lake Washington. The first electric streetcar was introduced the following year, and horses were phased out; Seattle became the first western city to have a fully electric streetcar system. By 1891 lines had reached Ballard, and within a few more years lines covered the city (Map 281, *overleaf*). An interurban line between Seattle and Tacoma, the Puget Sound Electric Railway, was opened in 1902, and a link to Everett in 1910 (Map 282, *overleaf*).

When the Northern Pacific purchased the Puget Sound Shore Railroad (see page 95), Seattle was offered parity in grain transport rates with Tacoma and Portland on the condition that the city provide space for at least half a million bushels. For this purpose the Seattle Terminal Railway & Elevator Company was incorporated in March 1890. As there was little suitable space on the Seattle waterfront, a site across Elliott Bay in West Seattle was purchased, where elevators were built. They were reached by a three-mile-long trestle across the bay, and passenger service was added to freight. The trestle formed a barrier to silt from the Duwamish River, and silting of the area south of the trestle increased (Map 283, *overleaf*).

Map 280 (*above*).
A superb bird's-eye map of Seattle in 1891, which emphasizes the downtown area as brick built following a great fire two years earlier. Alki Point, first settled by Arthur Denny in 1851 and then abandoned, is at left. This map illustrates the positions of *Lake Washington* and *Lake Union*. In 1917 (though not officially completed until 1934) a system of locks and channels joined the two lakes, permitting access to Lake Washington from Puget Sound. This map shows a narrow connection between the two lakes, the Montlake log canal, first cut in 1861 and enlarged in 1883. The shallow southern part of Elliott Bay would eventually be filled in, in part using fill from the regrading of Seattle's downtown hills. *Great Northern* railroad track runs to the harbor from the north; the company would complete its direct transcontinental link to Seattle in 1893. Also visible are the three different street orientations of the city center, the result of three different original owners.

In the 1880s plans had been made for a canal to be built from Lake Washington through Lake Union to Shilshole (see page 128) to transport coal and timber to Puget Sound. This plan met some serious competition in 1893 when a former territorial governor, Eugene Semple, managed to get a bill through the state legislature allowing private companies to construct public waterways and charge liens on reclaimed tideland to finance the project. Then Semple announced a canal scheme of his own—one that would cut through Beacon Hill south of downtown and sluice the fill into the tidelands, at the same time dredging channels into the Duwamish River to accommodate oceangoing ships. Semple raised half a million dollars to finance his grand scheme. Work began in July 1895, and within a year nearly a hundred acres of tidelands had been filled and sold.

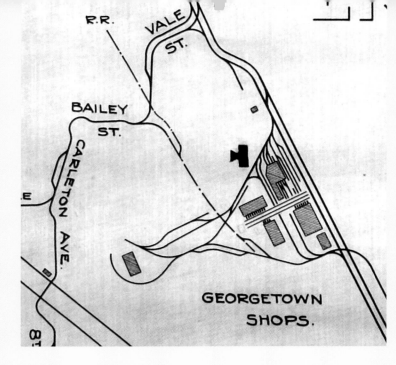

GEORGETOWN SHOPS.

MAP 281 (*left* and *above*).
Two parts of a map of the streetcar system in Seattle in 1917, the lines of the Puget Sound Traction, Light & Power Company. The system in North Seattle and detail of the *Georgetown Shops* in South Seattle are shown. The map is a very large finely drawn pen-and-ink creation on linen. The black symbols indicate the locations of telephones, the point of connection and control between streetcar crews and head office before radio had become practical for such communications. Single lines depict single tracks; double lines, double tracks. Dashed lines are competitive lines; here the *Municipal Car Line* is named. Lines with a long and short dash, only shown where they cross streetcar lines, are steam railroad tracks.

MAP 282 (*left, below*).
A map and timetable for the Puget Sound Electric Railway published in 1910, the year an extension northward to Everett was completed. Most Seattle-to-Tacoma services took one and a half hours; two trains each way were expresses taking only an hour and fifteen minutes. The service lasted until 1928; it was put out of business by automobiles and trucks.

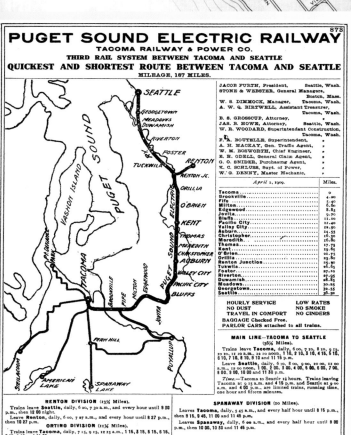

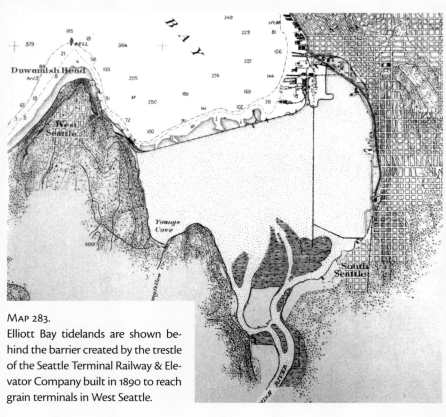

MAP 283.
Elliott Bay tidelands are shown behind the barrier created by the trestle of the Seattle Terminal Railway & Elevator Company built in 1890 to reach grain terminals in West Seattle.

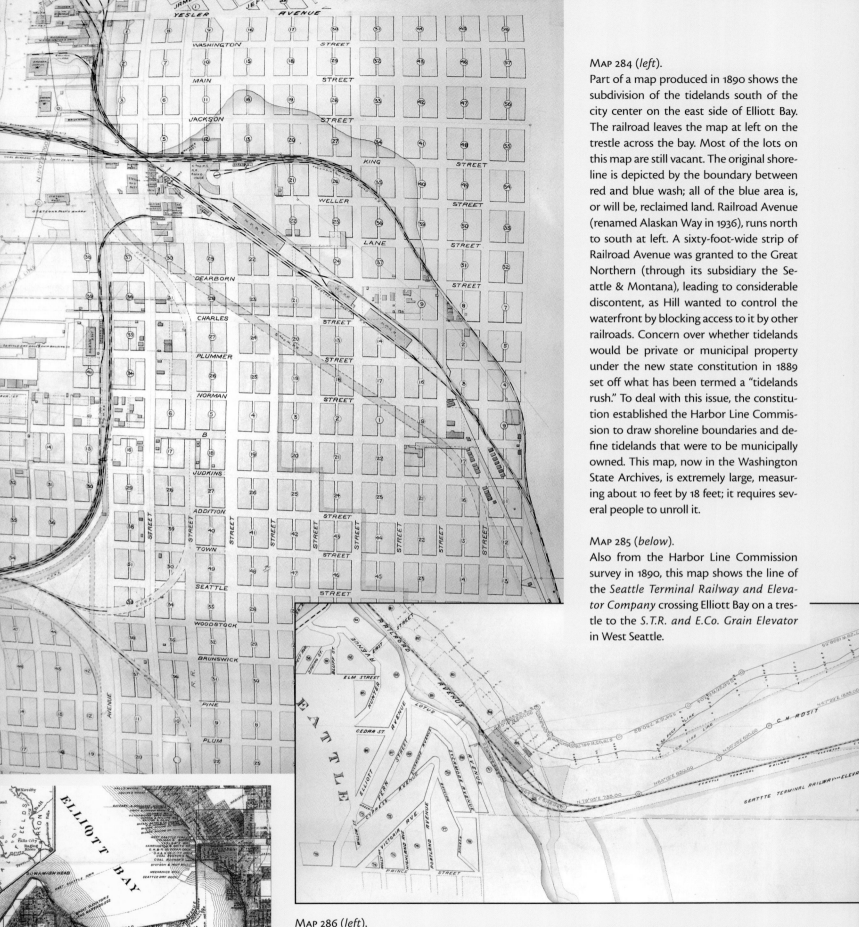

MAP 284 (left).
Part of a map produced in 1890 shows the subdivision of the tidelands south of the city center on the east side of Elliott Bay. The railroad leaves the map at left on the trestle across the bay. Most of the lots on this map are still vacant. The original shoreline is depicted by the boundary between red and blue wash; all of the blue area is, or will be, reclaimed land. Railroad Avenue (renamed Alaskan Way in 1936), runs north to south at left. A sixty-foot-wide strip of Railroad Avenue was granted to the Great Northern (through its subsidiary the Seattle & Montana), leading to considerable discontent, as Hill wanted to control the waterfront by blocking access to it by other railroads. Concern over whether tidelands would be private or municipal property under the new state constitution in 1889 set off what has been termed a "tidelands rush." To deal with this issue, the constitution established the Harbor Line Commission to draw shoreline boundaries and define tidelands that were to be municipally owned. This map, now in the Washington State Archives, is extremely large, measuring about 10 feet by 18 feet; it requires several people to unroll it.

MAP 285 (below).
Also from the Harbor Line Commission survey in 1890, this map shows the line of the Seattle Terminal Railway and Elevator Company crossing Elliott Bay on a trestle to the S.T.R. and E.Co. Grain Elevator in West Seattle.

MAP 286 (left).
In contrast to MAP 284, this map depicts more correctly the actual state of affairs in Elliott Bay in 1890, whereas MAP 284 represents the legal situation by showing subdivision on the tidelands, which, by definition, were only land at low tide. Some subdivision encroachment onto the bay is shown on this map, however. Seattle formally annexed the southern Elliott Bay tidelands in 1895, adding close to five square miles to the city's area. Note the meandering course of the Duwamish River prior to its straightening (see page 123). The railroad shown to the west of West Seattle is the Portland & Puget Sound, an unbuilt road by a Union Pacific subsidiary that would have connected to Railroad Avenue in West Seattle.

Semple's extravaganza did not fit in with the plans of James J. Hill and his Great Northern Railway to control the waterfront, and Hill's lawyer, Judge Thomas Burke, executed a series of legal maneuvers that brought construction to a halt. By the time all the legalities were sorted out in 1901, the proponents of a northern canal had put together financing and had begun work excavating the channel that would become the Lake Washington Ship Canal (see page 128), and Semple had lost much of his financing. Nonetheless, Semple's legacy is the reclamation of about 175 acres of the tidelands, and his canal excavation is used today by the Spokane Street interchange on Interstate 5.

Eventually, most of the tidal flats south of the railroad trestle would be reclaimed, such was their obvious utility to a city bounded by steep hills and hemmed in by Lake Washington. The biggest reclamation project of all was Harbor Island now some 406 acres in size, filled by the Puget Sound Bridge and Dredging Company and completed in 1909. At the time, Harbor Island was the largest artificial island in the world. Fill for the project came from 24 million cubic yards of material removed from the Jackson and Dearborn street regrades and dredged from the Duwamish River.

Much of the fill for Seattle's tidal flats has come from carving away at its confining hills. There have been, by one count, as many as sixty individual regrades. The first were privately done, on a small scale. The larger public regrades were mainly the work of Seattle's energetic city engineer, Reginald Heber Thomson, who had been appointed in 1892 following a search for competency after it was found that a sewer trench on First Avenue had been installed running uphill. Semple had begun the process of cutting through Beacon Hill for his canal; Thomson continued the process but cut farther north, along Jackson Street. The Jackson Regrade, which was carried out between 1907 and

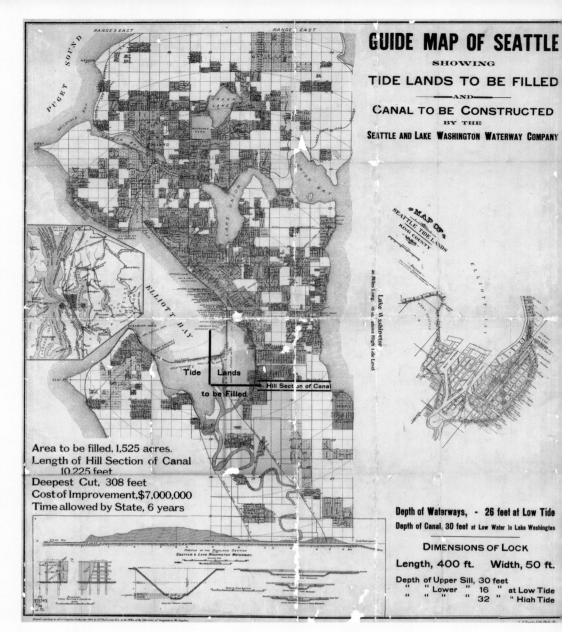

MAP 287 (*above*).

Using the same map as MAP 286 (*previous page*) as a base, this map was produced by Eugene Semple's Seattle and Lake Washington Waterway Company in 1895 to show the route of its proposed south Seattle canal, the tideland to be filled, and the waterway through the tidelands to be left for access to the *Hill Section of Canal*. Semple was a member of the Harbor Line Commission that produced MAPS 284 and 285. He resigned to lead his south canal company.

1910, lowered Beacon Hill by eighty-five feet and required the demolition of two large schools and many other buildings.

The Jackson Regrade was quickly followed by the regrading of Dearborn Street half a mile to the south. Here the land was lowered in some places by more than a hundred feet. Some 1.6 million cubic yards of earth were sluiced away to the tidal flats. The Dearborn Regrade was so deep that a bridge was required to carry 12th Avenue across Dearborn; the bridge is still there today, officially renamed the Jose P. Rizal Bridge after the Philippine hero.

To the north of the downtown area even more extensive regrading took place. Denny Hill was the subject of several regrades. Property owners

MAP 288 (*left*).

On 17 July 1897 the Steamer *Portland* arrived in Seattle with $700,000 worth of gold aboard, and the Klondike Gold Rush began. Seattle benefited immensely from the gold rush, becoming the main supply port for gold seekers departing for Alaska and the Yukon. This map, which was similar to one produced by competitor Tacoma, crudely depicts the routes to the Klondike via *Seattle the Gateway*: to Skagway and over the Chilkoot Pass and down the Yukon River; and to St. Michael, on the west coast of Alaska, and up the Yukon River.

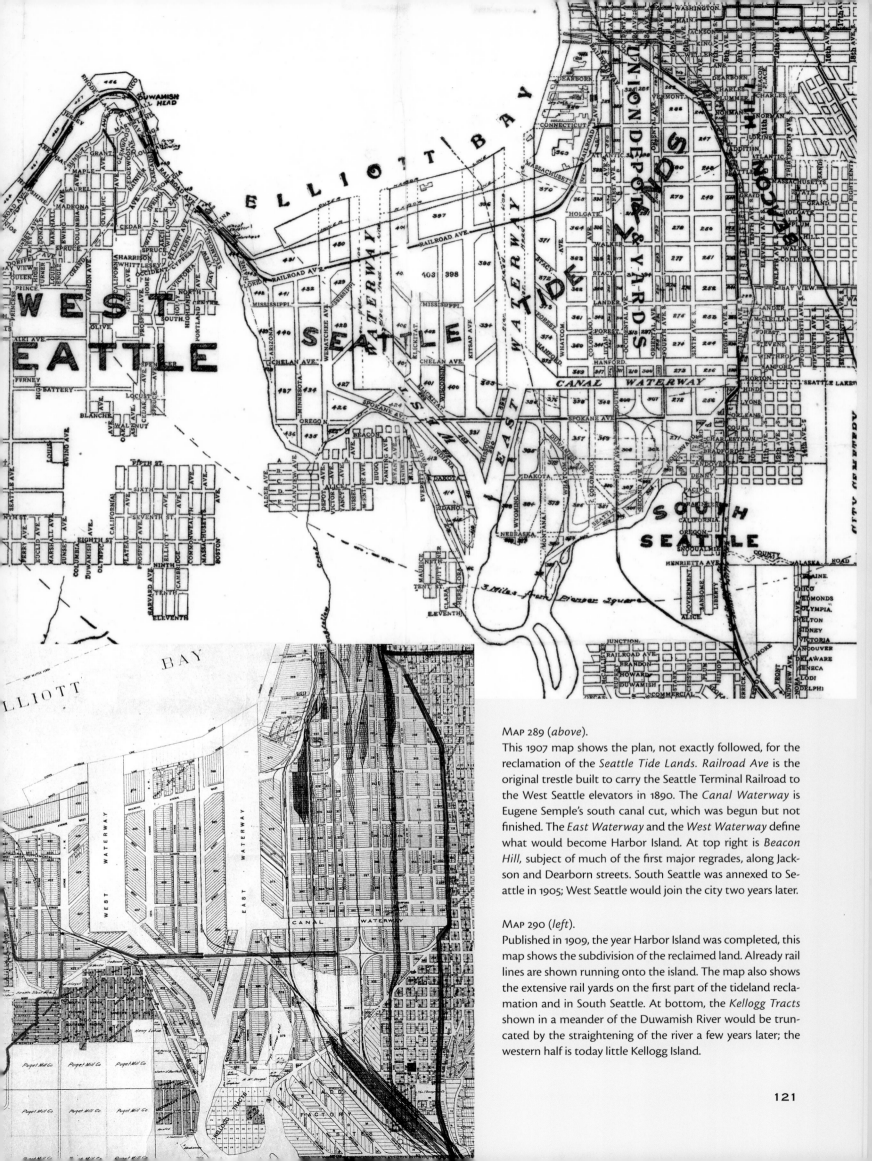

MAP 289 (*above*).
This 1907 map shows the plan, not exactly followed, for the reclamation of the *Seattle Tide Lands*. *Railroad Ave* is the original trestle built to carry the Seattle Terminal Railroad to the West Seattle elevators in 1890. The *Canal Waterway* is Eugene Semple's south canal cut, which was begun but not finished. The *East Waterway* and the *West Waterway* define what would become Harbor Island. At top right is *Beacon Hill*, subject of much of the first major regrades, along Jackson and Dearborn streets. South Seattle was annexed to Seattle in 1905; West Seattle would join the city two years later.

MAP 290 (*left*).
Published in 1909, the year Harbor Island was completed, this map shows the subdivision of the reclaimed land. Already rail lines are shown running onto the island. The map also shows the extensive rail yards on the first part of the tideland reclamation and in South Seattle. At bottom, the *Kellogg Tracts* shown in a meander of the Duwamish River would be truncated by the straightening of the river a few years later; the western half is today little Kellogg Island.

Above. Sluicing operations on Denny Hill about 1910.

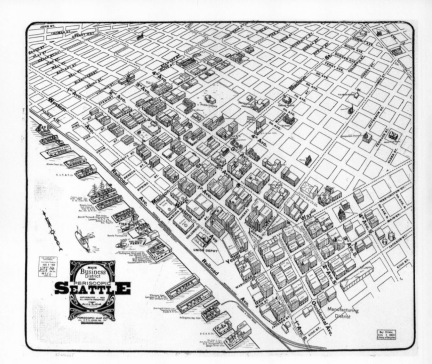

cut First Avenue through from Pike Street to Cedar Street about 1900. Thomson's engineers did the same to Second Avenue in 1904 and also regraded Pike and Pine streets between Second and Fifth avenues. The area can be seen on MAP 291 (*right*).

Thomson had more plans for this neighborhood, now called the Denny Regrade. Between 1908 and 1911 the half of Denny Hill nearest the waterfront was sluiced away, with the fill going by flume and tunnel into the tidelands. Some twenty-seven city blocks were leveled, the area bounded by Pine and Cedar streets and from Second to Fifth avenues. Five million gallons of water were pumped through hydraulic cannon, carrying away a total of six million cubic yards of earth. The leveled area is clearly shown on MAP 292 (*below*).

MAP 292 (*below*).
This promotional bird's-eye map depicts the Denny Hill area after the first major regrade in 1908–11 (the flat and largely unbuilt area bounded by Second and Fifth avenues and Pine and Cedar streets) but before the second regrade in 1929–31, which would level the east part of the hill, visible in the foreground, the blocks nearer to the viewer (northeast) from Fifth Avenue.

MAP 291 (*above*).
This bird's-eye or "periscopic" view of downtown Seattle in 1903 shows the area regraded by both private and city projects prior to the first major attack on Denny Hill beginning in 1908. *Below* are profiles of the Third-, Fourth-, and Fifth-Avenue 1908–11 regrades, from an engineering magazine.

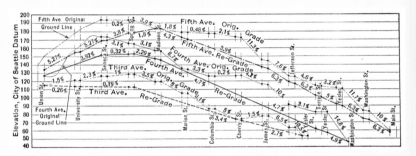

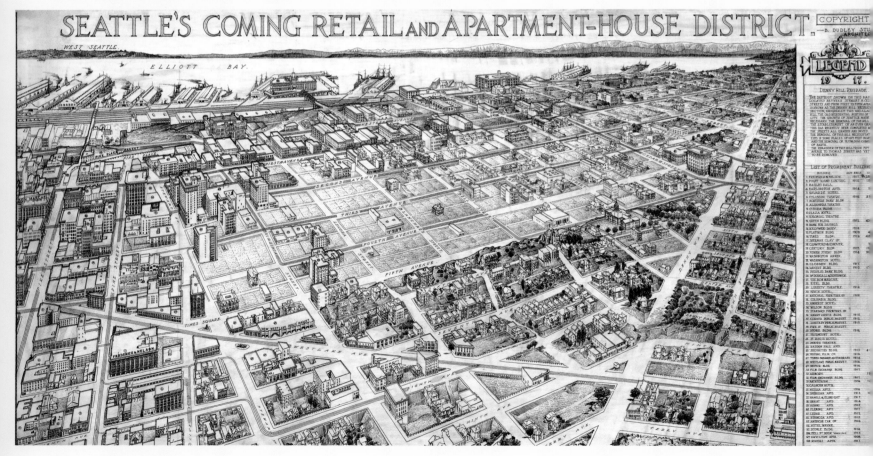

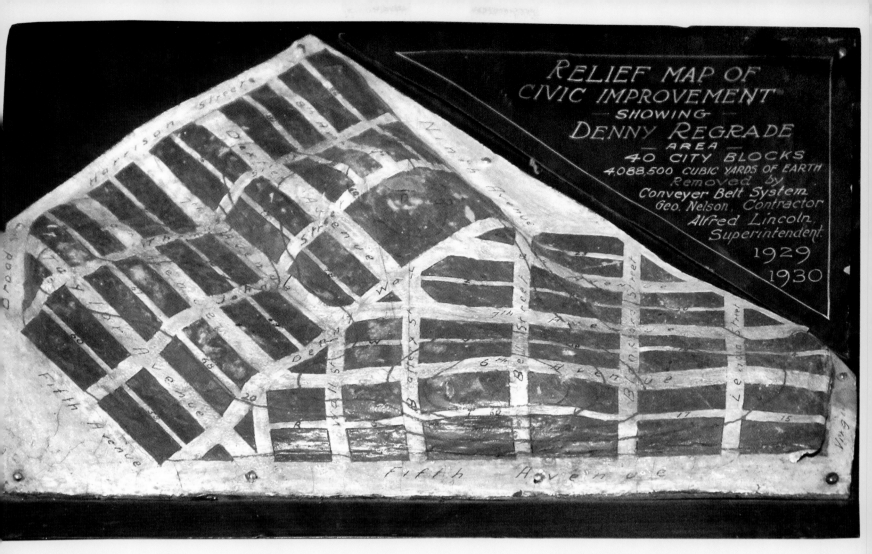

MAP 293 (*above*).
This three-dimensional map was made by the contractor for the second Denny regrade in 1929–31 and shows the original topography of the half of Denny Hill northeast of Fifth Avenue.

Thomson resigned in 1911, becoming chief engineer for the Port of Seattle, but the Denny Regrade was not complete; the northeastern half of the hill remained. A further forty city blocks were regraded between 1929 and 1931, the area shown on the contractor's model (MAP 293, *above*). Sluicing had by this time fallen into disfavor, and technology had improved, so excavators were used to load conveyor belts, which dumped their contents into special scows designed to capsize in a controlled fashion once towed out into Puget Sound.

These regrades fundamentally altered the topography of Seattle, allowing in modern times its downtown to expand and thrive.

Other major projects that altered the face of Seattle were the construction of two waterways: the Lake Washington Ship Canal (see page 128), and the straightening of the lower part of the Duwamish River to make it accessible to oceangoing ships. Once again, one of the principal proponents of the scheme was Thomson. The straightening, with accompanying dredging, would not only help sell reclaimed industrial land (MAP 294, *right*) but also alleviate the flooding that plagued the valley. By 1920 the Duwamish River had been straightened and deepened to fifty feet for four and a half miles, creating the Duwamish Waterway.

Seattle was soon one of the most topographically modified cities in the country. But aesthetics were not overlooked either. In 1903 the Olmsted Brothers, consulting landscape architects of some renown, were retained to produce a comprehensive parks plan for the city, along with one for Portland. In 1908 John Charles Olmsted produced a supplementary report extending the proposed park system into newly annexed areas (MAP 295, *overleaf*). This was followed in 1911 by a major comprehensive plan produced by Virgil Gay Bogue (discoverer of Stampede Pass in 1881; see page 85), who had been selected by Thomson to lead the Municipal Plans Commission, established in 1910. The following year Bogue produced a

MAP 294 (*below*).
In 1908 this advertisement appeared in newspapers offering factory sites on reclaimed land and showing the proposal to straighten the Duwamish River.

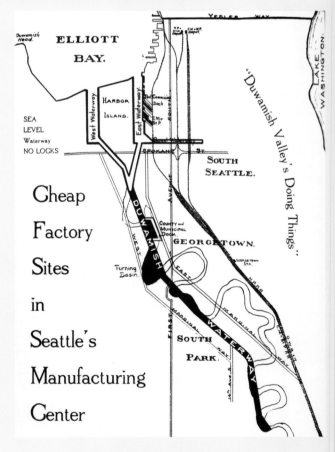

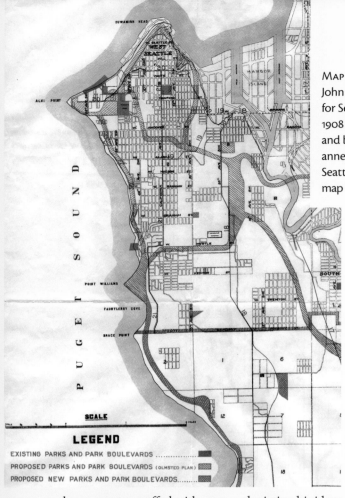

MAP 295 (*left*).
John Olmsted's parks plan for Seattle was revised in 1908 to extend his parks and boulevards into newly annexed areas such as West Seattle, shown here on a map from the Bogue report.

MAP 296 (*right*).
Plans for the straightening of the Duwamish River drawn up in 1911. The wide meanders of the old course of the river are shown.

MAP 297 (*below* and *below, center*).
Two details of a map from the Bogue plan showing proposed dock facilities. The first shows those at Renton, at the southern end of Lake Washington, at the mouth of the Cedar River. The straightened Duwamish is just visible at lower left. The bottom one shows docks at Bellevue, at the location of today's Newport Shores, with Mercer Island on the left.

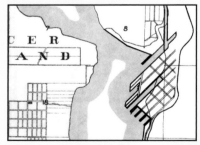

long report—stuffed with maps—depicting his ideas for everything from the parks system, in which he followed Olmsted's work quite closely, to railroad and transit planning, road building, and industrial lands, especially docks and waterfront facilities. He also drew up plans for a new civic center to be built on the newly available Denny Regrade. Some of his maps are shown on this page. Bogue's grand plan was not to be, however; the voters rejected it in 1912.

Despite this setback, many features of Bogue's plan were implemented later. And the site of the last Denny Regrade is now the location of a civic center—Seattle Center, built on the site of the 1962 World's Fair (see page 206).

MAP 298 (*below*).
A fine bird's-eye view of Virgil Bogue's proposed civic center for the Denny Regrade area. This is the red circle on MAP 299, *right*.

MAP 299 (*below*).
The location of Bogue's proposed civic center in the Denny Regrade area and a grand boulevard leading to it are shown in red. Dashed red lines are proposed new arterial highways; solid red lines existing arterials. More of Bogue's docks line *Lake Union*.

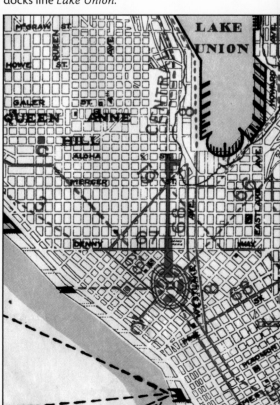

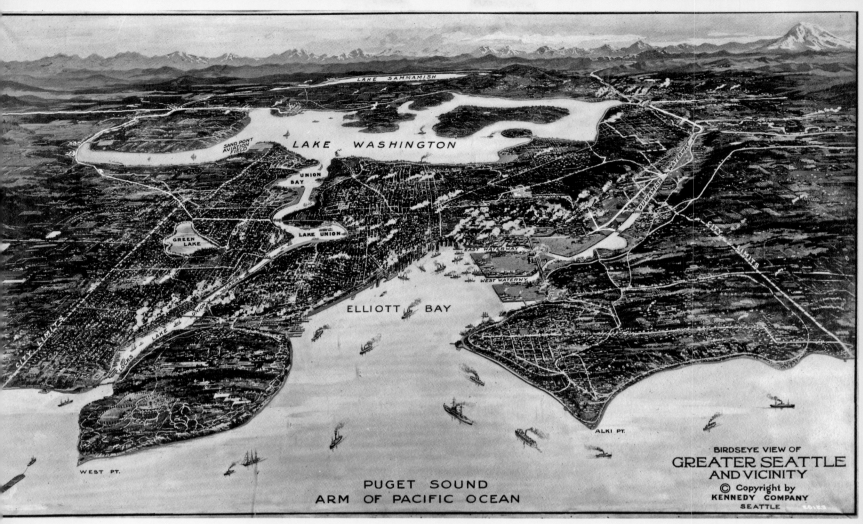

BIRDSEYE VIEW OF
GREATER SEATTLE
AND VICINITY
© Copyright by
KENNEDY COMPANY
SEATTLE

PUGET SOUND
ARM OF PACIFIC OCEAN

MAP 300 (*above*).

This superbly detailed bird's-eye map of Seattle is undated but appears to have been published about 1922, as it shows the *Sand Pont [Sand Point] Aviation Field,* which was established about 1920. The map is a grand summary of Seattle's achievement to that date: the Lake Washington Ship Canal, which began operation in 1917 though not officially completed until later; the newly straightened *Duwamish Waterway*; and the reclaimed tidelands, including Harbor Island. The Denny Regrade area, however, is shown crammed with buildings.

MAP 301 (*right*).

Fire insurance maps are one of the best sources historians have for determining the detailed use of buildings at a given date. This colorful map comes from a fire insurance atlas published in 1912 and shows the docks of Seattle's shoreline between *King St.* and *Atlantic Street* (now the location of Qwest Field). The significance of the railroads to Seattle is underlined by this map—there are rail lines everywhere. The key indicates building materials, all too important in determining the risk of fire.

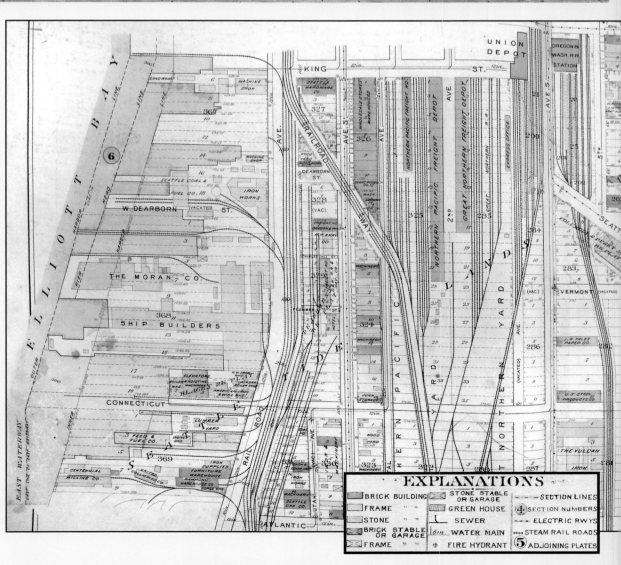

EXPLANATIONS

BRICK BUILDING	STONE STABLE OR GARAGE
FRAME	GREEN HOUSE
STONE	SEWER
BRICK STABLE OR GARAGE	WATER MAIN
FRAME	FIRE HYDRANT

SECTION LINES
SECTION NUMBERS
ELECTRIC RWYS
STEAM RAIL ROADS
ADJOINING PLATES

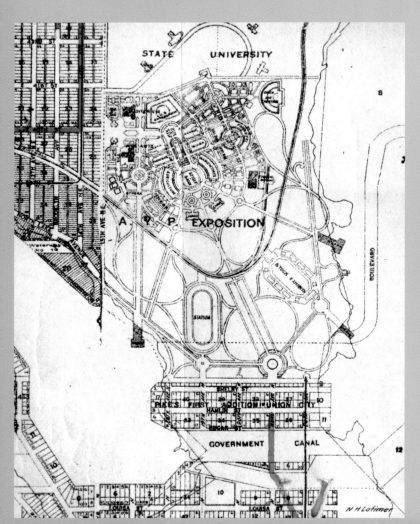

The Big Exposition

In 1893 Chicago hosted a massive fair called the World's Columbian Exposition. Its success and popularity led to many similar fairs in cities around the United States in the years that followed. Portland's Lewis and Clark Centennial Exposition (see page 110) was one of these, and in 1906 Seattle decided to hold its own fair. That year the grounds of the University of Washington were chosen as the fair's site.

Intended to promote Seattle's ties to Alaska, promote the importance of Alaska to the United States, and additionally promote the economic growth of Seattle and the region, the fair was originally scheduled for 1907, but delays led to the fair being held in 1909. It was probably a bigger success as a result, since 1907 had turned out to be a recession year, whereas 1909 was a boom time. The Panama Canal, then under construction, was thought likely to lead to booms in all West Coast port cities.

Landscape architect John C. Olmsted was hired to plan the site layout, and he produced an inspired design that was integrated into the visual environment surrounding the site. Of particular note was the grand boulevard aligned to a vista of a distant Mount Rainier (MAP 305, *right*).

The elaborately prepared and built exposition ran from June to October 1909 and was a great success, with about 3.75 million people visiting the fair. The exposition's legacy was of even greater benefit. The University of Washington, which had moved to the site from downtown in 1895, had all the land it would need for many years cleared, and the university requested that twenty of the neo-classical buildings be retained, providing an instant campus expansion. Several of the buildings remain to this day.

MAP 302 (*above*).
This fine advertising piece for the Alaska–Yukon–Pacific Exposition was produced in 1907. A classical Seattle holds ribbons reaching to the far corners of the earth, superimposed on a map of the Northwest Coast, where all roads lead to Seattle.

AUTHORIZED BIRDS EYE VIEW OF THE ALASKA-YUKON-PACIFIC EXPOSITION
SEATTLE, U.S.A. 1909
OPENS JUNE 1ST   CLOSES OCT. 16TH

MAP 303 (*opposite page, left*).

A map published in 1909 shows that Olmsted had to deal with the existing rail line curving round the main vista of the site. This was the line built by the Seattle, Lake Shore & Eastern in 1890, in 1909 belonging to the Northern Pacific. The rail line was accommodated by a bridge over the main boulevard walk. The site layout is an early version.

MAP 304 (*opposite page, right*).

This is John Olmsted's final plan for the exposition layout. Note that the Lake Washington Canal cut was slightly to the south (see next page).

MAP 306 (*left*).

Real estate investors and speculators were very interested in the exposition. This map advertised the event as an investment opportunity.

Right. One of a set of colorized postcards printed as souvenirs.

MAP 305 (*above*).

A superb advertising bird's-eye map of the Alaska–Yukon–Pacific Exposition site, looking south, no doubt inspired by the similar bird's-eye produced four years earlier for Portland's Lewis and Clark Centennial Exposition (MAP 264, *page 110*). The Olmsted Brothers landscape design company had designed both fairs, and both bird's-eyes are drawn in a very similar style. Beyond the fair site is the Lake Washington Ship Canal cut through Montlake (see next page), and Mount Rainier is shown prominently in the distance. Designer John Olmsted had selected the mountain as the fair's visual focus. An airship, representing the height of progress and modernity, hovers above.

Lake Washington Ship Canal

In 1854, Thomas Mercer suggested the name for Lake Union on the basis that the lake would one day unite Lake Washington with Puget Sound. And unite them it did, some sixty-three years later.

In the interim, there were many proposals for canals of one sort or another, and following various routes. Some attempted to build canals, and others actually succeeded. When the Union City plat was filed by Harvey Pike in 1869, he reserved a two-hundred-foot-wide strip for a canal between Lake Union and Lake Washington (MAP 308, *right, center*) and attempted to hack a channel before deciding it was too big a task. Two years later he incorporated the Lake Washington Canal Association and petitioned Congress to support it, but Washington Territory had little influence in the capital in those days, and the request was ignored.

In 1883 the Lake Washington Improvement Company was created by a group of luminaries that included David Denny and Thomas Burke. They hired Chinese workers to dig a three-quarter-mile-long channel north of Queen Anne Hill, and a small wooden lock was built. At the same time a channel was cut through the portage between Lake Union and Lake Washington. A canal of sorts was complete—but big enough only for the passage of logs, not for ships (MAP 309, *below*).

After Washington became a state in 1889, Congress passed an act authorizing the survey of a ship canal. The report, published in 1891, established the feasibility of the Shilshole Bay route, though it also examined a route terminating at Smith Cove (MAP 307, *right*).

Although this route seemed proven as the optimal one, in the 1890s it faced opposition from Eugene Semple's southern canal, which he actually began in 1895 (see page 117). The federal government authorized more surveys, as it was looking for a freshwater basin that the U.S. Navy could use. In 1908 the U.S. Army Corps of Engineers, now led in Seattle by General Hiram M. Chittenden, recommended a canal 75 feet wide and with a minimum depth of 25 feet, with 825-foot-long locks. The Cedar River was to be diverted into the southern end of Lake Washington to provide a constant water flow though the locks. Construction of a ship canal was finally authorized in June 1910, and construction began in November 1911. The gates of what would be named the Hiram Chittenden Locks were closed in July 1916, and the water of Salmon Bay began to rise. A few days later Lake Washington's waters were released into the Montlake Cut, and by October Lake Washington was at the same level as Lake Union. In July 1917 formal dedication of the Lake Washington Ship Canal took place, with a grand parade of ships from Puget Sound to Lake Washington and back.

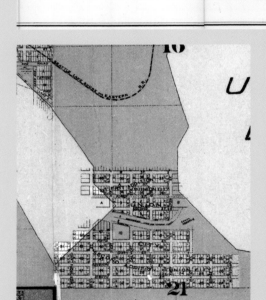

MAP 307.
The map from the report by the U.S. Army Corps of Engineers. Published in December 1891, the report estimated the cost of a canal to Shilshole Bay at $2.9 million and one to Smith Cove, also shown on this map, at $3.5 million.

MAP 308 (*above*).
The plat of *Union City* with the *Canal Reserve* left by Harvey Pike in 1869. This map dates from 1891.

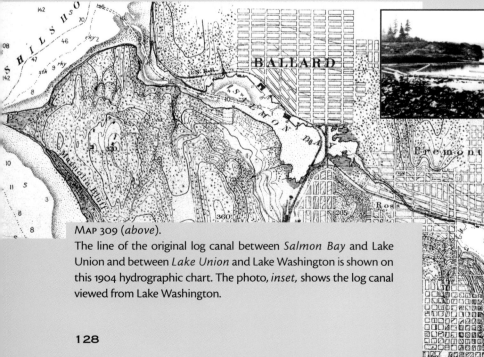

MAP 309 (*above*).
The line of the original log canal between *Salmon Bay* and Lake Union and between *Lake Union* and Lake Washington is shown on this 1904 hydrographic chart. The photo, *inset*, shows the log canal viewed from Lake Washington.

PROPOSED ROUTE OF CANAL
TO CONNECT
LAKES UNION and WASHINGTON
WITH
PUGET SOUND

SURVEYED UNDER THE DIRECTIONS OF
CAPT THOMAS W. SYMONS, CORPS OF ENGINEERS U.S.A.
BY
Philip G Eastwick Civil Engineer
Feb 1891.
To accompany report of Board of officers of the
Corps of Engineers, consisting of
Colonel George H.Mendell
Major Thomas H.Handbury
Captain Thomas W. Symons
Appointed by Special orders N° 99 Dated Sept.30th 1890.

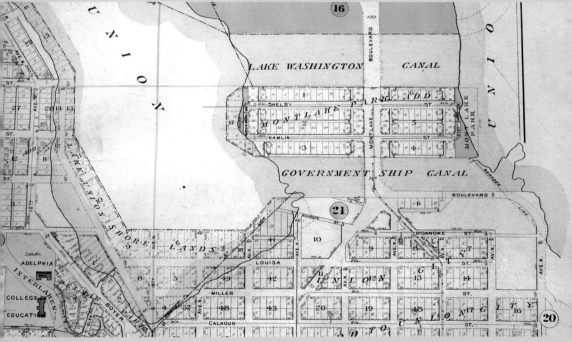

Map 312 (*below*).
The U.S. Army Corps of Engineers issued this informational brochure, complete with a map, in 1949.

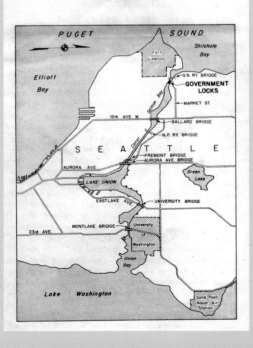

LAKE WASHINGTON SHIP CANAL

Seattle , Washington

Constructed by

**SEATTLE DISTRICT
CORPS OF ENGINEERS
U. S. ARMY**

Map 310 (*above*).
A 1912 map from a fire insurance atlas shows both the original reserve, marked *Government Ship Canal*, and the route actually cut, labeled *Lake Washington Canal*. The *Union City* subdivision is at bottom, the *Montlake Park Add[ition]* at top. *Far left, top,* is a photo of the Montlake Cut under construction about 1914.

Map 311 (*below*).
A 1925 bird's-eye map shows the route of the completed canal.

COMMUNITIES OF THE NORTHWEST

TACOMA

Tacoma, child of the Northern Pacific, was quickly a contender for preeminence in the Northwest, yet its early promise soon faded as its rivals acquired their own rail links to the East. By 1880 the city had a third of the population of Seattle and a decade later came close to exceeding its rival's size—36,000 persons when Seattle had 43,000—yet by 1900, seven years after the Great Northern came to Seattle and three after the Klondike Gold Rush, Tacoma could not boast even half the souls of Seattle. By 1910 Tacoma had fallen even further behind: the city's population was 42,000 that census year compared with Seattle's 237,000.

Nevertheless, great things were expected of Tacoma, as demonstrated by the wealth of promotional advertising for the city, including quite a few bird's-eye maps, only some of which are shown on these pages—not to mention grand plans that came to nothing, such as the Olmsted plan for the city (MAP 313, *below*) or a harbor development plan drawn up by Virgil Bogue in 1912 (MAP 319, *overleaf*).

What became Old Tacoma had been claimed in 1864 by Job Carr, and most was then sold to Morton McCarver, a real estate speculator who correctly guessed that the Northern Pacific would select Commencement Bay as its Puget Sound terminus. What McCarver had not bargained for was the selection of a depot and townsite deep within the mudflats of the bay, two miles south of his claim. New Tacoma, as it was called, was incorporated in 1875. Both townsites grew, however, and merged as Tacoma in 1883.

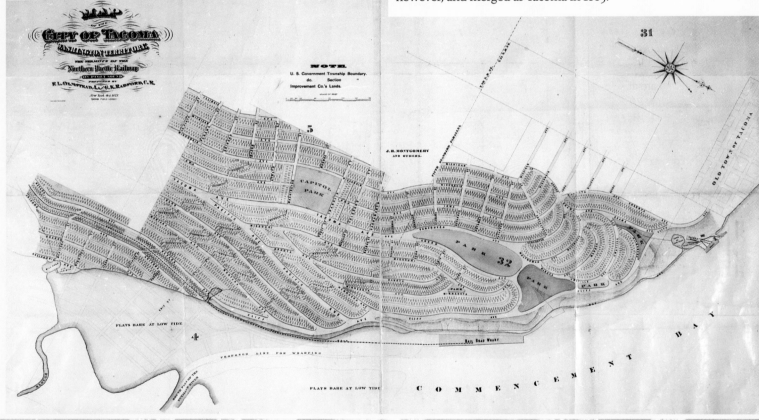

MAP 313 (*left, top*).

Frederick Law Olmsted was a brilliant but controversial landscape architect, designer of New York's Central Park. When the Northern Pacific selected Tacoma as its terminus in 1873, it assigned its subsidiary, the Tacoma Land Company, to sell lots in the townsite to raise money. Olmsted was hired to plan the city. Olmsted's design principles were such that his subdivisions, far from being the usual rectangles common at the time, tried to conform to the topography. His innovative plan for Tacoma, shown here, demonstrates this sympathy with the difficult, hilly site. Not only do his streets curve around the hillsides, but the steep slopes are ascended via diagonals rather than by lines at right angles to the slope. Alas, his design was too far ahead of its time and was rejected by the Northern Pacific board, which wanted something more traditional and—board members thought—more saleable. The result was the normal rectangular grid shown in MAP 315—and the steep, unforgiving inclined streets of today. Olmsted's company, with his nephew and adopted son John Charles Olmsted, was to play a part in the design of rival Seattle (see page 123).

MAP 314 (*left, bottom*).

Tacoma in 1884, a year after its connection—through Portland—to the transcontinental railroad. Old Tacoma is at far right.

MAP 315 (*right*).

More factually accurate than the bird's-eye maps, this is a map by the U.S. Coast and Geodetic Survey showing *Tacoma* in 1889. *Old Tacoma* is at top. The mudflats that would one day become the Port of Tacoma are at right. What had until 1887 been the transcontinental connection is the *Kalama Branch N.P.R.R.*; the new transcontinental link, via Stampede Pass, is the *Puyallup Branch,* crossing the tideland on a trestle.

MAP 316 (*below*).

This superb bird's-eye map depicts Tacoma in 1890, at the peak of its challenge to Seattle. The city is shown with few vacant lots and with the hustle and bustle of economic activity at every turn, typical of such promotional maps.

TACOMA.

Western Terminus of the Northern Pacific Railroad.

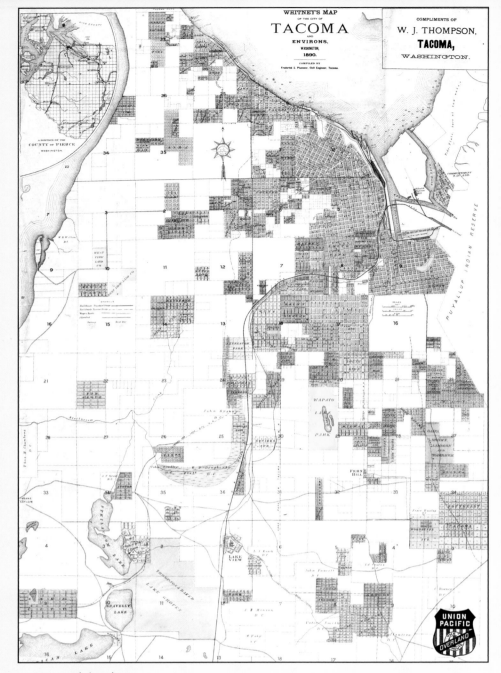

MAP 317 (above).

This is essentially a real estate development map from 1890, showing Tacoma and the area to the south. The colored blocks are subdivisions, mostly offered for sale. Cities like Tacoma often grew somewhat piecemeal, driven by rail lines and then streetcar systems servicing developments where the owners had the inclination to try to create communities of their own—and hopefully make a handsome profit doing it. This map illustrates the barrier created by the *Puyallup Indian Reserve* just to the east of the original townsite and Northern Pacific depot. On this map, the land right up to the boundary of the reserve has been platted. In 1894 the subdivision would cross into the reserve here with a new development called the Indian Addition. Note that this map (like others in the Pacific Northwest; see, for example, the map of Port Townsend, MAP 343, *page 141*) has a Union Pacific logo on it. Although that railroad long had designs on Puget Sound, it would not reach Tacoma until 1909, following a difficultly extracted agreement to share the Northern Pacific line from Portland.

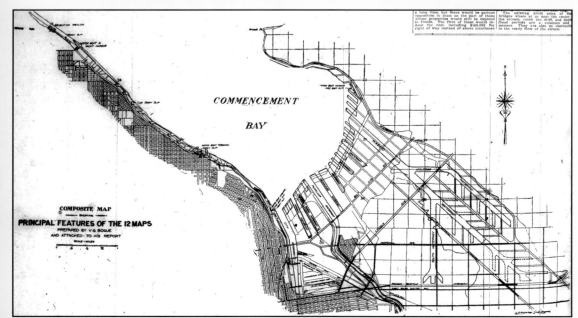

MAP 319 (left).
Virgil Bogue, fresh from his city plan for Seattle, in 1911 was hired by the Tacoma Commercial Club and Chamber of Commerce to prepare a plan for Tacoma Harbor. This map is a summary of a number of maps in the report; it appeared in the *Tacoma Daily Ledger* in January 1912. The plan's main feature was an extensive reclamation scheme, with multiple broad waterways to allow dock access—a major port serviced by railroads. Many western ports had the same idea at this time because of the imminent opening of the Panama Canal. Bogue's plan was not adopted, but similar reclamation schemes would create today's port.

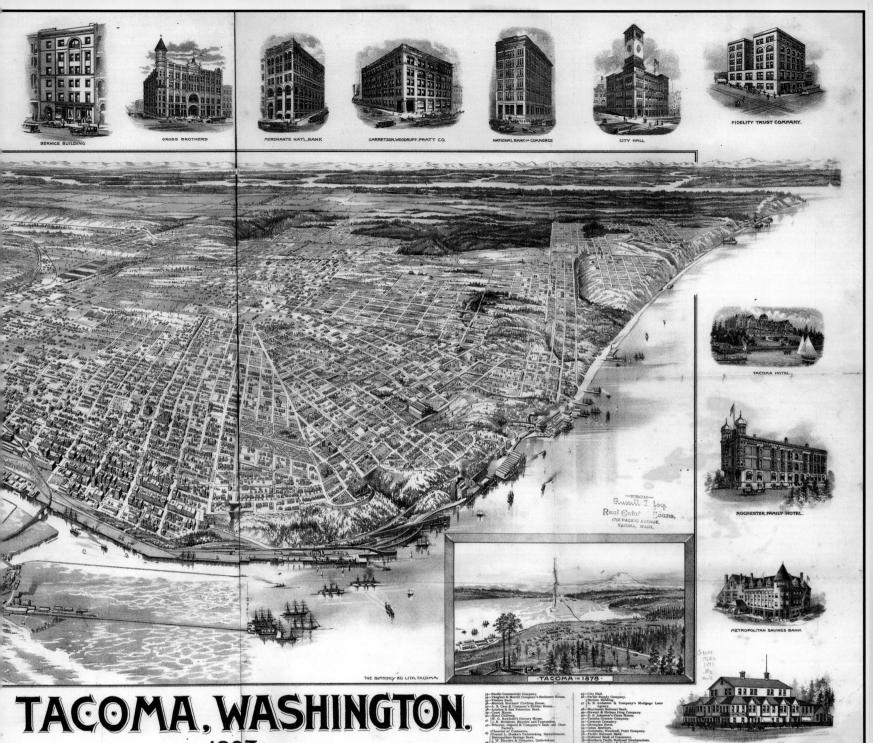

BERNICE BUILDING   GROSS BROTHERS   MERCHANTS NAT'L BANK   GARRETSON, WOODRUFF, PRATT CO.   NATIONAL BANK OF COMMERCE   CITY HALL   FIDELITY TRUST COMPANY.

TACOMA HOTEL.

ROCHESTER FAMILY HOTEL.

METROPOLITAN SAVINGS BANK

CRESCENT CREAMERY COMPANY.

TACOMA IN 1878

TACOMA, WASHINGTON.
1893.

MAP 318 (*above*).
Perhaps the most magnificent of the bird's-eye maps of Tacoma is this one published in 1893, complete with views of advertisers' buildings around the map edges. Note the extensive tidal mudflats, yet to be reclaimed. Inset at right is another bird's-eye view, this one of Tacoma in 1878, a view from the west with Mount Rainier prominent on the horizon.

MAP 320 (*right*).
The plan for the reclamation of the *Tacoma Tide Lands* of *Commencement Bay* is well shown on this detailed but more mundane map dated 1907, though it seems to show information from slightly later. The lines of both the *Union Pacific Ry.* and *C.M.&St.P. Ry.* (Chicago, Milwaukee, St. Paul & Pacific) are shown; Union Pacific's line was not completed once a common carrier agreement was forced from Northern Pacific, and the Milwaukee did not arrive until 1911. South of downtown the Union Pacific line is shown dashed (beside the subdivision named *Prescott's*); this is where the tunnel was to be, but construction was abandoned following the agreement. *Pacific Traction Co.* was an interurban that began service in 1909.

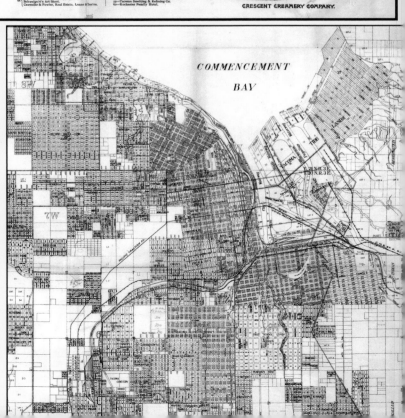

COMMENCEMENT BAY

The Union Pacific wanted a Puget Sound terminus for an extension northward from Portland. In 1905 the company began building a new line in the Tacoma area, including a tunnel to downtown, but this seems to have been part of a ploy to force the Northern Pacific to agree to joint use of its line. This agreement was reached in 1909, and a new Union Station was built two years later. The same year the rival Chicago, Milwaukee, St. Paul & Pacific also reached Tacoma (see page 166).

PUYALLUP

Puyallup, ten miles southeast of Tacoma, was platted by Ezra Meeker on his claim on the Puyallup River in 1877. Meeker had traveled west on the Oregon Trail in 1852. In 1906, thinking that the historical importance of the trail had been unfairly forgotten, he set about promoting it. That year he rode the trail in an ox-drawn covered wagon and continued to Washington, D.C., to meet President Theodore Roosevelt. In 1924, at the age of ninety-four, he flew over the trail in an open-cockpit plane, this time continuing to Washington to meet President Calvin Coolidge.

Puyallup grew substantially during the first decade of its existence and was incorporated in 1890. The city's first mayor was Ezra Meeker. The location of an important railroad junction, the city was connected to Tacoma by an interurban line in the early twentieth century and began the process of suburbanization, which has continued to this day.

MAP 321 (*below*).
Puyallup, as it appeared in an atlas in 1889, the year before incorporation. Note *Meeker Street*.

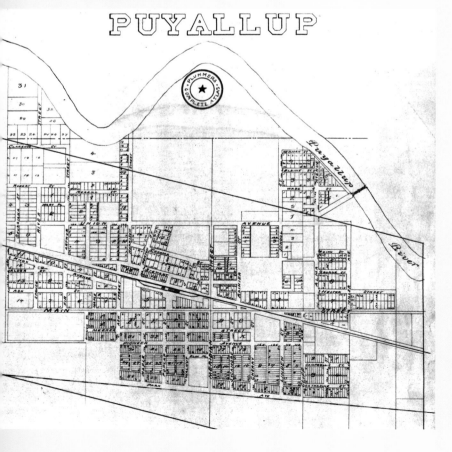

OLYMPIA

Olympia had been named as the territorial capital in 1853 (see page 70), yet in 1873 the Northern Pacific unexpectedly bypassed the city for Tacoma. To be connected at all, the citizens of Olympia had to build their own line—which was narrow gauge—to connect to the Northern Pacific main line at Tenino. It was not until 1891 that the Northern Pacific deigned it worthwhile to build its own line into Olympia.

Olympia continued as state capital after Washington achieved statehood in 1889. A design competition for the capitol was held in 1911, with the result being impressive buildings with fine vistas across Capitol Lake. The new domed legislature was completed in 1928.

BREMERTON

In 1888, the U.S. Navy decided that Point Turner (MAP 323, *below*) would be the best place to establish a shipyard to service the Pacific Fleet. Bremerton was founded by a German immigrant, William Bremer, in 1891 after he bought the land at an inflated price, thinking it worthwhile given the demand that the new shipyard would create. The rival plat of Charleston, to the west, was created at the same time.

Later that year Bremer sold some 190 acres to the Navy. Bremerton was incorporated in 1901, just as the city temporarily lost its position as the primary Pacific naval shipyard to Mare Island, in San Francisco Bay, because of criminal elements that were robbing sailors. After the city shut down all its saloons, the shipyard was reinstated.

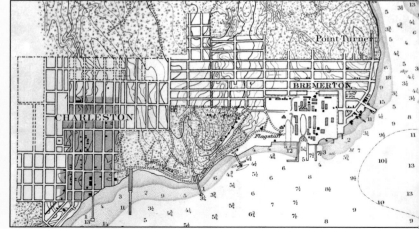

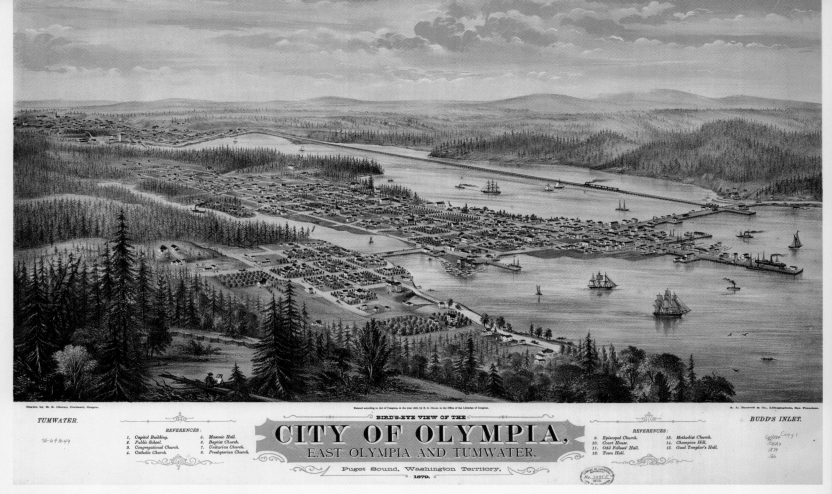

TUMWATER.

REFERENCES:
1. Capitol Building.   5. Masonic Hall.
2. Public School.   6. Baptist Church.
3. Congregational Church. 7. Unitarian Church.
4. Catholic Church.   8. Presbyterian Church.

BIRD'S-EYE VIEW OF THE
CITY OF OLYMPIA,
EAST OLYMPIA AND TUMWATER,
Puget Sound, Washington Territory.
1879.

REFERENCES:
9. Episcopal Church.   13. Methodist Church.
10. Court House.   14. Champion Hill.
11. Odd Fellows' Hall. 15. Good Templar's Hall.
12. Town Hall.

BUDD'S INLET.

MAP 322 (*left, top*).
A pencil-and-ink bird's-eye view of the proposed new legislative buildings in Olympia, the winner of a 1911 design competition by architects Walter Wilder and Harry White.

MAP 323 (*left, bottom*).
This 1912 map shows *Bremerton* with *Point Turner* and the relatively small naval yard then operational. To its left is the land onto which the shipyard would expand. Beyond that is *Charleston,* a neighboring and rival city annexed by Bremerton in 1927.

MAP 324 (*above*).
A fine bird's-eye map of Olympia and Budd Inlet in 1879 by Eli Glover. In the foreground is an artist drawing the scene, thought to be Glover's representation of himself. At *1* is the capitol building, a wooden affair on the edge of the forest.

MAP 326 (*below*).
Bremerton business district in 1909. North is to the right. The U.S. Navy Yard is at left. *Smith's Bay* is now the location of Evergreen Park.

MAP 325 (*below*).
A later postcard bird's-eye view of the Bremerton Navy Yard, undated but probably published about 1930.

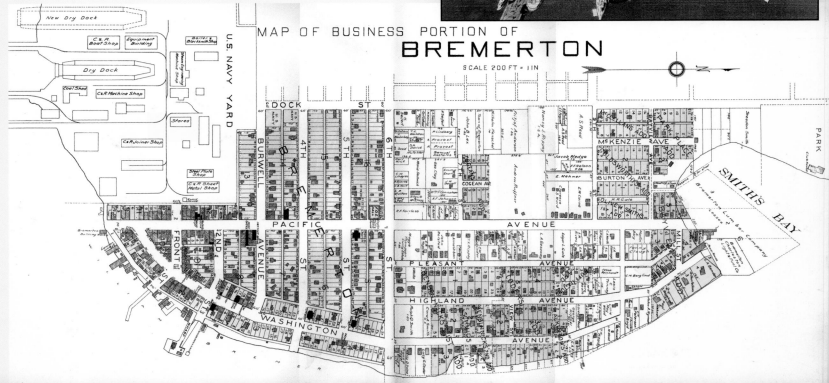

MAP OF BUSINESS PORTION OF
BREMERTON
SCALE 200 FT = 1IN

EVERETT

In July 1890 Henry Hewitt, Jr., a Wisconsin lumber merchant, took a cruise to Alaska. On board he met Charles L. Colby, a Seattle investor, and the two hatched a plan for what was to be a major industrial city at Port Gardner, on Possession Sound, north of Seattle. Others had the same idea. In August the Rucker brothers, Wyatt and Bethel, entrepreneur migrants from Ohio, filed a plat for a city, beating Hewitt and Colby to the punch. The reason for all the sudden haste was, of course, the Great Northern Railway; both groups had correctly guessed that James Hill would build his new transcontinental through to the Snohomish Valley. What they had not bargained for is that he would continue the line south to Seattle, making that city his terminus.

The investors quickly decided to cooperate. In November the Ruckers transferred more than half of their property to Hewitt in return for an undertaking to develop the site into an industrial city. The same month Hewitt and Colby incorporated the Everett Land Company to handle the development. And they chose a name for their city—Everett—after Everett Colby, Charles Colby's fifteen-year-old son.

MAP 327 (*right, top*).
This 1890 map shows the location where Everett would spring up, but at this date the principal city is Snohomish. Note the newly completed *Nor[thern] Br[anch] of S.L.S.&E. Ry.* (Seattle, Lake Shore & Eastern).

MAP 328 (*below*).
Henry Hewitt's Everett Land Company created this detailed illustrated map in 1892, complete with a bird's-eye and three other maps showing the location of the new city at various scales. The *Bird's-Eye View of the Town-site of Everett* is reproduced larger at *right*.

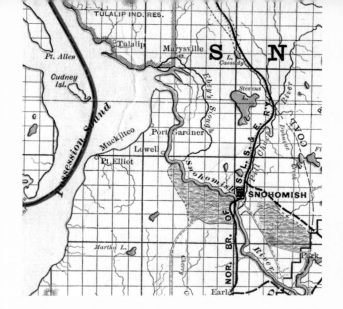

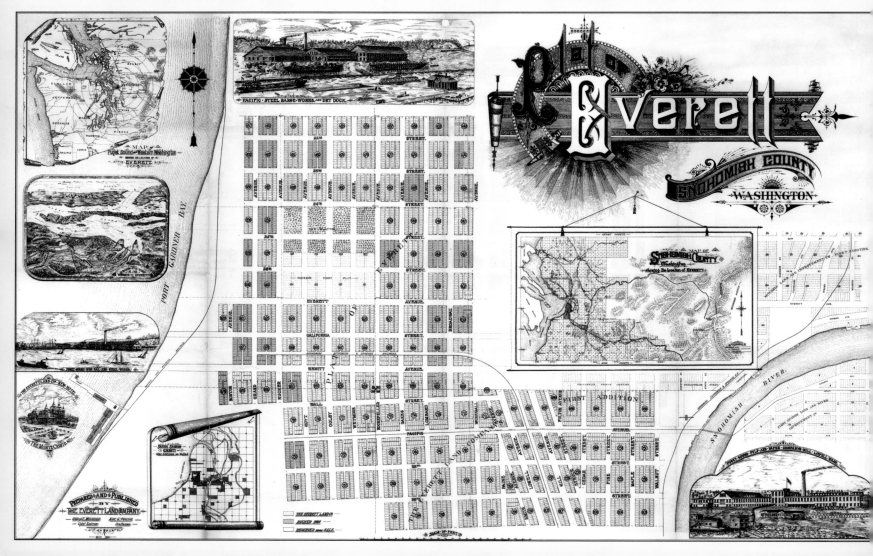

The Everett Land Company succeeded in building its city, but it was principally dependent on the processing and shipping of lumber—in line with most other Puget Sound cities of the period—rather than being the industrial metropolis Hewitt and Colby had envisaged.

The city of Everett was incorporated in May 1893, just as the Great Northern completed its transcontinental line. The city, while not the terminus, was still well positioned, but the Everett Land Company did not make its investors much money. The economic downturn of 1893 made it difficult to interest buyers in Everett land, and the Everett Land Company was sold to a company controlled by James Hill. The city now forms the vibrant northern end of the Seattle metropolitan area and is the home of several Boeing aircraft manufacturing facilities.

The early efforts to market Everett land has left us with some fine examples of real estate promotional maps, such as those illustrated here, documenting the infant settlement in a remarkable way.

MAP 329 (above).
A superb bird's-eye map of Everett published in 1893, the year the city was incorporated. The all-important railroad lines of the *Great Northern Rail Road* are prominently displayed; the *Seattle and Montana Rail Road*, also part of James Hill's railroad empire, runs along the waterfront. The informative text touts Everett as *the future industrial city of the Pacific Coast.*

MAP 330 (right).
Snohomish, founded in 1859, was eclipsed by the establishment of Everett at the mouth of the Snohomish River once the Great Northern arrived. This bird's-eye was published in 1890, just before the rise of Everett. The railroad shown is the earlier Seattle, Lake Shore & Eastern (see page 96).

A TALE OF FOUR CITIES

The attractions of Bellingham Bay—a good harbor in close proximity to the Pacific, supplies of coal surrounded by lumber just waiting to be cut—created no less than four separate communities around the bay. In the 1880s the Bellingham Bay Improvement Company and railroad builder–turned–real estate investor Nelson Bennett promoted Fairhaven (MAP 335, *right, center*) as the likely terminus of the Great Northern's transcontinental line. That railroad went to Seattle, but in the 1890s three regional railroads connected with the bay and so prosperity seemed assured.

MAP 332 (*above*).
All four of Bellingham's towns are shown on this detail of a map published in 1890.

Whatcom had been platted in 1853. Sehome, now the downtown area of Bellingham, was platted in 1858 (MAP 158, *page 73*). Fairhaven, to the south, was not platted until 1883, and Bellingham was a small area purchased and incorporated by a Fairhaven investor in 1890 (MAP 332, *above*).

In 1891, Whatcom and Sehome, whose residents could already see the value in being bigger, voted to merge, becoming New Whatcom; the name was changed once more, to simply Whatcom, in 1901. By this time the cities were closely integrated and shared an electric streetcar system. In 1903, Fairhaven and Whatcom voted to amalgamate, and a name was chosen so as to be neutral to both—Bellingham. From a practical point of view, doing this required only half the voters—all men—to approve, whereas adopting the name of either Fairhaven or Whatcom would have required two-thirds. As it happened the merger passed with a large majority, 2,163 to 596. The new and enlarged city was soon advertising itself as a first-class city with a population of 25,000 (MAP 336, *right*).

MAP 333 (*below*).
This bird's-eye view of Whatcom (in the foreground) and Sehome (in the background) with Fairhaven (in the distance) was drawn in 1889 by James T. Pickett, son of Captain George Pickett, who in 1856–58 had commanded Fort Bellingham (see MAP 157, *page 73*). The artist has taken a number of liberties with his work: although the view is to the south, both Lake Whatcom and Mount Baker—both to the east—can be seen.

MAP 331 (*above*).
This intriguing newspaper advertisement from 1891 touts the virtues of land at *Fairhaven*, and from the map one would be led to believe that Fairhaven was the only city between *Seattle* and *Vancouver*, B.C. In fact the *Great Northern Railway* would two years later meet Puget Sound at Everett, not Fairhaven.
Below. A real estate ad from 1890 directed at Scandinavian immigrants.

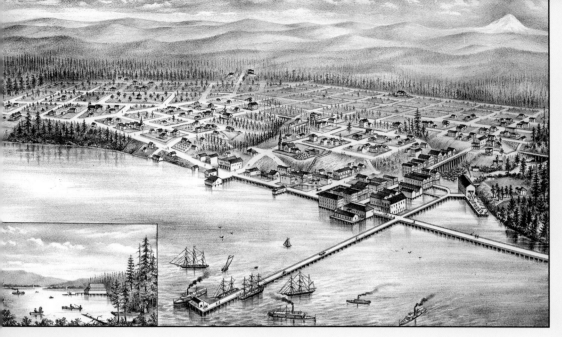

MAP 334 (*left*).
A bird's-eye view of Whatcom in 1888, from a history book published in 1889. A number of buildings along the shoreline are on pilings.

MAP 335 (*left, center*).
Not to be outdone, Fairhaven produced this bird's-eye map in 1890, considerably overstating the industrial and shipping activity, as was typical with these maps, which were for promotional purposes. The artist has been unable to exclude the competing Sehome and Whatcom, visible at right, but has managed to depict them far smaller than the sprawling—but not by any means built out—Fairhaven. The Great Northern track shown along the waterfront is the subsidiary Fairhaven & Southern, which would connect with New Westminster, B.C., in 1891.

MAP 336 (*below*).
This map was made instantly up-to-date in 1903 when it was overprinted in red across the names of *Fairhaven* and *Whatcom* after the vote to amalgamate the cities of Bellingham Bay into one city—*Bellingham*.

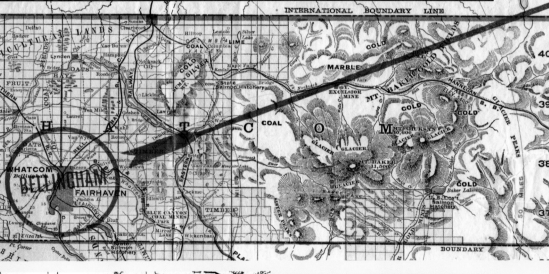

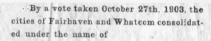

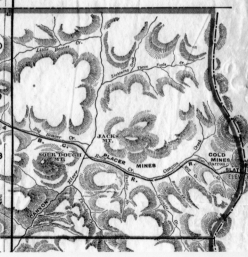

By a vote taken October 27th, 1903, the cities of Fairhaven and Whatcom consolidated under the name of

BELLINGHAM

now a city of the first class, with a population of 25,000.

BELLINGHAM, on BELLINGHAM BAY, has the best Harbor on the Pacific Coast.

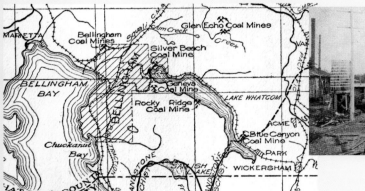

MAP 337 (*left*).
This 1923 map shows the coal mines then scattered around *Bellingham*. The *inset* photo is of Bellingham Coal Mine No. 1 (to the north of the city), taken in 1921. The mine produced between 500 and 1,000 tons of coal per day.

ANACORTES

Anacortes was of interest to real estate speculators in the 1870s because it was thought a likely terminus for the Northern Pacific's Cascade Branch, and there are maps that show precisely that (two maps on this page). The site was considered a probable terminus because of its superior access to the open Pacific as compared with anywhere else within Puget Sound. By 1873 much of the land in what was called Ship Harbor had been acquired by Hazard Stevens, a Northern Pacific lawyer and coincidentally the son of Washington's first governor, Isaac I. Stevens. His supposed inside track came to naught when the railroad chose Tacoma as its terminus, and Stevens sold his holdings to Anna Curtis Bowman. Her husband, Amos, named a city he platted in 1877 after his wife, and the post office changed it slightly—to Anacortes.

Railroad fever returned toward the end of the 1880s with the pending arrival of the Seattle & Northern, Northern Pacific sponsored until James Hill obtained control about 1890 (complete purchase was not until 1901). For a while Anacortes and the Northern Pacific vied with Fairhaven and the Great Northern as the Pacific terminus to watch. The Seattle & Northern built up the Skagit Valley with the goal of connecting with the Northern Pacific main line east of the Cascades but never made it farther than Rockport, fifty miles from Anacortes (see MAP 235, *page 99*). Nonetheless, the speculative fever boosted Anacortes's development.

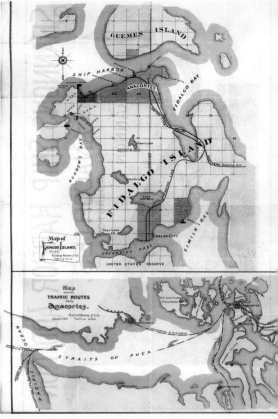

MAP 338 (*below*) and MAP 339 (*right, top*).
An 1893 real estate broker's map of a planned *Northern Pacific Addition to Anacortes*, just east of Ship Harbor, with, on the reverse (MAP 339) two smaller-scale maps showing the city's position on Fidalgo Island and explaining the advantageous position of Anacortes relative to the Pacific Ocean.

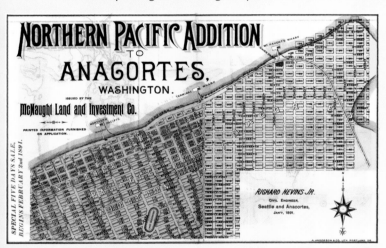

MAP 340 (*right*).
Detail from an atlas map published in 1898 still shows Anacortes as the terminus of the Northern Pacific transcontinental line.

MAP 341 (*below*).
A booklet published in 1890 by the Oregon Improvement Company, part of the Northern Pacific group, extolled, with much hyperbole, the virtues of Anacortes. This interesting map was folded into the front of the booklet. It shows the city as the terminus of the Northern Pacific transcontinental line. Anacortes is, the booklet boasted, "directly on this American line to India . . . further set in a locality of unequalled resources to render the population of such a city prosperous."

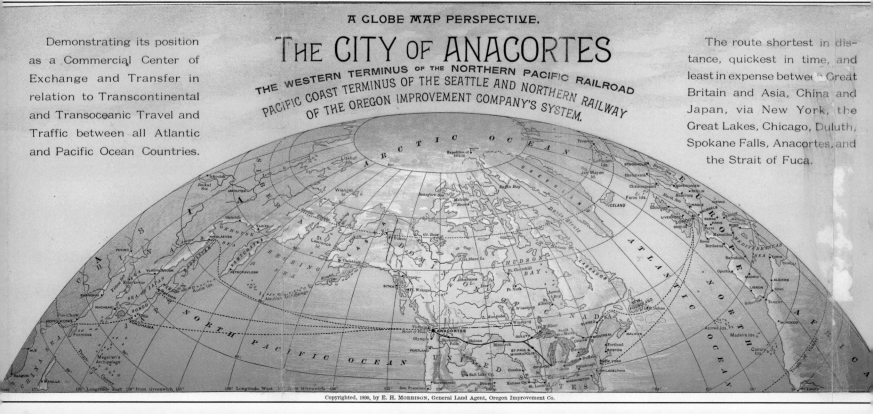

A GLOBE MAP PERSPECTIVE.

THE CITY OF ANACORTES

THE WESTERN TERMINUS OF THE NORTHERN PACIFIC RAILROAD
PACIFIC COAST TERMINUS OF THE SEATTLE AND NORTHERN RAILWAY
OF THE OREGON IMPROVEMENT COMPANY'S SYSTEM.

Demonstrating its position as a Commercial Center of Exchange and Transfer in relation to Transcontinental and Transoceanic Travel and Traffic between all Atlantic and Pacific Ocean Countries.

The route shortest in distance, quickest in time, and least in expense between Great Britain and Asia, China and Japan, via New York, the Great Lakes, Chicago, Duluth, Spokane Falls, Anacortes, and the Strait of Fuca.

Copyrighted, 1890, by E. H. MORRISON, General Land Agent, Oregon Improvement Co.

Port Angeles

At the top of the Olympic Peninsula, on the Strait of Juan de Fuca, Port Angeles sits within the shelter of a long spit, in a natural harbor first mapped by Spanish explorers in 1790 (see Map 43, *page 22*).

Port Angeles has a unique status as America's second "national city." In 1863 the federal government created a military and townsite reserve with the intention that it would one day become the western equivalent of Washington, D.C., and a center for federal activities in the West. Some waterfront lots were sold in 1864 but not enough even to defray the government's survey and auction costs.

By the late 1880s more and more settlers arrived but found no land available within the city, yet found themselves staring at a three-thousand-acre federal reserve surrounding the town. In 1890 some would-be settlers began a virtual land rush, moving onto the reserve, clearing homesites, and squatting on the land. The following year the federal government allowed the legal claiming of land on the reserve, and in 1894 the remaining lots were auctioned off; the reserve was no more.

Port Townsend

The first settlers at Port Townsend arrived in 1852 and included none other than Francis Pettygrove, who had previously been responsible for naming Portland (see page 61). Throughout the 1880s the city grew, on speculation that it would become the premier port of Puget Sound owing to its good harbor and position near the entrance. But the Northern Pacific failed to connect the city with its line to Tacoma, and ports such as Seattle, on the more accessible east shore of the sound, grew rapidly. In the 1890s growth faltered at Port Townsend, and the town rapidly declined; but no one tore down existing structures when they left, thus leaving the legacy of the considerable collection of Victorian buildings we see today.

MAP 344 (*below*).
A fine bird's-eye view of Port Townsend from its growth period; this was published in 1878.

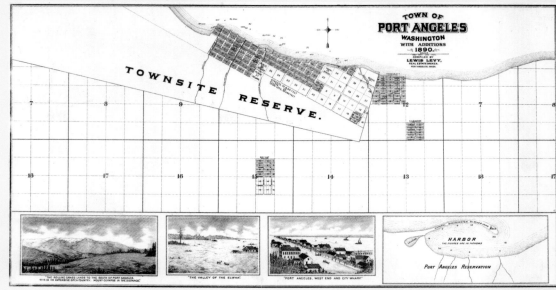

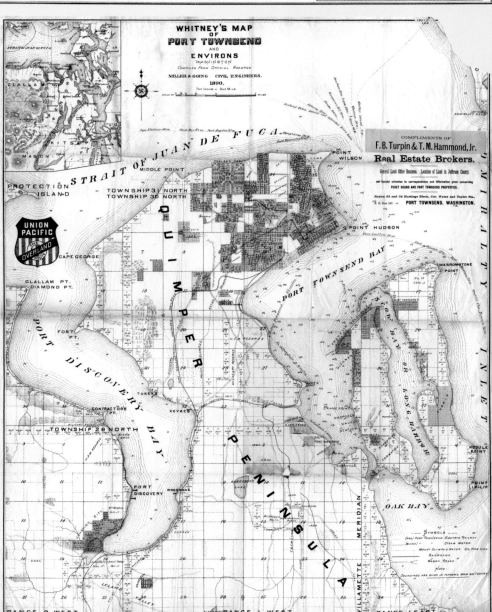

MAP 342 (*above, top*).
The federal *Townsite Reserve*, now partly subdivided, surrounds the original waterfront lots of the town of Port Angeles on this 1890 map.

MAP 343 (*above*).
A fine real estate map of Port Townsend and area from 1890. Many of the lots for sale were never developed, and the rail lines shown likewise were stillborn. Note the Union Pacific sponsorship (see MAP 317, *page 132*).

SPOKANE

EuroAmerican settlement of Eastern Washington began in the 1870s as the threat of Indian attack slowly diminished. Camp Spokane, set up in 1880 by the U.S. Army fifty miles northwest of today's city, helped to secure the region. In 1878 James Nettle Glover filed a plat for an area adjacent to Spokane Falls, on the Spokane River. He had arrived here five years earlier, and although a few others were already living along the river, it seems Glover was the one who appreciated the potential of the falls for powering a large sawmill, which he built, along with two partners.

A boom of sorts began after Glover sold half of his holdings to two other settlers and it seemed the Northern Pacific line would pass through. In 1881 the village of Spokan Falls was incorporated, just in time for the arrival of the Northern Pacific, connecting the infant city to the coast by late 1882 and to the East with the completion of the transcontinental the following year.

The railroad, the falls, and the 1883 discovery of gold in the Coeur d'Alene district of Northern Idaho close by assured the city of a preeminent position in Eastern Washington. The rise of Spokane has been described as miraculous, with the population increasing from 350 persons in 1880 to nearly 20,000 ten years later. By 1900, more than 36,000 lived there, and by 1910 the population stood at an astonishing 104,000.

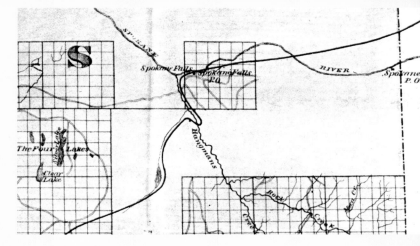

MAP 345 (*above*).
A planned route for the Northern Pacific is shown on this 1878 map. Although the route as indicated was not followed, the line did pass through Spokane Falls.

Fire claimed some thirty-two blocks of the city in August 1889, with a failure in water pressure hampering all efforts to extinguish it. Much of the city's rebuilding afterward was financed by Dutch investors, who, with the Northern Pacific in place and other lines promised, could see the potential for growth. A second transcontinental, the Great Northern, arrived in the city in June 1892 after citizens gave the railroad a free right-of-way, and Hillyard became the site of vast rail yards and maintenance facilities. The city eventually became a major railroad hub (MAP 350, *right*).

Spokan Falls changed its name to Spokane Falls in 1883, and to Spokane in 1891.

BIRD'S EYE VIEW OF
SPOKANE FALLS, W.T.
1884
TOTAL FALL OF WATER FROM A TO B: 158 FEET.

SPOKANE
⇥ WASHINGTON ⇤

MAP 346 (*left, bottom*).
This elegant bird's-eye map of Spokane Falls, viewed from the northwest, emphasizes the falls on the Spokane River and the mills utilizing their power, by showing them in the foreground. The total drop in water level from *A* to *B* is noted—158 feet. As can be seen from the list below the map, a wide variety of industries and businesses operate in the city. The Northern Pacific passes through the center.

MAP 347 (*above*).
Spokane is beginning to look like the city of over one hundred thousand inhabitants that it would be just five years after this bird's-eye map was published in 1905. In this view, from the southwest, the Northern Pacific line runs into the map from the foreground, while the *Great Northern R.R.* is on the north side of the river in the foreground, crossing to the south side over a bridge in the center of the city. Note that the downtown area had been rebuilt mainly in brick after the 1889 fire.

MAP 348 (*below*) and **MAP 349** (*right*).
The Spokane & Inland Empire Railroad was an interurban that promoted the growth of Spokane by feeding people and freight to the city from the rich agricultural Palouse to the south—the *Rolling Wheat Belt*—and the *Spokane Valley* to the east. **MAP 348** is from a 1910 timetable; **MAP 349** is a detail of a 1911 map.

MAP 350 (*right*).
The position of *Spokane* as a major railroad hub is emphasized here to advertise the *Power City* on the falls. The map dates from 1910.

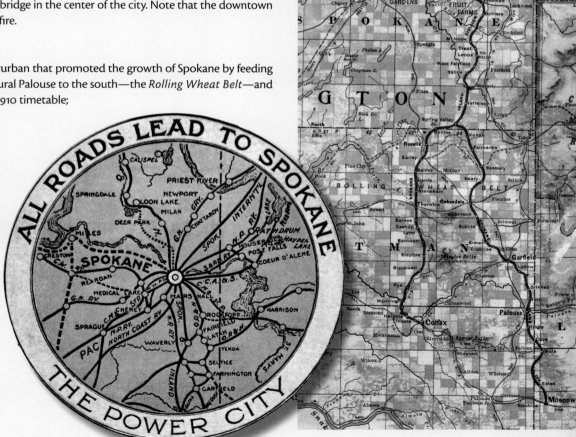

YAKIMA

Yakima in 1884, as the Northern Pacific's Cascade Branch approached, was a thriving town of six hundred that had been incorporated the previous year. Then, in December, the railroad made a startling announcement—its depot would be built four and a half miles west of the town. Here the railroad would own the land rather than the citizens of Yakima, who had filed claims and purchased lots in the town in anticipation of the railroad's coming.

Not surprisingly, the townspeople were outraged; Washington Territory even brought legal action against the railroad to compel it to establish a depot at Yakima. Yet despite the success of this action, the site the railroad had chosen, called North Yakima, was actually a superior one that would allow more growth, and by April 1885 the entire town of over a hundred buildings was physically moved—with screw jacks, rollers, and teams of horses—to the new site.

MAP 351 (*above*).
The Northern Pacific's creation, North Yakima, has a lot of well-ordered space to fill on this bird's-eye map from 1889. The railroad runs through the center of the city, and the depot that started it all is shown.

North Yakima was incorporated in 1886, and in 1918 changed its name to Yakima. The old city was renamed Union Gap and is today effectively a southern suburb of Yakima.

WALLA WALLA

Fort Walla Walla, distinct from the Hudson's Bay Company post of the same name, had been established in 1858 by the U.S. Army as part of its campaign to control Indian depredations east of the Cascades (see MAP 143, *page 65*). The city was incorporated in 1862 and grew rapidly after gold was discovered on the Clearwater River in Idaho in 1860. In the 1860s the city became the largest in Washington Territory and was thought likely to become its capital. By 1875 the city had a railroad link with the Columbia River, where it connected with steamboats. The construction of the Walla Walla & Columbia River Railroad was largely due to the efforts of city resident Dorsey Baker (see page 84). Walla Walla's governmental aspirations came to nothing, but it thrived as the center of a rich agricultural region.

MAP 352 (*below*).
An 1884 bird's-eye map of Walla Walla. An Oregon Railway & Navigation Company train is in the foreground.

Eugene & Springfield

Eugene Franklin Skinner built a cabin in 1846 on the Middle Fork of the Willamette, which served as a trading post for the area. He and a judge, David Risdon, filed the first plat six years later. The site proved to be too near the river and became known as Skinner's Mud Hole; a new plat was laid out on higher ground in 1853. By this time Skinner was operating a ferry across the river (Map 354, *below*). The settlement became known as Eugene City (Map 353, *right*), incorporated in 1862 under that name; two years later it became the City of Eugene.

Springfield, by contrast, was not named after an individual. Elias Briggs chose a location for his house close by a spring, an area that became known as the spring field. A plat was filed in 1856, and the city was incorporated in 1885.

Map 355 (*below*).
By 1902, the date of this map, there had been extensive subdivision into city lots and four localities are named: *Eugene, Fairmount, Glenwood,* and *Springfield*. The railroad is the *O. & C. R.R.* (Oregon & California), completed to Eugene in 1871, and a *depot* has been built just south of *Skinner Butte*. There is now a bridge at the location of Eugene Skinner's ferry. Just upstream the Willamette has cut off a meander, which the river did frequently throughout its course.

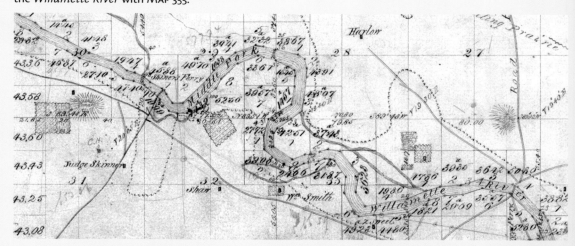

Map 353 (*above*).
In 1866 this General Land Office map showed both *Eugene City* and *Springfield* on the *Middle Fork* of the Willamette.

Map 354 (*below*).
This 1853 survey map shows the location of Eugene *Skiner's [Skinner's] Ferry*, the beginnings of the city of Eugene. There is no settlement yet at what will become Springfield. Compare the course of the *Willamette River* with Map 355.

MORE CITIES OF OREGON

The majority of the cities of the Willamette Valley have origins in the decade after the international boundary was defined in 1846. Many changed positions because of flooding. Slowly settlement reached beyond the Willamette, prompted by gold rushes or transportation opportunities, but on an 1866 map of Oregon published by the federal General Land Office (Map 356, *right*), the only settlements noted east of the Willamette Valley were Dallas (The Dalles), incorporated in 1857; the steamboat wharves at Umatilla, incorporated as Umatilla Landing in 1864; and the mining supply and agricultural town of La Grande, incorporated in 1865.

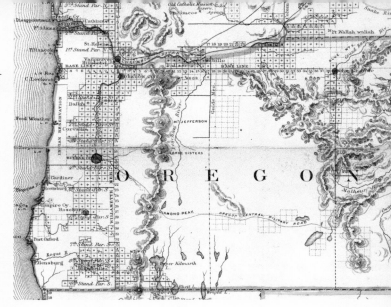

Map 357 (*left*).

An 1852 survey map shows *Cincinatti*, an early city of the Willamette with great hopes for the future; unlike some, Cincinatti did not disappear but became the village of Eola, incorporated as such in 1856. The town was originally named after a presumed resemblance to the Ohio city.

Map 356 (*above*).

A General Land Office map of Oregon in 1866. The green circle, representing the location of the surveyor general's office, is *Eugene City*; the red circles are other land offices. Gold finds are marked in yellow. Few Oregon cities exist east of the Willamette Valley.

Map 358 (*below*).

A bird's-eye map of Salem in 1876. At *A* is the state capitol building, which was completed that year, though it received a distinctive copper dome in 1893. This structure was on the same site as an earlier one built in 1855 and destroyed by fire only two months later. The building depicted here also burned down, in 1935, and was replaced three years later by the current art deco design. Where *State St.* meets the Willamette is an interesting and unusual aerial bridge. In the background, Mount Hood towers over the valley a little larger than life.

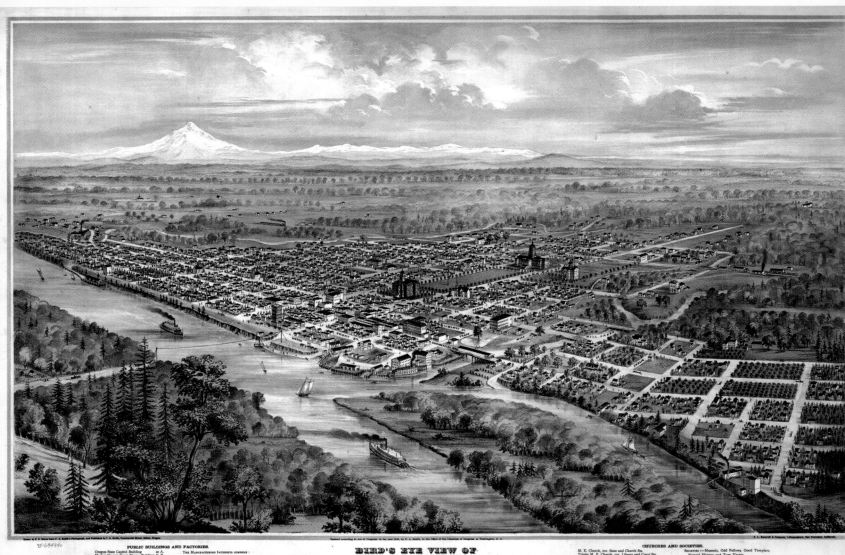

PUBLIC BUILDINGS AND FACTORIES.

Oregon State Capitol Building   A.
Marion County Court House Building   B.
Willamette University Building   C.
Academy of the Sacred Heart   D.
Salem Public Schools   E.
Water Works at G.   Gas Works   H.
State School for the Deaf   I.
Oregon Home for the Sick   J.
State Penitentiary   two miles distant.
County Fair Grounds   S.

THE MANUFACTURING INTERESTS COMPRISE:

Flouring Mills, Saw Mills, Wagon Factory,
Pump Factory, Woolen Mills, Oil Mills,
Machine Shops and Foundries,
Sash and Door Factories, Chair Factory,
Agricultural Implement Works,
Plow Factory, Bag Factory, &c.
Mount Hood, sixty miles distant.

CHURCHES AND SOCIETIES.

M. E. Church, cor. State and Church Sts.
Trinity M. E. Church, cor. Liberty and Court Sts.
First Baptist Church, cor. Liberty and Marion Sts.
First Presbyterian Church, St. Chemeketa and Center.
Congregational Church, E. E. cor. Liberty and Center.
Evangelical Church, N. E. cor. Liberty and Center.
St. John's Catholic Church, cor. Cottage and Chemeketa Sts.
Christian Church, cor. High and Center Sts.
St. Paul's Episcopal Church, cor. Church and Chemeketa Sts.
Cumberland Presbyterian Church, High Street, near Union.

Societies—Masonic, Odd Fellows, Good Templars.
Natural History and Turn Verein.
Opera House.
Three Newspapers, (Two Daily.)
Several First-class Hotels.
Banks and numerous Mercantile Institutions.
Situated on the Willamette River, Forty Miles above
Portland, on the Oregon and California Railway.

BIRD'S EYE VIEW OF SALEM, OREGON
FROM THE WEST, LOOKING EAST.
1876.

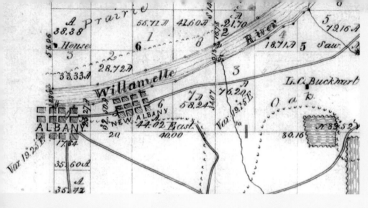

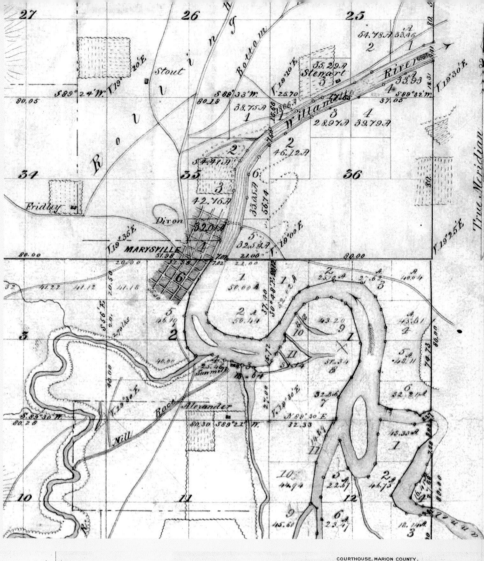

Map 359 (*above*).
Albany was founded in 1848 by the Monteith brothers, Walter and Thomas, who built the following year what was reputed to be the finest house in all of Oregon. Just east of the Monteiths' Albany (named after the city in New York) a rival town was platted in 1849 called Takenah. The name of the post office was changed to *New Albany*, as shown on this 1852 survey, in 1850; the territorial legislature combined both towns under the name Takenah in 1854 but the following year reversed itself and named them both Albany—an altogether confusing situation.

Map 360 (*right*).
Marysville is shown on the banks of the *Willamette River*, just below where Marys River enters, on these two survey map sheets (the top one is from 1852, the bottom one 1853). The town's name was changed to Corvallis in 1853.

Map 361 (*below*).
Another rather splendid bird's-eye map of Salem, this one from 1905 touting its population—14,768. The aerial bridge over the Willamette has been replaced by a steel structure.

STATE HOUSE OF OREGON.

COURTHOUSE, MARION COUNTY.

Capital City of Oregon.
POPULATION IN 1905   14,768.

SALEM

MUTUAL L. & LITH. CO. PORTLAND, OR.   COPYRIGHTED BY E. KOPPE & CH. FROMM.

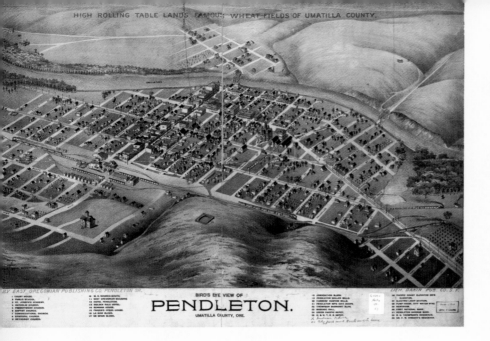

BIRD'S EYE VIEW OF

PENDLETON.

UMATILLA COUNTY, ORE.

Map 362 (*above*).

A bird's-eye map of Pendleton, in the deep valley of the Umatilla River in eastern Oregon, then as now a commercial center for the surrounding rich agricultural area. This map was published about 1895. The city grew from a trading post established in 1851 and was at first named Marshall, being renamed in 1868 for the Democratic candidate for vice-president in 1864, George Hunt Pendleton. The city was incorporated in 1880.

Map 363 (*below*).

The text accompanying this stunning bird's-eye map of the Rogue River Valley touts *Grants Pass* as "the coming inland metropolis of Southern Oregon." The map was created by a landscape artist, Gibson Catlett, and published in 1910 by the Grants Pass & Rogue River Railroad to promote its new development, *South Grants Pass*, across the river in the location of today's Harbeck-Fruitdale. The railroad never lived up to the promise of this superb map. It changed its name to the California & Oregon Coast Railroad and was intended to connect with Crescent City, on the California coast, but was only constructed for fourteen miles west of Grants Pass. The railroad investors had hoped that the Oregon & California Railroad would purchase the new line, but it did not. The caption to the map notes points to "seventy thousand acres of unsurpassed apple, peach, and diversified farming lands" in this region of the Rogue River, Applegate, and Williams valleys; adjacent is two billion standing feet of saw timber—a rich region indeed. Note *Medford* in the distance.

Map 364 (*right*).

A classic 1890 bird's-eye map of Astoria, now grown from the original Pacific Fur Company post established in 1811 (see page 30) into a city of some size. This clearly promotional illustration has not one but two bird's-eye views in addition to an array of fine residences and businesses—some with interior views—around the edge.

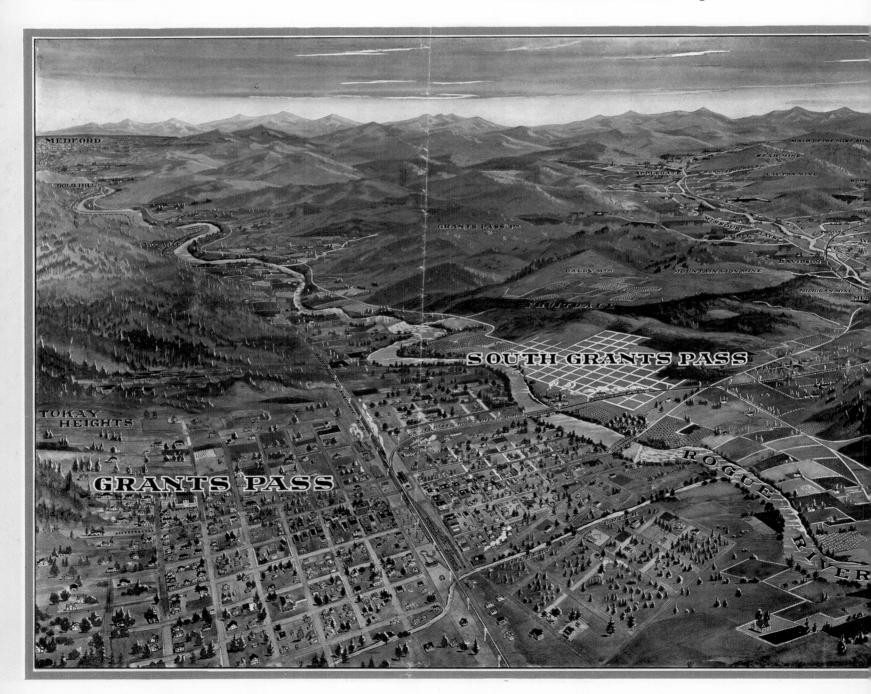

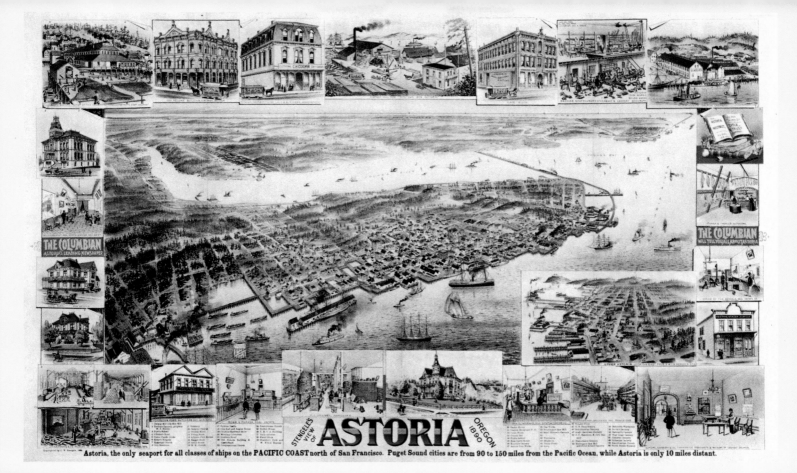

STENGELE'S VIEW OF **ASTORIA** OREGON 1890

"Astoria, the only seaport for all classes of ships on the PACIFIC COAST north of San Francisco. Puget Sound cities are from 90 to 150 miles from the Pacific Ocean, while Astoria is only 10 miles distant."

THE COLUMBIAN ASTORIA'S LEADING NEWSPAPER

THE COLUMBIAN WILL TELL YOU ALL ABOUT ASTORIA

TWO BILLION FEET SAW TIMBER

Cities That Never Were

The Northwest is littered with ghost towns—settlements that sprang up to house and service mineral seekers of one kind or another. But there were also hundreds of cities in the Northwest that never progressed much beyond a gleam in their founders' eyes. Three that seem particularly interesting have been selected for inclusion here.

The earliest is Semiahmoo City. Generally speaking, a city site is selected with future growth in mind, but not Semiahmoo City, which was platted on the long and thin spit that divides Drayton Harbor from Semiahmoo Bay in northwest Washington, close by the international boundary at today's Blaine. The spit had been used by Native peoples for centuries before.

The map shown here, a contemporary copy made before photocopiers made the task so much easier, is for an addition to an original plat and was filed in 1889. There are no details other than the street grid and surveyed lots, so the map has the appearance of a major city; the thin peninsula is nowhere to be seen.

Semiahmoo Spit was later the site of a salmon cannery and is today developed with a resort hotel partially converted from the old cannery.

A much later attempt at a city was on the tidal flats of the delta of the Nisqually River, just west of the original site of the Hudson's Bay Company's Fort Nisqually (see MAP 84, *page 37*). The latter site became an explosives factory, belonging to E.I. du Pont de Nemours & Company, in 1906 and is today the city of Dupont. The fort itself was moved to Point Defiance Park in Tacoma.

In the 1940s there was a proposal, shown on MAP 368, *right,* for the creation of a foreign trade zone (where import duties were not applied until goods from abroad were carried out of the zone), together with an accompanying city, all on the rich tidal flats of the delta.

Our third non-city is an example of the stop-at-nothing attempts at marketing a townsite in the days before the regulation of advertising.

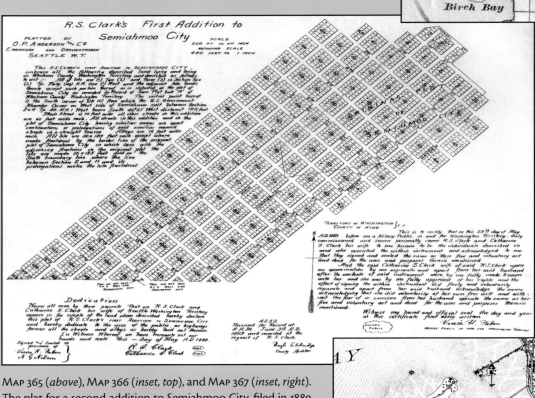

MAP 365 (*above*), MAP 366 (*inset, top*), and MAP 367 (*inset, right*). The plat for a second addition to Semiahmoo City, filed in 1889 by R.S. Clark and his wife, Catherine Clark. MAP 366 shows *Semiahmoo* on an 1890 map; MAP 367 is a 1908 map of *Drayton Harbor*, showing the spit, with cannery buildings, across from Blaine.

MAP 368 (*below*).
The proposed foreign trade zone and city on the Nisqually tidal flats in Puget Sound. Thought to have been proposed in the 1940s, foreign trade zones exist today at some ports, such as nearby Olympia. By today's standards a more improbable site could hardly have been found, yet port facilities at Olympia, Tacoma, Seattle, and elsewhere were created on what were originally tidal flats. Today this area is a state wildlife recreation area and a national wildlife refuge, 2,800 acres of fresh- and saltwater marsh, prairie, and mudflats.

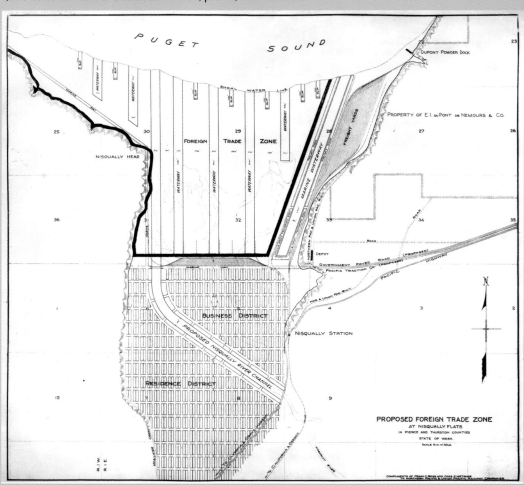

A Little Over One Year Ago the Site of Lakeport Was Just as Nature Made it. Now Lakeport is a Thriving Town of Two Hundred Souls, and This Year (1911) Will See Its Population Increased by Many Hundreds

MAP 369 (*above*).

From a promotional booklet published in 1911 comes this little bird's-eye map of the extensive town of Lakeport, with *Floras Lake* and the canal *under construction* that was to create the port city worthy of the interest of an investor or potential resident. *Cape Blanco* is in the distance. The map is at least truthful in that it shows no railroad, although the accompanying text states that surveys have been made and that "a complete coast line will be in operation before the fair at San Francisco in 1915," a reference to the Panama-Pacific International Exposition of that year.

Left. The "substantial" hotel at Lakeport, from the 1911 promotional booklet.

Below, left center; below, left bottom; below, right. The cover and excerpts from the same booklet.

The city of Lakeport was promoted in 1910–11 for a site on the shores of Floras Lake, on the southern Oregon coast, about four miles north of Cape Blanco. A canal had been planned in 1894 to link the lake with the ocean to allow export of timber. When it was found that the lake was about six to ten feet higher than the sea, the scheme was abandoned. This did not prevent its revival about 1908, this time accompanied by plans for a city of some size on the east shore of the lake (MAP 369, *above*). A post office was established and a few buildings were constructed, including a "substantial" hotel. The scheme collapsed, of course, when it finally became known that the canal could not be built owing to the water level difference. It is possible that the promoter did not know of this difficulty, but of course he would not likely have been able to sell his townsite lots had it become widely known. Truth in advertising would be some time coming yet. Many port schemes were hatched for the Pacific Coast at this time, intended to benefit from the opening of the Panama Canal in 1913, and many failed when soon after the world sank into war and real estate prices collapsed.

CANAL AND HARBOR

THE CANAL to connect Floras Like with the Pacific Ocean is now under construction. The construction is approved by the U. S. Engineer's office and the War Department. There is a sand spit about 700 feet wide between the lake and the ocean and it is through this sand spit the canal is being built. The canal will be 300 feet wide, and the contract is let for its excavation 20 feet below low tide line; this will give a channel 20 feet deep at low tide and about 28 feet at high tide. The channel will admit most of the vessels plying the Pacific waters.

Cities in the Northwest are building rapidly; the eyes of the world are centered on the Pacific Northwest, because the world is coming to know its resources and possibilities.

LAKEPORT courts investigation and will grow on its merits.

LAKEPORT began building March 1, 1910. It now has a sawmill, three-story hotel, newspaper, post office, laundry, livery stable, butcher, barber and blacksmith shops, general, grocery, confectionery, cigar and hardware stores; telephone connections with the main line and the wireless station at Cape Blanco. A bank will be established this summer and other lines of business represented. Other mills and factories will soon be built and in operation.

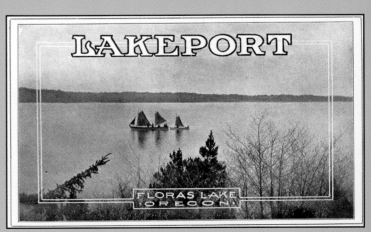

NATIONAL PARKS AND NATIONAL FORESTS

The idea of reserving land for recreation or conservation purposes slowly gained traction in the West throughout the second half of the nineteenth century since Abraham Lincoln's protection of the California sequoias in 1864. The Northwest's first national park was created at Mount Rainier in 1899, and Crater Lake National Park followed three years later. The Olympic Forest Reserve, created in 1897, became a national monument in 1909 and was transformed in 1938 into Olympic National Park. Today, 95 percent is designated wilderness. In 1981 it became a World Heritage Site. The Northwest's newest national park, the 1,069-square-mile North Cascades National Park, was created from reserves in 1968; this park is also principally wilderness.

Some forest reserves were created in 1891. More were created by President Theodore Roosevelt in 1897 and are known as the "birthday reserves" because they were reserved on George Washington's birthday. Chelan Forest Reserve (MAP 371, *below, right*) was one such reserve. Forest reserves were renamed national forests in 1908.

In August 1910, the largest forest fire the United States has ever seen swept from northeast Washington across the Idaho panhandle into western Montana, in three days burning nearly three million acres, destroying seven towns, and killing eighty-six people, including a team of twenty-eight firefighters trapped by the inferno. The fire had begun as a series of smaller fires caused by trains, lightning, and fires of homesteaders, miners, and hoboes, and

Left. Crater Lake National Park, created in 1902, was close to the route of the Natron Cutoff, completed by the Southern Pacific–controlled Central Pacific as an alternate route to California in 1926. This brochure dates from about that time.

MAP 371 (*below, right*). A 1908 map from an early U.S. Forest Service brochure for Chelan National Forest.

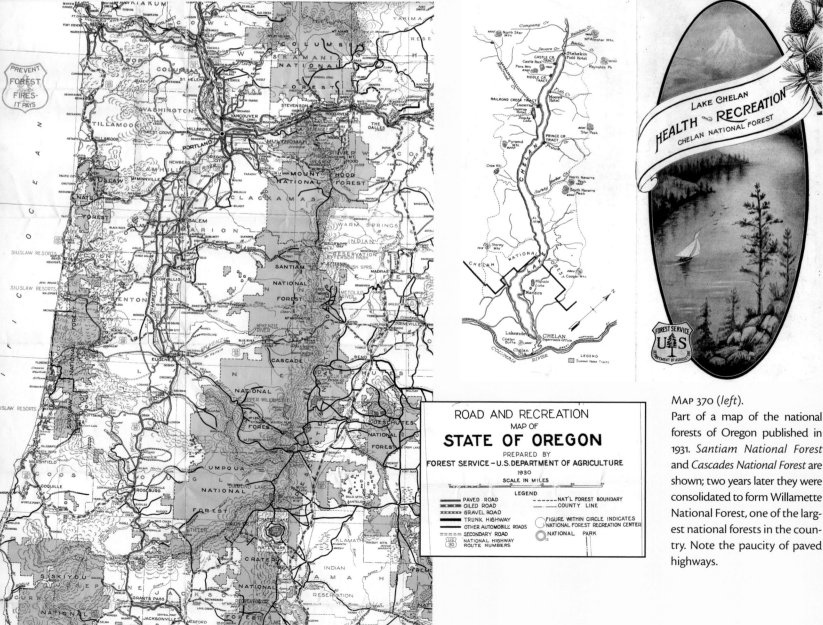

MAP 370 (*left*).
Part of a map of the national forests of Oregon published in 1931. *Santiam National Forest* and *Cascades National Forest* are shown; two years later they were consolidated to form Willamette National Forest, one of the largest national forests in the country. Note the paucity of paved highways.

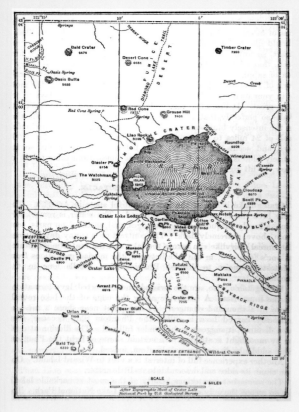

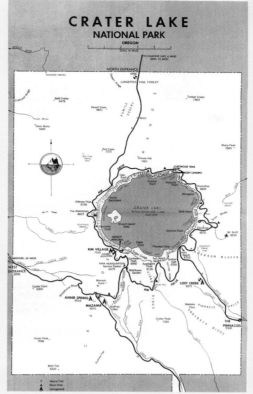

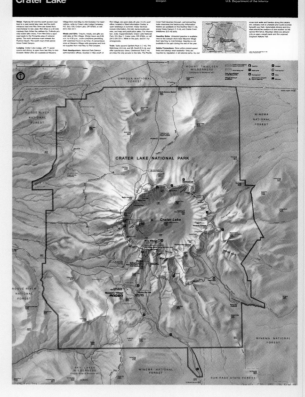

they had all come together under the driest of conditions. It was a matter of debate before this fire as to whether it was best to let forest fires burn or to try to stop them; following this fire Congress gave the U.S. Forest Service much more extensive powers to deal with the threat of fire.

National parks and forests now cover some 46,400 square miles in Washington and Oregon, equal to 29 percent of the total land area.

MAP 372 (*top left*), MAP 373 (*top center*), MAP 374 (*top right*). An interesting sequence showing the evolution of cartographic presentation over the years. This is Crater Lake National Park, established by President Theodore Roosevelt in 1902. MAP 372 was published by the U.S. Geological Survey in 1916, the year the U.S. National Park Service (NPS) was created. MAP 373 was published by the NPS in 1965, and MAP 374 was published by the NPS in 1996. The latter two maps show Rim Drive circling the lake, a road completed in 1918. The nearly two-thousand-foot-deep lake was discovered by gold seekers in 1853 but not properly mapped until geologist Clarence Dutton and his men carried a half-ton survey boat up to the lake in 1886.

MAP 375 (*right, center*).
In 1899 Mount Rainier National Park became the fifth U.S. national park. This detailed survey was published in 1907 and shows existing roads and those under construction. The boundary was later refined slightly and in particular was extended east to the *Summit of the Cascade Mountains*.

MAP 376 (*right*) and MAP 377 (*far right*).
Not everyone agrees with national parks. These two maps are of unknown origin but perhaps were published by logging interests. They appeared in 1955 and are part of a booklet demanding a reduction in the size of the Olympic National Park, and they use excellent graphics to convey their message. One compares the park's size to the state of Rhode Island, and in the other a giant hand demonstrates the park's size by lifting it skyward from the Olympic Peninsula.

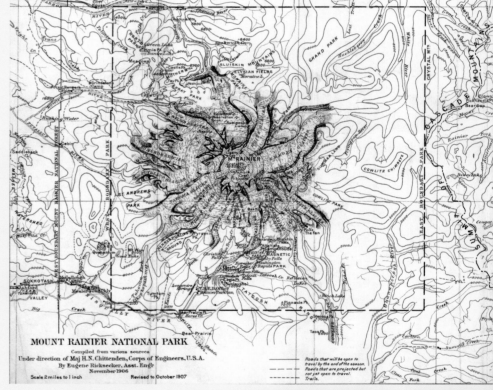

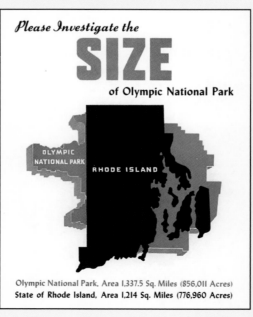

WHEAT AND LUMBER

Wood and agriculture were important to the Northwest's economy from the beginning. By 1910 Washington had the largest timber production of any state. Much of the wood came from the forests adjacent to Puget Sound and was shipped out via that waterway—a convenient channel while it lasted. All around Puget Sound sawmills and wharves sprang up, much like that of Samuel Hadlock at Port Hadlock, at the southern end of Port Townsend (MAP 378, *above, right*). Later, once all the easy wood was gone, logging progressed farther away from the water and began to rely on steam power: ephemeral logging railroads that could be easily relocated once that section of wood was exhausted (MAP 379, *below*). Today, much more carefully regulated, the forest industry is still a backbone of the Northwest economy.

The extensive and fertile Palouse region of Eastern Washington, and usually considered to extend into northeast Oregon, owes it fecundity to loess, wind-blown soil of glacial derivation, which accounts for the random rolling hills. In the 1880s this agricultural area briefly exceeded that of Puget Sound in population as the railroad and other land grants were taken up, and wheat became the agricultural staple, as it was ideally suited to the climate. The ability to mechanize the entire process led to large-scale farming. The old photos of multi-horse harvesting teams still impress today (see the postcard image, *above, left*).

MAP 378 (*above*).
Port Hadlock, on Puget Sound, had been founded by Samuel Hadlock in 1870 when he built a sawmill and wharves to cut and export nearby timber. This bird's-eye view is from 1889.

MAP 379 (*left*).
Hundreds of short and tortuously laid-out logging railroads dotted the Northwest. A detail of a 1931 map shows two logging railroads, one on each side of the Columbia River, about ten miles east of Gresham. The southern bank line is completely isolated, probably instead hauling logs to creeks, while the northern bank line connects with the river, crossing the main line of the Spokane, Portland & Seattle to get there. Logging railroads usually did not last long, as they were moved with the wood supply.

MAP 380 (*below*).
This unusual perspective advertises *Pullman*, Washington, and the *Palouse Country* that it serves. Trains ply a network of rail lines, with the Columbia River on the left, *Spokane* on the right, and *Seattle* and Puget Sound in the distance. The *State College of Washington*, which until 1890 was the state agricultural college and which would become Washington State University in 1959, is at right. The map comes from a 1911 Pullman Chamber of Commerce brochure.

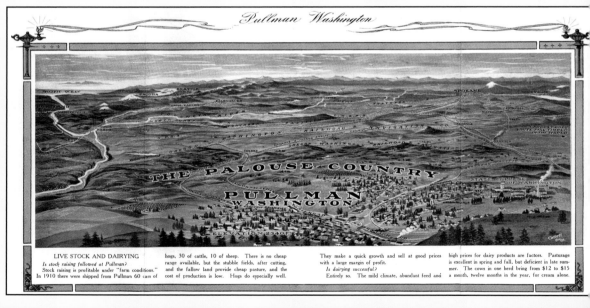

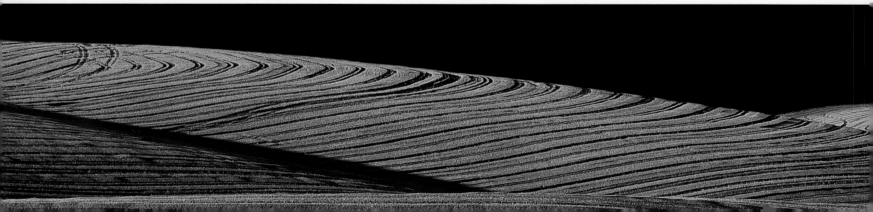

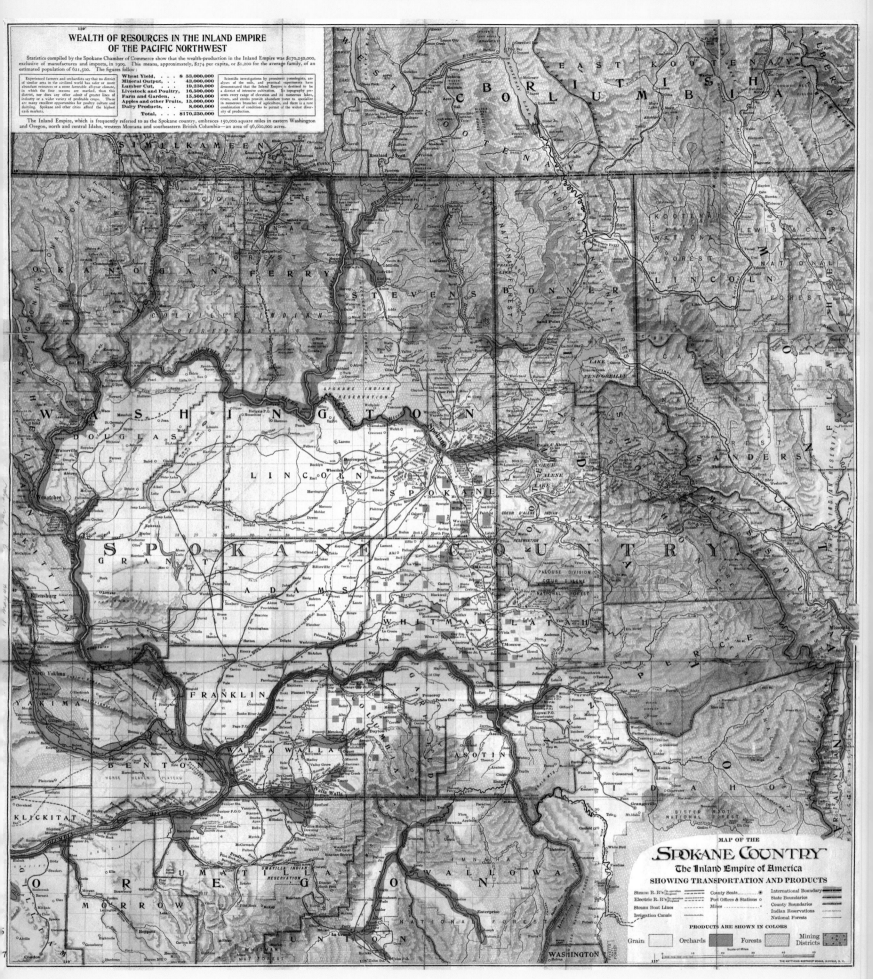

MAP OF THE
SPOKANE COUNTRY
The Inland Empire of America
SHOWING TRANSPORTATION AND PRODUCTS

Steam R. R's { In operation / Proposed } — County Seats ⊙ — International Boundary
Electric R. R's { In operation / Proposed } — Post Offices & Stations ○ — State Boundaries
Steam Boat Lines — Mines ✕ — County Boundaries
Irrigation Canals — Indian Reservations — National Forests

PRODUCTS ARE SHOWN IN COLORS

Grain | Orchards | Forests | Mining Districts

MAP 381.

The yellow-colored areas are those that grow grain on this 1911 map of the Inland Empire. Also shown are orchards (in red), forest (in green), and mining districts (pink with red dots). With the by now extensive transportation network of both electric interurbans and steam-powered lines, and some early irrigation canals, it all adds up to one prosperous-looking region.

Left. A storm approaches late-summer rolling wheat fields near Walla Walla.

SELLING OREGON

The lure of cheap land, and perhaps a chance to start life anew, drew many a settler to the Northwest. And where there was a demand, supply would be found. All manner of promoters, ranging from the honest to the get-rich-quick type, plied their trade. Two fine examples are shown here.

A real estate company called the Rogue River Valley Orchards Company advertised land for fruit growing in 1910 (MAP 383, *below*) and especially promoted the growing of apples and pears. The land was all on the north side of the river and required irrigation for it to be successful. Today that irrigation has been applied, but there are few orchards.

An interesting land sales case is that of the Oregon & Western Colonization Company, based in St. Paul, Minnesota, which in 1908 purchased 1,250 square miles of land grants associated with the Willamette Valley and Cascade Mountain Military Wagon Road, also known as the Santiam Wagon Road, which had been completed with the aid of such grants in 1868 (see MAPS 172 and 173, *page 78*). By 1915, the sale of these lands was in full swing (MAP 384, *right*), with the map back having illustrations and text showing "a prosperous Oregon rancher," "apples grow[ing] thick in Oregon," and "stock raising and dairying pay[ing] big cash dividends." Despite the obvious hyperbole of such promotions, many settlers did buy land and create a fine life for themselves, especially where they could benefit from irrigation. However, today all but 6 percent of Malheur County is rangeland.

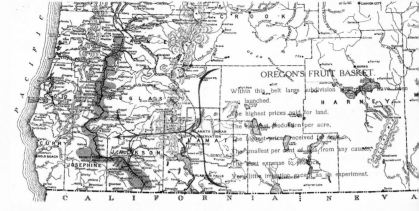

MAP 382 (*above*).
From another agricultural real estate promotion comes this overprinted map touting the valleys from *Eugene* to the California boundary as *Oregon's Fruit Basket*—"highest prices paid for land."

MAP 383 (*below*) with cover (*right*).
Paralleling the city of Ashland along the side of the valley, the *Ashland Orchard Tracts* stretch for four miles. Without irrigation they would have been of limited agricultural utility. This map and promotional brochure were published in 1910 and may be considered quite typical of this genre of promotional material for the period. The cover shows predictably fruit-laden trees, and panels from the map's back (*left*) extol the "large and certain profits" and give a multiplicity of reasons why the land is such a good investment. By such means were settlers enticed to the Northwest. Ashland is here called the "Italy of America," but the brochure is careful to note the "population of 6000 *Americans*."

LARGE AND CERTAIN PROFITS

Fruit growing is a science, from selecting the land and trees to preparing the fruit for market, and there are certain and big profits to the skilled orchardist.

The peculiar soil conditions of this section, aided by proper altitude, the right aridity of air, days of bright sunshine, nights of cool crisp air and a temperature that is always mild, have the quality of producing fruits of rare flavor and beautiful color, juicy, crisp and delicious, yet having a firmness, even in peaches and grapes, that enables them to be shipped in perfect condition to markets so distant as New York, London and Vladivostok. Charles V. Galloway, Oregon Commissioner to the St. Louis World's Fair, reported that of the peaches for exhibit coming from long distance those from the Rogue River Valley arrived in better condition and stood up longer than did the peaches from any other section of the United States.

Bartlett pears, that growers in many fruit districts have difficulty in transporting to local markets, are shipped from the Rogue River Valley to London with no difficulty whatever.

REASONS WHY

The following are some of the more concise and explicit reasons why you should invest in this land:

1 Largest Profits
2 Greatest Security
3 Productive Soil
4 Beautiful View
5 Healthful Climate
6 Nearness to City
7 Social Advantages
8 Educational Advantages
9 City Accommodations
10 Values Sure to Increase
11 Comfortable Home
12 Best Saving Proposition
13 Provides for Old Age
14 Springs and Resorts
15 Transportation Facilities
16 Perfect Fruits
17 Agreeable Occupation
18 Long List of Successes
19 Improvements, Roads, Bridges and Boulevards
20 Easy Plan of Purchase

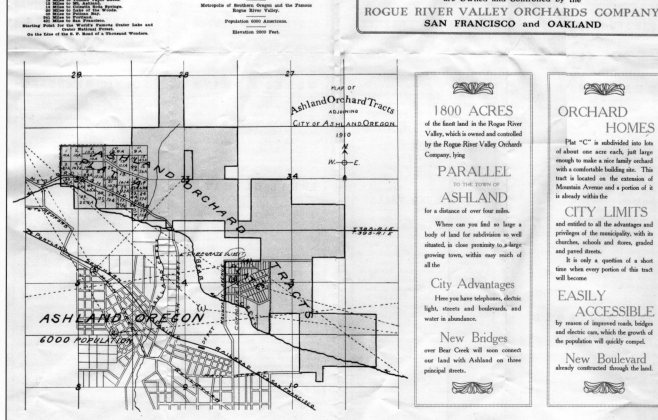

800,000 ACRES OF LAND IN CENTRAL OREGON

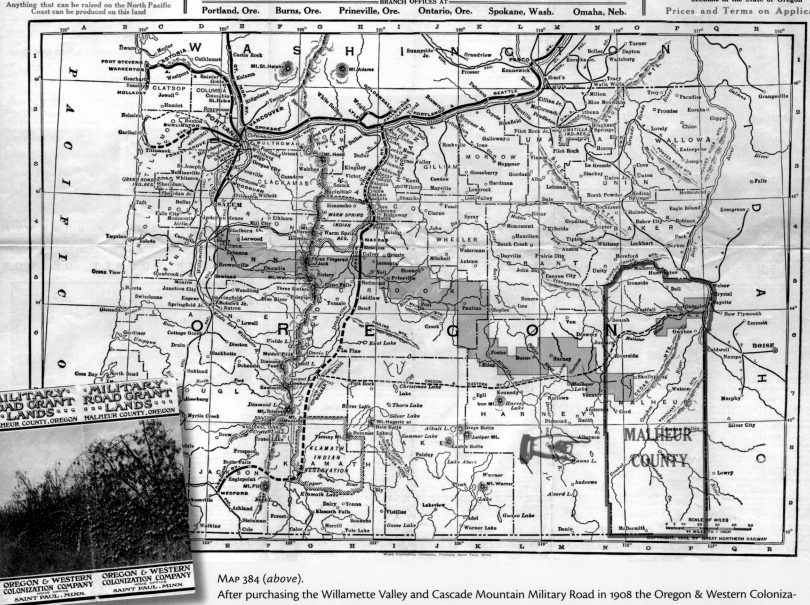

MAP 384 (*above*).

After purchasing the Willamette Valley and Cascade Mountain Military Road in 1908 the Oregon & Western Coloniza-tion Company set about marketing its lands; here it promotes those in Malheur County. This map was published in 1915.

Below is an illustration from the back of the map titled "Plowing Sixty Acres a Day in Oregon." The map's back was crammed with promotional letters from existing settlers. "Italian prunes if properly handled will make from $350 to $500 per acre net per year," enthused one. "This beats any place I ever saw for the dairy and hog business," wrote another. "I am no booster and don't believe in it, but these are the facts."

MAP 385 (*left*).

This bird's-eye view with multiple illustrations pro-moted the Ashland area in 1909.

EARLY IRRIGATION PROJECTS

The halcyon years of Northwest irrigation came once the large Columbia Basin Project got under way in the 1930s (see page 189), yet, long before that, the region was covered with smaller-scale projects. The 1890 census recorded their extent for us (Map 389, *right*). Projects typically diverted water upstream into canals for which the drop was carefully controlled so as to retain gravity flow that could easily be distributed further via a system of lesser channels. Nowhere is this better illustrated than in the West Okanogan project (Map 390, *far right*), where canals run along the valley sides. Railroads often promoted irrigation schemes, for they stood ready to benefit from transportation of agricultural produce, not to mention the general increase in freight levels that came with a higher population. Many areas of the drier Northwest are prosperous today because of these early irrigation efforts.

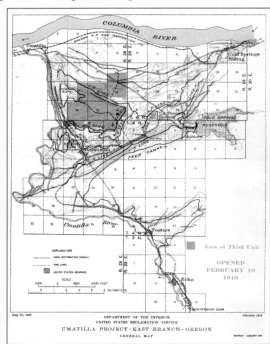

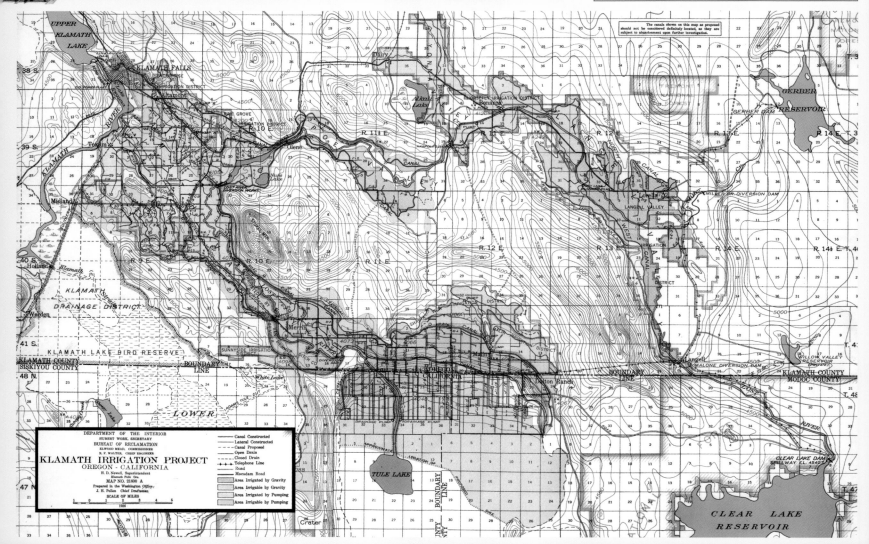

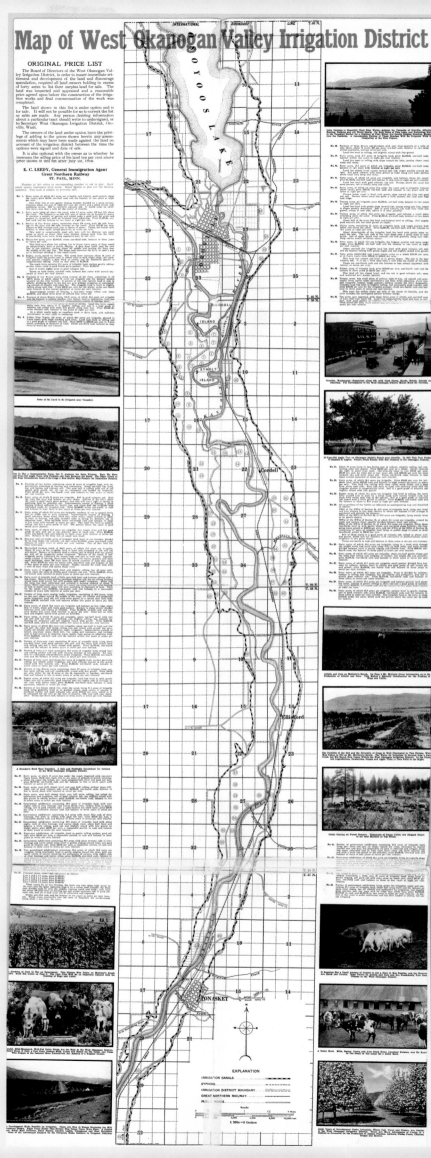

MAP 386 (*far left*), MAP 387 (*left*), and cover illustration (*far left, top*).
The Umatilla Project, on the Umatilla River at Hermiston, Oregon, reclaimed 36,000 acres between 1903 and 1927. This bird's-eye map of the project, a more formal U.S. Reclamation Service map, and a superb illustration, come from a booklet published by the Umatilla Project Development League, a local booster group at least partly funded by the Oregon Railroad & Navigation Company, by this time part of Union Pacific. The name of Hermiston was taken from a then-popular Robert Louis Stevenson novel, *Weir of Hermiston*.

MAP 388 (*left, bottom*).
The Klamath Irrigation Project, straddling the Oregon–California boundary line, in 1926. A series of seven dams were constructed between 1906 and 1921, converting 350 square miles of rangeland to more intensive use. The project also involved draining lakes and wetlands to allow them to be farmed. In 2001 water was cut off because of concerns about endangered fish species, but public outcry ensured water returned the following year. It now seems probable that four dams will be dismantled.

MAP 389 (*above*).
The extent of irrigated areas in the Northwest can be seen on this census statistical map produced in 1889 for the 1890 census but published in 1894. The green-colored areas denote irrigation. Although irrigation was limited in area at this time, it was still widely practiced on a small scale.

MAP 390 (*right*), with location map, MAP 391 (*below*).
The West Okanogan Valley Irrigation Project, sponsored by the Great Northern Railway, whose line ran up the valley, irrigated the area between Tonasket and the international boundary near Oroville, at Osoyoos Lake. This large fold-out map was published in 1915 to list properties over forty acres for sale in the new irrigation district and contains the usual promotional photographs of the agricultural wonders worked by applying water. Irrigation canals are shown as the black-and-white dashed lines and run along both valley sides, from where water could be applied to the fields below by gravity feed.

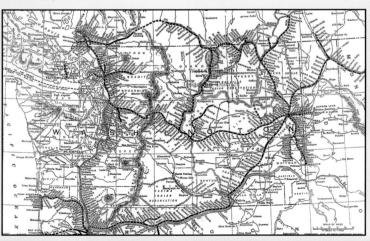

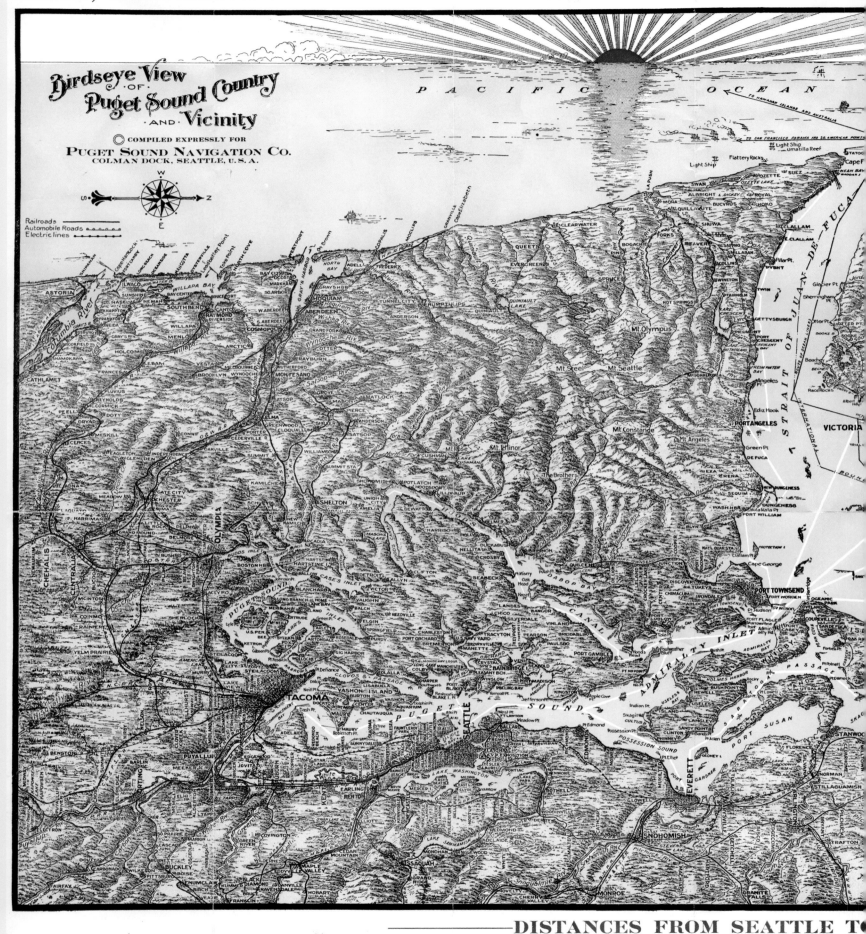

Birdseye View
OF
Puget Sound Country
AND Vicinity

◎ COMPILED EXPRESSLY FOR

PUGET SOUND NAVIGATION CO.
COLMAN DOCK, SEATTLE, U.S.A.

Railroads
Automobile Roads
Electric lines

DISTANCES FROM SEATTLE TO

	Miles		Miles		Miles		Miles		Miles	
Anacortes, via outside	74	Blaine, via inside	112	Deer Harbor	117	Fort Flagler	38	Langley	38	Pleasant Harbor
Anacortes, via inside	80	Bremerton	15	Diamond Point	52	Fort Worden	42	Lilliwaup	71	Point Roberts, via out
Argyle	84	Brinnon	54	Dungeness	64	Friday Harbor	94	Lopez	89	Point Roberts, via ins
Beach	109	Brown's Point	35	East Clallam	128	Gettysburg	105	Neah Bay	147	Port Angeles
Bellingham, via outside passage	92	Charleston	17	East Sound	143	Hoodsport	74	Oak Harbor	54	Port Crescent
Bellingham, via inside passage	99	Camano	42	Edmonds	18	Holly	62	Olga	154	Port Gamble
Bellingham, via San Juan Island	176	Clinton	32	Everett	33	Kingston	18	Orcas	126	Port Ludlow
Blaine, via outside	105	Coupeville	54	Fotr Casey	39	La Conner	64	Pleasant Beach	11	

PUGET SOUND NAVIGATION COMPANY, COLMAN DOCK, SEATT.

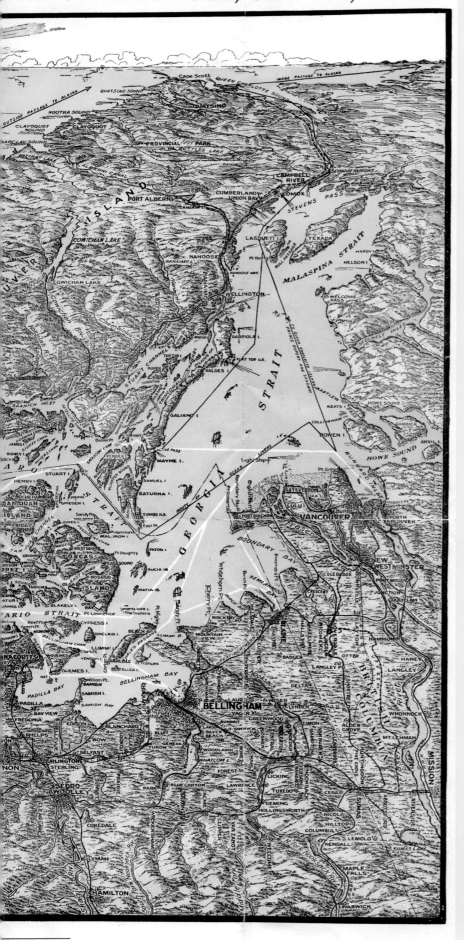

NAVIGATING PUGET SOUND

In a land with no roads but naturally endowed with a large and complex body of water, it was probably inevitable that the Puget Sound country would quickly become dependent on shipping. A considerable number of competing freighters and ferries seemed to buzz around the sound, earning the moniker "the Mosquito Fleet."

In 1898 Charles Peabody, whose family had begun the first scheduled sailings from New York to Liverpool flying a black ball flag in 1818, saw an opportunity: he founded Puget Sound Navigation Company and began building a fleet of boats that became known as the Black Ball Line.

In 1921 Peabody created the first auto ferry, and freight-hauling coastal steamers gave way to ferries linked to highways. After 1937, Peabody's son Alexander purchased fourteen ferries idled by the completion of the Bay Bridge in San Francisco. Peabody had bought the Kitsap County Transportation Company and its ferries in 1935, but a series of strikes and increasing public backlash against Black Ball's then-monopoly position led successive governors to investigate a publicly run system.

The urgent need for ferries during World War II—six were needed to service the Bremerton Navy Yard—delayed any action, but in 1951, after much maneuvering on both sides, including a complete ferry shutdown for nine days, Governor Arthur Langlie engineered a state takeover of the ferry system, and Washington State Ferries was born.

Intended only as a stopgap measure until cross-sound bridges could be built (see page 212), the ferries continued after the legislature later rejected plans for bridges, and today they form one of the largest ferry fleets in the world.

MAP 392 (*left*).
This magnificent bird's-eye map of Puget Sound was published by the Puget Sound Navigation Company in 1910 to illustrate its *short, inexpensive steamer trips*. Here Puget Sound is viewed from an unfamiliar angle—from the east.

MAP 393 (*below*).
Colman Dock, the Seattle base of the Puget Sound Navigation Company, is shown on this 1912 fire insurance plan.

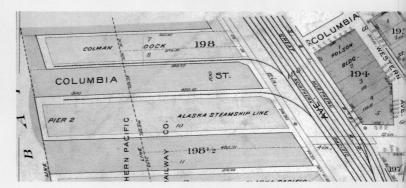

	Miles		Miles		Miles
Port Madison	13	Roche Harbor	105	Twin	111
Port Townsend	42	Rosario	148	Union City	78
Port Williams	58	San de Fuca	54	Utsalady	59
Potlatch	77	Seabeck	53	Vancouver, B. C.	149
Pysht	118	Sidney	16	Victoria, B. C. (Inside Harbor)	84
Quilcene	61	Smtih's Island	53	West Clallam	129
Richardson	72	Tacoma	30	West Sound	123

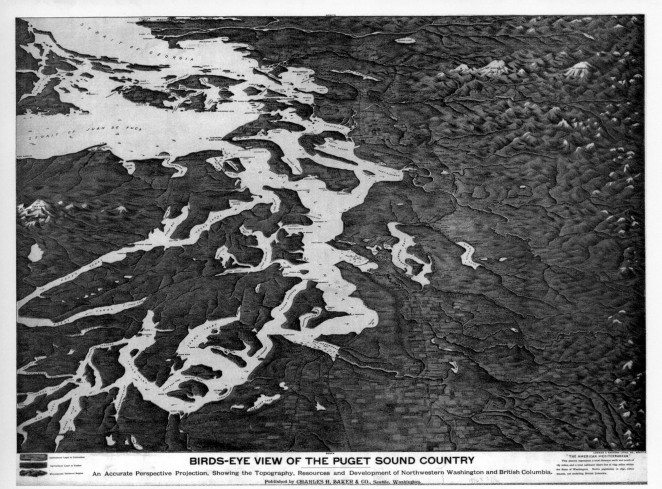

BIRDS-EYE VIEW OF THE PUGET SOUND COUNTRY

An Accurate Perspective Projection, Showing the Topography, Resources and Development of Northwestern Washington and British Columbia.

Published by CHARLES H. BAKER & CO., Seattle, Washington.

MAP 394 (left).
This bird's-eye map of Puget Sound, published in 1891, uses a high perspective that makes it look like a satellite photo of today. Covering the region from Olympia in the south to Vancouver, B.C., in the north, this map would have appeared very unusual in its day. At bottom right reference is made to Puget Sound as "The American Mediterranean."

MAP 396 (below).
A strange subject, perhaps, for a postcard, but a postcard it was, with this map showing the shipping routes of the world—here more pompously *the World's Highway of Commerce*—feeding into *Puget Sound Ports* (shown as a red circle), many through the new Panama Canal. The postcard is undated but seems to have been published about 1923.

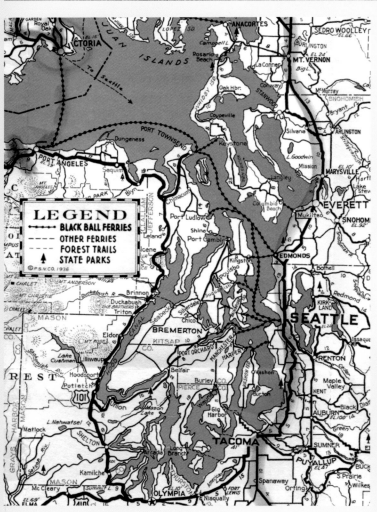

LEGEND
— BLACK BALL FERRIES
···· OTHER FERRIES
--- FOREST TRAILS
▲ STATE PARKS
© P.S.N.CO. 1936

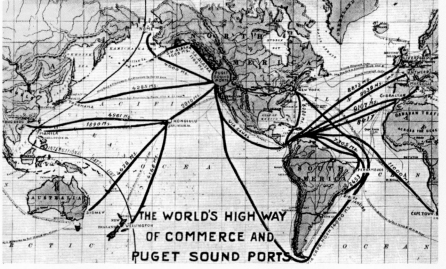

THE WORLD'S HIGHWAY OF COMMERCE AND PUGET SOUND PORTS

MAP 397 (below).
Another postcard promoting the port of Seattle, probably from the 1920s. A bird's-eye view of the city is complemented by a map of the world with shipping routes to Seattle.

MAP 395 (above).
Puget Sound Navigation's Black Ball ferry routes in 1936 stretched to *Port Angeles* and *Victoria*, and to the San Juans, as well as the heavily used cross-sound routes.

162

A Pacific Harbor

The paucity of harbors on the Pacific Coast north of San Francisco was one of the factors that led the United States to press for an international boundary that would include Puget Sound in American territory—the one deepwater port in all the Northwest. Nonetheless, the relatively shallow harbors available elsewhere often had men dreaming of vast new ports—if only they could be dredged. Willapa Bay seemed the best bet, and with the imminent opening of the Panama Canal in 1913, railroads were building to Willapa Bay and businesspeople and investors anticipated great things. In 1912 the Commercial Club in South Bend, the community that stood to gain the most if Willapa Bay was developed, published a newsletter that included MAP 399, a fine bird's-eye view of the harbors of southwest Washington that pointed out the natural entrance depths of Willapa Bay. The newsletter was unashamedly addressed to "persons seeking a good investment on the Pacific Coast prior to the opening of the Panama Canal." As it happened, the canal's opening had much less impact than people had anticipated, as it was followed a year later by war in Europe, which dramatically reduced shipping volumes.

Two decades later the governor of Washington struck a special committee to examine the possibility of a canal connecting Puget Sound with Grays Harbor, Willapa Bay, and the mouth of the Columbia (MAP 398, *right*). But it was not a good time to be examining such things; while the canal was found to be feasible, the Depression was at its height, and money was needed more urgently for many other things.

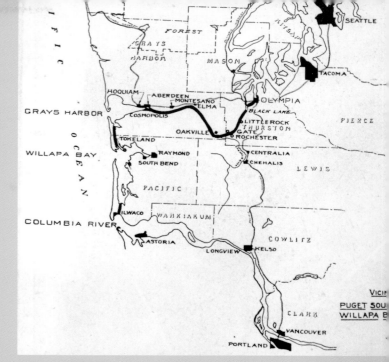

MAP 398 (*above*).
The 1933 plan to connect the southern end of Puget Sound with the Columbia River via Grays Harbor and Willapa Bay.

MAP 399 (*below*).
Published by the South Bend Commercial Club in anticipation of the opening of the Panama Canal the following year, this fine bird's-eye map of Willapa Bay, with Grays Harbor on one side and the Columbia on the other, was to advertise the superiority of Willapa Bay over its competitor havens.

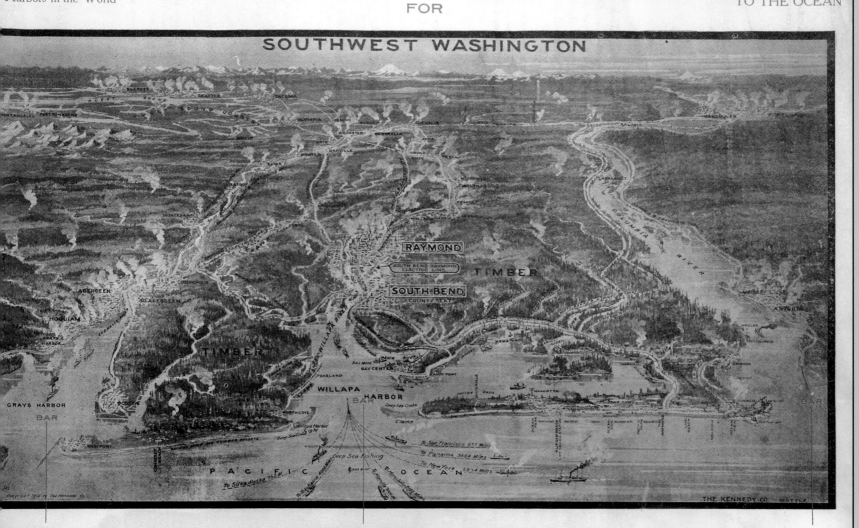

WILLAPA Harbor is One of the Best Natural Harbors in the World

WHAT NATURE HAS DONE
FOR

WATER GRADE FROM SPOKANE TO THE OCEAN

SOUTHWEST WASHINGTON

$1,500,000 Expended
Depth 19 Feet

Not One Penny Expended
Depth **26** Feet

$20,000,000 Expended
Depth **21** Feet

OPENING THE OREGON COAST

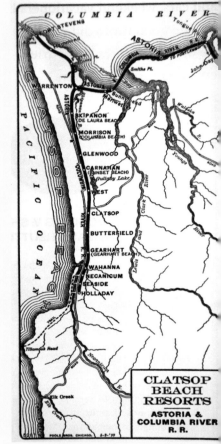

Parts of the scenic Oregon Coast were made accessible by road after a defense road, the Roosevelt Coast Military Highway, was begun in 1914. The highway was improved in the 1930s, and a series of bridges was built to prevent a constant reliance on ferries. But not everywhere was accessible by road.

Herbert Logan built a toll road to Elk Creek (which became Ecola in 1910 and Cannon Beach in 1922) to bring tourists to his sixteen-room hotel. By 1912 this was a public road, but it was

Above. Cannon Beach's cannon, from which the city gets its name, was found on the beach in 1898. It came from the *Shark*, a ship wrecked on the Columbia Bar in 1846.

MAP 400 (*right*).
The only way to get to the northern Oregon beaches in 1910 was by railroad from Astoria.

MAP 401 (*left*).
The *Wolf Creek State Hwy* is shown as (*under construction*) crossing the Nehalem River on this 1934 blueprint drawn on a 1930 base map.

MAP 402 (*left, center*) and
MAP 403 (*left, bottom*).
These 1932 and 1940 highway maps show the Wolf Creek Highway (Route 2) as unimproved in 1932 and nearly complete in 1940. The Wilson River Highway (later Route 6) is shown incomplete on MAP 403 but both were opened in July 1939. Wolf Creek Highway involved a tunnel (photo, *right*) and a bridge over the Nehalem River. The drawing of this bridge, *below*, was done in 1937 by State Highway Commission artist Frank Hutchinson (see page 113). The road created a direct link between Portland and the Oregon Coast. The two highways might have succumbed to the Depression, given their timing, but instead became projects of the Works Progress Administration

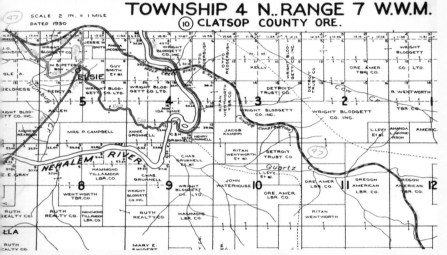

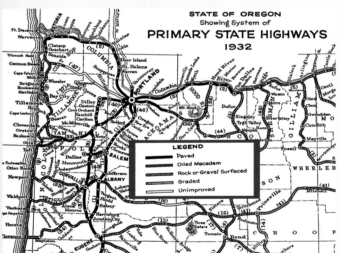

(WPA), a New Deal agency, and employed some 1,500 workers for its construction. Nevertheless, work was slow, beginning in 1932 and lasting until 1939. The Wolf Creek Highway was renamed the Sunset Highway in 1946.

·WOLF·CREEK·HIGHWAY·NEHALEM·RIVER·BRIDGE·

MAP 404 (*above*), with cover (*left, top*). Bus service to the Oregon Coast in 1939 was able to take advantage of the new coast road (Highway 101), which is shown complete except for a few details.

Left. The cover of an Oregon Coast Highway map from the late 1930s shows the quality of the road at that time.

Right. The Siuslaw River Bridge, at Florence, opened in March 1936. The drawing is by Frank Hutchinson.

necessary to drive on the sand at many points, such as Arch Cape, south of Cannon Beach, where a tunnel was constructed in 1937 to avoid this inconvenience. By the 1920s there was a core of perhaps two hundred year-round residents at Cannon Beach, and "Summer Housing By the Sea" was offered for sale, with lots at $100 each. This was the beginning of the tourist mecca we see today, and access to the coast was improved in the 1930s with the building of the Wolf Creek and Wilson River highways (MAP 403, *far left*).

Considerable engineering work was necessary to complete a coastal road, notably so at points where rock jutted out into the Pacific, such as at Neahkahnie Mountain (MAPS 405 and 406, *below*); at 1,661 feet, it is one of the highest points on the coast. In 1927 the state took over the private ferries that had operated across the rivers up to that time, but a longer-term solution required bridges. These were built during the 1930s with the assistance of loans from the Public Works Administration (PWA), a federal New Deal agency. Five bridges, which had the added bonus of being beautifully designed art-deco affairs, were completed in 1936, making the Oregon Coast Highway continuous—other than a few minor diversions—for the length of the state. Tourism immediately increased—by 72 percent the following year.

MAP 405 (*above, left*) and MAP 406 (*above*).
One of the most difficult sections of the road along the Oregon Coast was the one around Neahkahnie Mountain, just north of Manzanita. Here, at two scales, are the 1939 engineering drawings for the road, together with a photograph of the section as it looks today (*left*). The photo was taken from the parking area at right on MAP 406; the *Rock Pinnacle (to be retained)* can be seen behind the car.

RAILROAD EXPANSION AND DECLINE

The first two decades of the twentieth century saw a last great surge of railroad construction in the Northwest, including the building of one of the last American transcontinentals, the Chicago, Milwaukee & St. Paul, or Milwaukee Road, which extended its line to Seattle in 1909. Although this latecomer built what was arguably the best grade, it was the first to fold in the postwar-era competition from automobiles, trucks, and planes. Bad business decisions killed the line, once the longest railroad of any in the United States, but the bottom line was that there were simply too many railroads at a time when fewer were becoming supportable.

The Milwaukee was trying to ensure its continued competitiveness against the Northern Pacific and the Great Northern when it decided to extend its line west from the Missouri. Its Puget Sound Extension was expensive, for the railroad had no land grant and had to buy all its land. The railroad chose Snoqualmie Pass for its crossing of the Cascades, a pass previously rejected by the Northern Pacific in favor of Stampede Pass a little to the south.

Built from both west and east, the two sections met near Garrison, Montana, where a last spike was driven on 14 May 1909. Through passenger service to the Pacific began on 10 July. A final link, the 2¼-mile-long Snoqualmie Tunnel, was not completed until 1915.

Soon after its completion, the railroad decided to electrify two mountains sections: one crossing the Continental Divide in Montana, and the other crossing the Cascades, a 216-mile section from Othello, Idaho, into Tacoma, where electrified service began in November 1919; the ten miles from Black River junction into Seattle was electrified in 1926 (Map 409, *right*). It

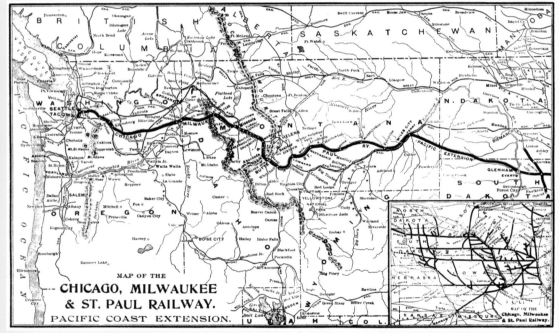

MAP 407 (*left*).
The *Pacific Coast Extension* (Puget Sound Extension) of the Chicago, Milwaukee & St. Paul, published in 1909, the year the line was completed.

MAP 408 (*below*).
The crack passenger train from Chicago to Seattle and Tacoma, the *Olympian*, in its original steam locomotive–powered form, together with a map of its route—shown somewhat straighter than reality. Also shown, marked *T.E.*, are the lines of the Tacoma Eastern Railroad, from 1909 an operating subsidiary of the Milwaukee and wholly owned after 1919.

Above, left.
The *Olympian*, now shown pulled by a powerful bipolar electric locomotive in this advertising poster from the Milwaukee about 1930. Electrification was intended to control operating costs on the mountain sections.

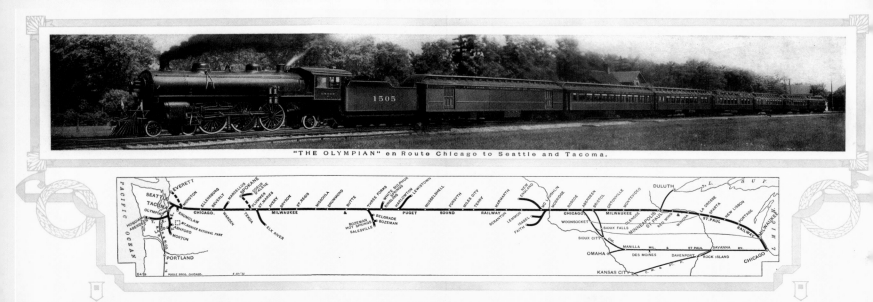

"THE OLYMPIAN" en Route Chicago to Seattle and Tacoma.

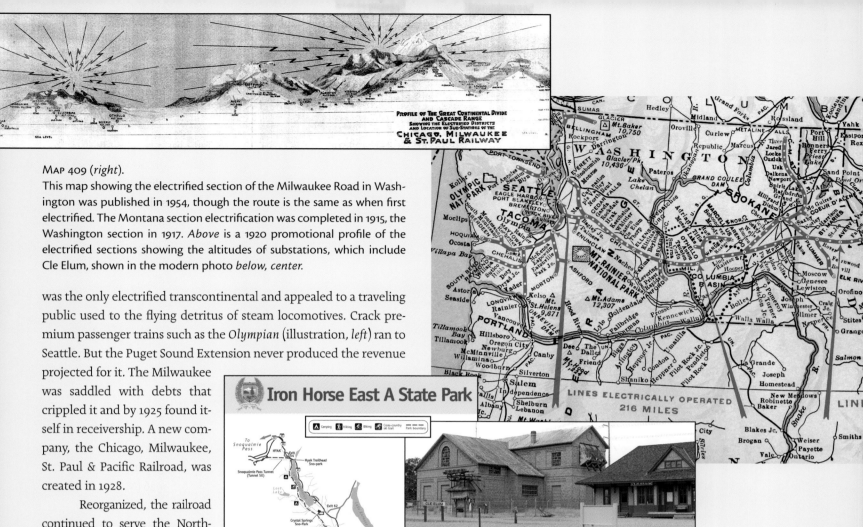

PROFILE OF THE GREAT CONTINENTAL DIVIDE
AND CASCADE RANGE
SHOWING THE ELECTRIFIED DISTRICTS
AND LOCATION OF SUB-STATIONS OF THE
CHICAGO, MILWAUKEE
& ST. PAUL RAILWAY

Map 409 (*right*).
This map showing the electrified section of the Milwaukee Road in Washington was published in 1954, though the route is the same as when first electrified. The Montana section electrification was completed in 1915, the Washington section in 1917. *Above* is a 1920 promotional profile of the electrified sections showing the altitudes of substations, which include Cle Elum, shown in the modern photo *below, center*.

was the only electrified transcontinental and appealed to a traveling public used to the flying detritus of steam locomotives. Crack premium passenger trains such as the *Olympian* (illustration, *left*) ran to Seattle. But the Puget Sound Extension never produced the revenue projected for it. The Milwaukee was saddled with debts that crippled it and by 1925 found itself in receivership. A new company, the Chicago, Milwaukee, St. Paul & Pacific Railroad, was created in 1928.

Reorganized, the railroad continued to serve the Northwest, but in 1973 its management made a decision to scrap the electrification in favor of diesels, a decision that, with rising oil prices, did not make economic sense. The railroad spiraled rapidly downhill, and in 1980 it abandoned its line to the Pacific. Parts of the Snoqualmie right-of-way are now a Washington state park (Map 410, *above*).

LINES ELECTRICALLY OPERATED
216 MILES

Iron Horse East A State Park

Map 410 (*left*).
The right-of-way of the Milwaukee Road has been maintained in case it might ever be required in the future. In the meantime, it is a state park and includes the Snoqualmie Tunnel. *Above* is the electrical substation and restored depot at Cle Elum.

Map 411 (*right*).
The Union Pacific system in Washington and Oregon in 1911. Much of the system consists of track built by the Oregon Railway & Navigation Company (the Oregon Railroad & Navigation after 1896) and purchased by Union Pacific in 1898. Union Pacific was bankrupt between 1893 and 1897 and had just reorganized at that time. The railroad became the Oregon–Washington Railroad & Navigation Company in 1910 and was operated as a subsidiary of the Oregon Short Line, itself a Union Pacific subsidiary. All were formally absorbed into Union Pacific in 1936.

The line from *Portland* to Puget Sound was jointly operated with the Hill lines as a result of an agreement reached in 1909 (see page 134). The map also shows the Deschutes Railroad line south from the *Columbia River* to *Bend* and an ex–Columbia Southern line from the river to sheep-raising center and now virtual ghost town *Shaniko*, a line completed in 1900.

THE PACIFIC NORTHWEST
IDAHO,
OREGON, WASHINGTON,
BRITISH COLUMBIA.
SCALE
0 10 20 30 40 50   100 Miles
Union Pacific System
Railroads

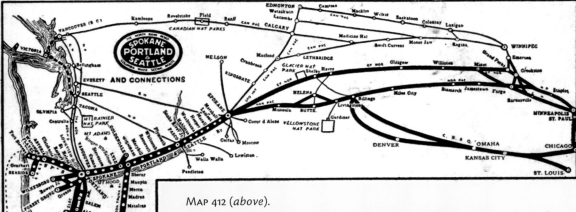

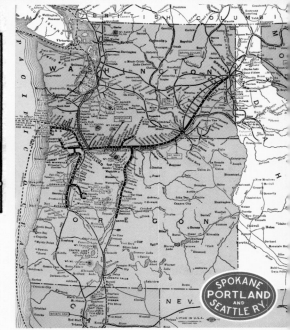

MAP 412 (*above*).
This strange-shaped map, which originally surrounded some text, shows the Spokane, Portland & Seattle Railway with its connections in 1938. The branch to Eugene (the Oregon Electric) and the one to Bend (the Oregon Trunk) are shown. East of Spokane the lines of the railroad's owners, the Northern Pacific and Great Northern, are depicted thicker than those of competitors, yet west of the city both of those companies' lines to Puget Sound are downplayed.

MAP 413 (*above, right*).
A similar map of the Spokane, Portland & Seattle Railway in the 1950s, which also shows many of the other main lines of the Pacific Northwest at that time. Just south of Eugene, the Natron Cutoff, from Natron to Black Butte, was built by Southern Pacific subsidiaries Oregon Eastern and then Central Pacific to provide an alternate line south to California, avoiding the gradients and curves of the Siskiyou Mountains (see MAP 223, page 94). Also shown is a Great Northern extension of the Oregon Trunk to Bieber, California, to connect with the Western Pacific. This route, dubbed the Inside Gateway, was completed in 1931 and allowed these railroads to compete with the Southern Pacific.

MAP 414 (*below*).
A timetable map for the Oregon Electric in 1910. The interurban began service between Portland and Salem in 1907. The Spokane, Portland & Seattle purchased the line in 1910 and extended service to Eugene in 1912, but the intended extension to Roseburg, shown on this map with a dashed line, was never built.

MAP 415 (*right*).
The Southern Pacific system in Oregon in 1927. In addition to the lines in the Willamette Valley, which included those of the Oregon Electric, a second main line, the Natron Cutoff, leaves the first main line near Eugene. Three branches now reach the coast: to Tillamook Bay; to Yaquina; and to Reedsport, Marshfield (now Coos Bay), and south to Powers. These lines were principally intended to serve the forest industry of the coastal region.

For a while, two railroad titans battled each other for supremacy in the Northwest. James J. Hill, who now effectively controlled the Northern Pacific as well as his Great Northern, wanted to keep out Edward H. Harriman of the Union Pacific. The latter gained control of the Southern Pacific in 1901, thus obtaining a second entrance into Portland, though Union Pacific was ordered to sell its Southern Pacific interest in 1913 for antitrust violation. In 1898 the Union Pacific had acquired control of the Oregon Railroad & Navigation Company (see page 90; in 1896 the name changed from Oregon *Railway* & Navigation). Harriman's empire thus appeared to threaten that of James Hill, and in 1905 Hill incorporated the Spokane, Portland & Seattle Railway (SP&S) to connect his Northern Pacific and Great Northern lines independently to Portland via the north bank of the Columbia. He also wished to tap the lumber industry of Oregon, at that time dominated by Union Pacific.

There followed what has been termed a railroad war, when both sides raced each other down the steep valley of the Deschutes River toward Bend, in central Oregon. The Oregon

Trunk Railroad, which branched from the sp&s via a bridge across the Columbia, fought the Union Pacific–controlled Deschutes Railroad. When it became obvious neither could gain the upper hand, a track-sharing agreement was worked out for part of the route. The first train pulled into Bend in November 1911.

Railroad competition became greater than what could be supported by the available traffic. The onset of the Depression in the 1930s, and the long-term rise of the automobile, the truck, and later the plane, led to the closing of many lines, leaving the two principal types we have today: the passenger commuter interurban and the heavy freight railroad, still the most economically viable method of transporting bulk items like coal.

MAP 416 (*below*).
The central part of a 1928 official railroad map of Washington. The trans-Cascades lines of the Great Northern, the Northern Pacific, and the Milwaukee are shown, the latter diverging from the Yakima Valley to take Snoqualmie Pass while the Northern Pacific takes Stampede Pass. Note the Northern Pacific coal town of Roslyn a little farther down the valley. Farther south is another intended trans-Cascade route—that of the Cowlitz, Chehalis & Cascade Railway (dashed blue line). This little line, originally thirty-two miles in length, was a joint venture between the Northern Pacific, the Great Northern, the Milwaukee, and the Union Pacific! Intended to haul logs to the other railroads' main lines, it is nonetheless shown as a proposed trans-Cascade line on this 1928 map. The line would have continued up the Cowlitz Valley, crossed the Cascade summit at White Pass, continued down into the Tieton River Valley, and connected with the Northern Pacific at Naches, a few miles west of Yakima. No more seems to have been heard of this interesting proposal, however. Along the Columbia east of Portland the Spokane, Portland & Seattle is on the north bank and the Oregon–Washington Railroad & Navigation (owrn) line runs along the south bank. At Wishram, the sp&s crosses the river to become the Oregon Trunk on the west bank of the Deschutes River; the owrn line branches south at Sherman for its east bank. At Portland the sp&s crosses the Columbia on a bridge built in 1908 (see MAP 260, page 108).

MAP 417 (*right*). An innovative bird's-eye map, with historical annotations of the Lower Columbia, that the sp&s gave to its passengers. The *Oregon Trunk Ry.* is at top. The map was originally produced in 1916 but updated from time to time; this is the 1938 edition.

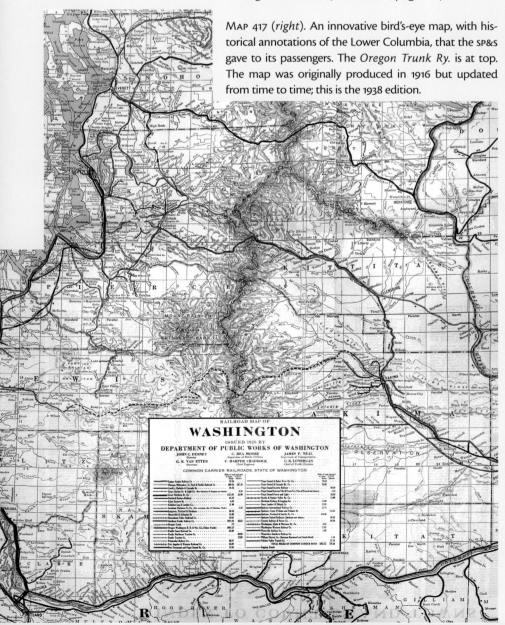

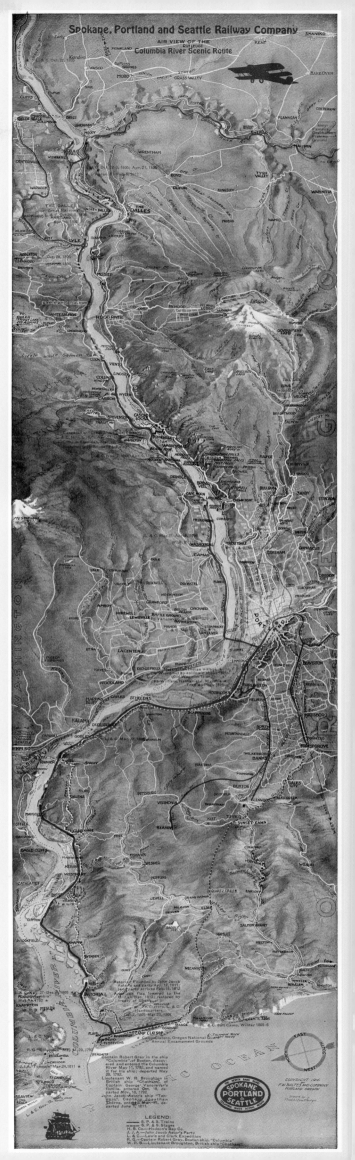

The Longest Tunnel

The Great Northern's first crossing of the Cascades had been by switchback; the second by a 2.63-mile-long tunnel (see page 100). Smoke in such a long tunnel became a serious issue. This was solved by electrification—for exactly three miles. Exchanging locomotives delayed trains. In addition, the slopes and curvatures of the mountain route were excessive. In summer there was a problem with fires caused by steam locomotives, and in winter there was a problem with snow—which the railroad built almost twelve miles of snowsheds to combat—which again created a smoke problem. The old route also required two short (and non-electrified) tunnels, making the total distance run in tunnels 3.66 miles.

The solution seemed to lie in a two-fold approach: a longer main tunnel and a longer stretch of electrified track. The new tunnel was 7.79 miles long, bored between Berne and Scenic. The tunnel was five hundred feet lower than the original one and thus eliminated much of the need for snowsheds on either side. Finally, the electrified section was extended to run from Skykomish through to Wenatchee, a distance of seventy-two miles. In addition the electrical system and the locomotives were upgraded to a more robust system that would allow heavier trains.

Drilling the tunnel was a major engineering feat, and 923,000 cubic yards of rock were removed in slightly less than three years. But it was expensive: the whole project cost $25.6 million, and to justify the expense the railroad needed increased traffic. Unfortunately for the company, when the tunnel was opened by President-elect Herbert Hoover on 12 January 1929, the Depression was only months away.

Above.
The new Cascade Tunnel was considered such a feat of engineering—it remains the longest railroad tunnel in the United States—that a special issue of *Railway and Marine News* featured it on its cover with this superb graphic showing a new electric locomotive posed by the tunnel. The use of modern electro-diesels instead of steam locomotives led to the removal of the electrification in 1956 in favor of a system of doors that open and close to allow trains to pass and huge fans to purge the air within the tunnel. This has now proved to be a problem for BNSF, Great Northern's successor, because of the limits it places on traffic volumes through the single-track tunnel.

Map 418 (*right*).
This late-1920s map from the Great Northern Railway shows the railroad's system in the Pacific Northwest.

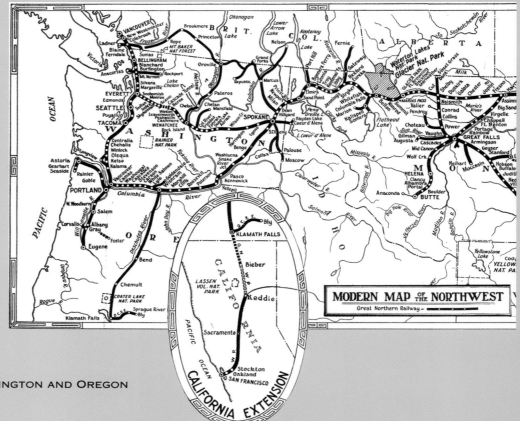

MODERN MAP OF THE NORTHWEST
Great Northern Railway

Map 419 (*right*).

This bird's-eye and sectional view shows the position of the new Cascade Tunnel in relation to the older one, much higher. A conventional map below indicates the positions of the two tunnels. Note the position of the old switchbacks.

Map 420 (*below*).

From a 1929 engineering magazine comes this map and profile of the entire new route up Chumstick Creek instead of the Wenatchee River. The new route is in red. *Inset* is a proposal that shows the snowsheds covering much of the west slope of the line to the old tunnel.

Map 421 (*bottom left*).

This triangulation for the tunnel was an illustration in a 1929 issue of *Railway and Marine News*, shown at left.

Map 422 (*bottom right*).

The Great Northern Railway inaugurated its new premium *Empire Builder* service from Chicago to Seattle following the completion of the new tunnel. This ad appeared in 1930.

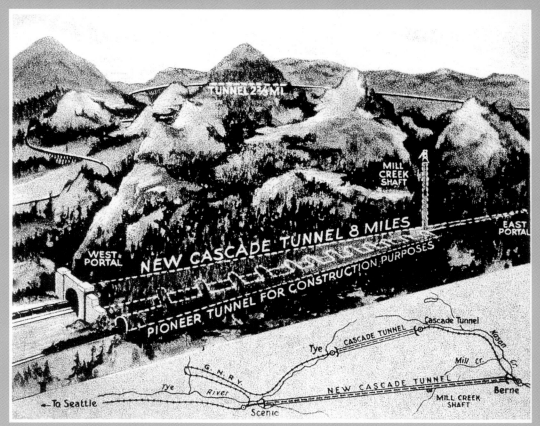

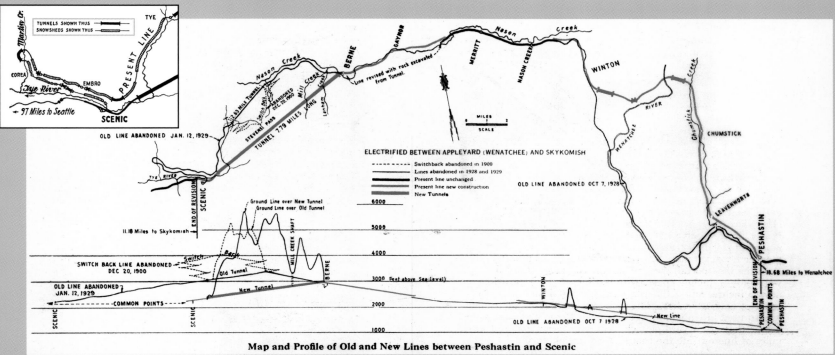

Map and Profile of Old and New Lines between Peshastin and Scenic

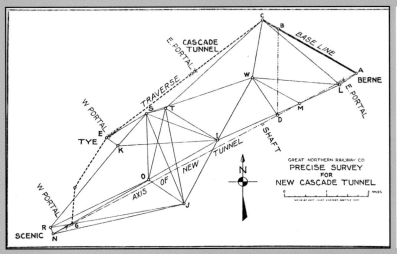

DEVELOPING A ROAD SYSTEM

There were always roads, but except for a few long-distance routes, such as the Oregon Trail, they provided only local access. The idea of having a state-sponsored road network, one that could be used to travel long distances in various directions at will, originated in the Northwest in 1913. That year Washington, which created a state highway department in 1905, enacted a state highway act creating a highway network across the state. Oregon did the same that year, creating its state highway department to "get Oregon out of the mud."

And get it out of the mud it would have to, in many cases, for the roads and trails adopted by the state were often hardly passable in the Northwest's frequent rains. The first asphalt and concrete roads were constructed in 1912. Existing roads were adopted and new routes proposed, as shown in the maps on these pages, and work began on a number of roads—the Oregon Coast Highway, for example, in 1914 (see page 164), and the Columbia Highway, which was completed—unpaved—from Portland to Hood River and to Astoria in 1915.

(see page 164)

The 1921 Federal Highway Act created a national system of highways, and in 1927 a national road-numbering system was created, aiding interstate travelers.

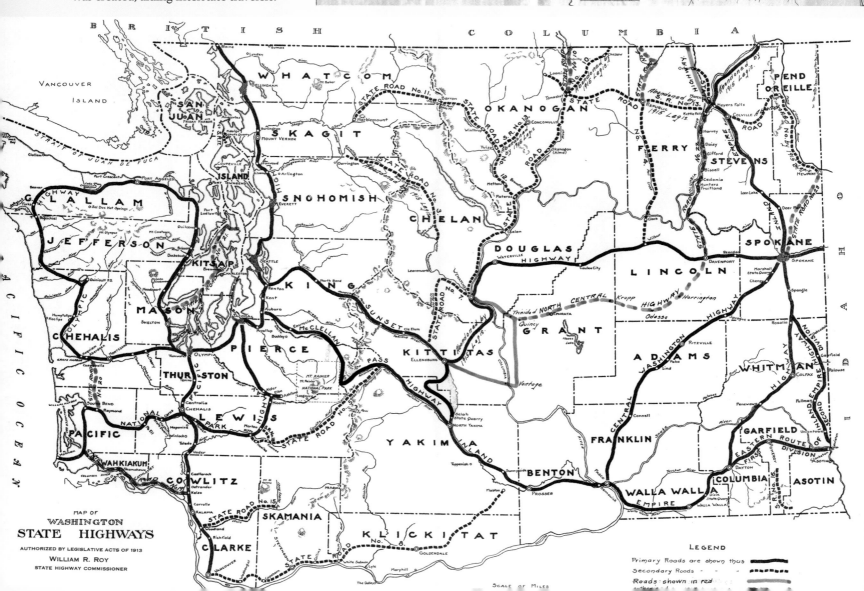

MAP OF
WASHINGTON
STATE HIGHWAYS

AUTHORIZED BY LEGISLATIVE ACTS OF 1913

WILLIAM R. ROY
STATE HIGHWAY COMMISSIONER

LEGEND

Primary Roads are shown thus ▬▬▬
Secondary Roads - - - -
Roads shown in red

SCALE OF MILES

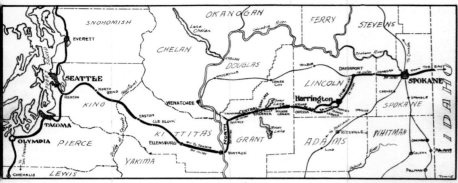

NORTH CENTRAL HIGHWAY
SHORTEST and BEST ROUTE BETWEEN SPOKANE and SEATTLE

Distance between Spokane and Seattle over North Central Highway, via Harrington, Soap Lake and Ellensburg, 347 Miles

Southern Route, via Walla Walla, 465 Miles ————— Sunset Route, via Waterville and Wenatchee, 408 Miles

Extreme Elevation above sea level in North Central Highway 2500 feet, and the only route north of North Central Highway 4700 feet.
Ruling Grade on West Side 3%, and East Side 3% for only three-fourths of a mile. Balance of way any car can make it on high.

A GOOD POWER FERRY AT THE COLUMBIA RIVER   NO SAND OR STONE ON NORTH CENTRAL HIGHWAY

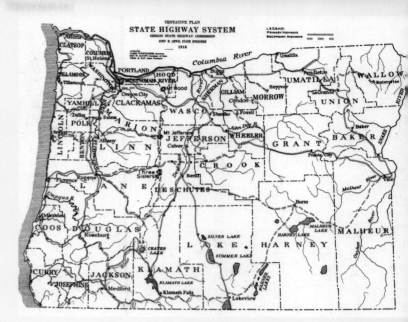

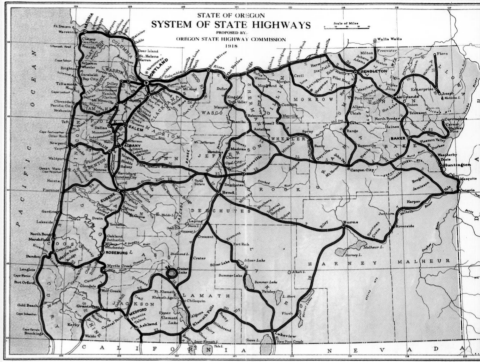

Left, top.
Somewhere in deepest Oregon, about 1910, an automobilist is in trouble, having slid off a narrow dirt road, a road seemingly in good condition for the period.

Map 423 (left).
Ironically, perhaps, the first Washington state road was designated where one of the last modern roads would be built. The approximate route of the North Cascades Highway (Highway 20) was adopted as a state highway in 1893 following this survey from *Marblemount* east across the North Cascades. A road, really more a trail, was constructed three years later.

Map 424 (left, bottom).
A state highway system was authorized by the Washington legislature in 1913. This map shows that system together with amendments passed in 1915, indicated in red.

Map 425 (above).
One of the state roads authorized by the 1915 amendments was the North Central Highway (now Highway 28), south from *Davenport* to reconnect with the Sunset Highway (now Highway 2) beyond *Ephrata*. This map was likely produced by businesspeople from communities along the new route, which they hoped would bring them customers. The text notes the distance drivable "on high"—that is, in top gear.

Map 426 (right, top), **Map 427** (right, center), and **Map 428** (right, bottom).
Proposals for a state highway system in Oregon in 1916 and 1918, together with an actual system in 1922. Careful reference should be made to the key, as only the red lines denote paved roads.

Below. This road in central Oregon in 1927 would likely have been classified as "graded"; the danger of getting stuck is clearly very real.

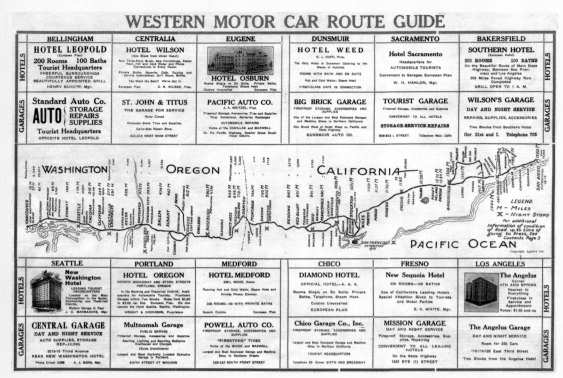

Map 429 (*above*).
An interstate road map from before the numbered highway: this was produced for the Pacific Highway, from Vancouver, B.C., to San Diego. As can be deduced from the ads, the trip took many days.

Map 430 (*left*).
Oregon developed or improved a system of secondary roads known as market roads. This 1925 map shows the proposed market roads of Marion County (in orange), showing how they connect with roads of adjacent counties (in yellow). State highways are labeled in black.

Map 431 (*below, left*).
Oregon's paved highways are still limited to the Willamette and Columbia valleys on this 1926 highway map. Note the addition of *National Highways Route Numbers*.

Map 432 (*below, center*).
Finding one's way through cities could be a problem for motorists. This map of Portland was contained in a 1926 Oregon State Highway Department map. Note the *Auto Camp Grounds*.

Map 433 (*right*).
Auto clubs and others issued little maps on cards that could be easily referred to by motorists on the move. This one showing the route from *Chehalis* to *Tacoma* via the Pacific Highway was published in 1919.

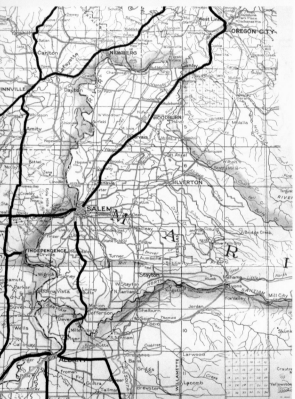

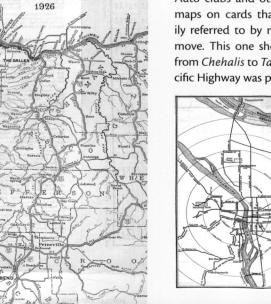

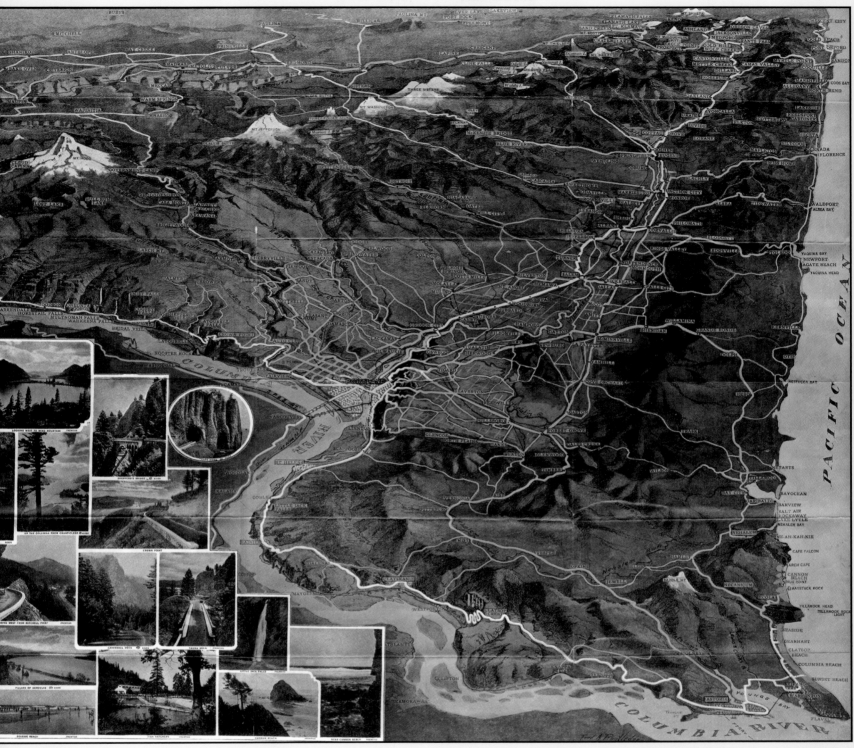

Map 434 (*above*).

This stunning bird's-eye-type map was published by the Portland Chamber of Commerce in 1923 and was directed at tourists in automobiles. Multiple views are carefully positioned to obscure the area of Washington, across the river, that the Portland businesses did not care to promote. The result is a magnificent and unique view of northwest Oregon and its roads. The illustration, *below*, comes from the cover of the map.

Right.
The cover of the 1936 Oregon State Highway Commission map shows one of the smaller bridges on the newly completed Oregon Coast Highway.

OREGON
1936 ROAD MAP

OREGON
1936 ROAD MAP

Oregon State Highway Commission
HENRY F. CABELL, Chairman
E. B. ALDRICH
F. L. TOU VELLE
Commissioners

R. H. BALDOCK, State Highway Engineer
H. B. GLAISTER, Secretary

Drive
OREGON
Highways

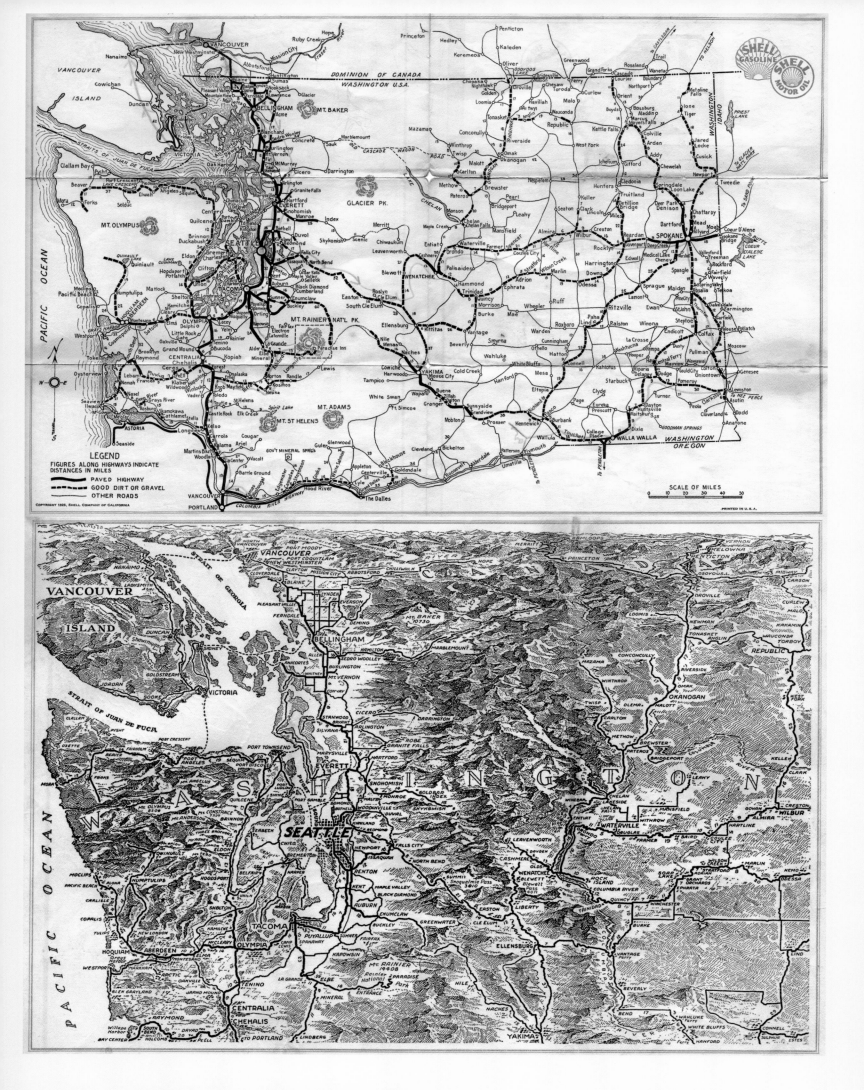

SEATTLE AND VICINITY

Map 435 (*left, top*) and Map 436 (*left, bottom*). Oil companies began supplying free maps to motorists as an inexpensive way to promote their wares as early as 1905. These two 1926 road maps of Washington were from Shell Oil; both were part of the same fold-out but show two completely different takes. Map 437 turns the road map into a bird's-eye map, an interesting perspective but one that is perhaps harder to read for practical wayfinding. These maps have a parallel in today's GPS-based map screens, many of which also have a perspective option.

SPOKANE AND VICINITY

Map 437 (*above*) and Map 438 (*left*).

Since road networks became denser in and around urban areas, mapmakers soon began printing more detailed maps for the vicinities of cities. These are 1927 maps for southern Puget Sound and for Spokane. Note the still-limited mileage of paved trunk roads evident here.

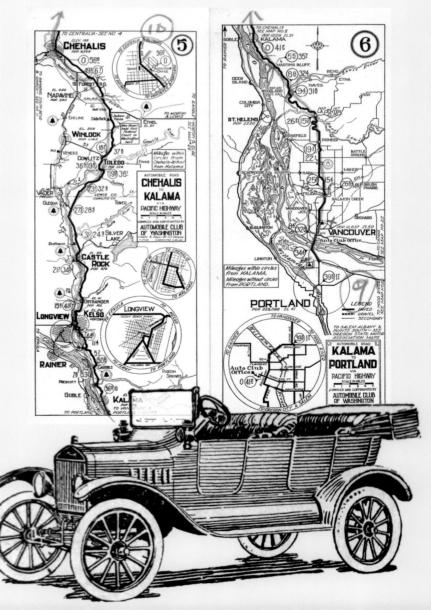

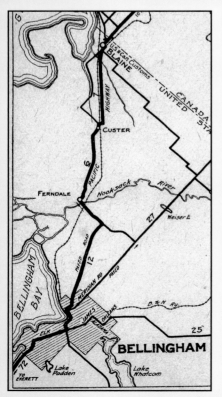

Map 439 (*left*) and Map 440 (*right*). The Automobile Club of Seattle had been founded by forty-six automobilists in 1904, and others followed. In 1917 all the clubs west of the Cascades came together as the Automobile Club of Western Washington, which published this 1917 Bellingham-to-Blaine card. In 1923 the club was renamed the Automobile Club of Washington and issued this Portland-to-Chehalis map pair (here obviously used) in 1927. The car is an illustration from the 1917 Bellingham city directory.

A Floating Bridge

The Washington Toll Bridge Authority was created in 1938 and immediately set about planning two new bridges— a floating one across Lake Washington and a suspension bridge across the Tacoma Narrows. The first was successful despite a pioneering design and remains in service to this day; the second was dramatically unsuccessful, a victim of an unforeseen design error.

The Lake Washington Floating Bridge was designed by engineer Homer Hadley (after whom the duplicate span, completed in 1989, is named) and utilized innovative floating concrete pontoons. It was the largest floating concrete structure in the world. Construction began in December 1938, and the bridge was officially opened on 2 July 1940. Among the toll charges was one of 35 cents for "wagons drawn by one or two horses."

The bridge was renamed the Lacey V. Murrow Memorial Bridge in 1967, honoring the director of the State Highway Department and the Toll Bridge Authority instrumental in its creation.

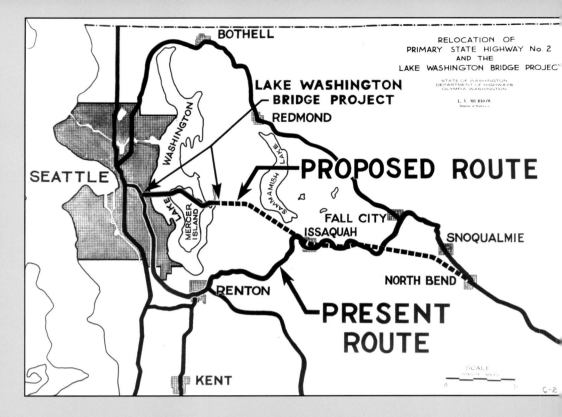

During repair and expansion work in 1990, the bridge sank. Engineers had diverted polluted water into one of the pontoons for temporary storage, and it sank in a storm, dragging the rest down with it. A new bridge was completed three years later.

Washington was a world leader in floating bridge technology, now with three such bridges over Lake Washington and another (which also sank) over Hood Canal (see page 212).

MAP 441 (*above*).
A planning report for the new bridge produced about 1938 compared the existing route to Seattle from the east with the new route. The bridge, initiated during the latter part of the Depression years, seems to have garnered widespread support. At the same time an organization called the Lake Washington Bridge and Highway Association published supporting ads promoting the route. This organization wanted direct links from the agricultural areas of Eastern Washington to the ports of Puget Sound.

MAP 442 (*left*).
An alternate map from the same report shows the new bridge and highway in red on a topographic base.

MAP 443 (*below*).
The plan for the elegant East Tunnel entrance at 35th Avenue, on the west side of the lake. Also shown is an aerial photo of the bridge under construction in early 1940.

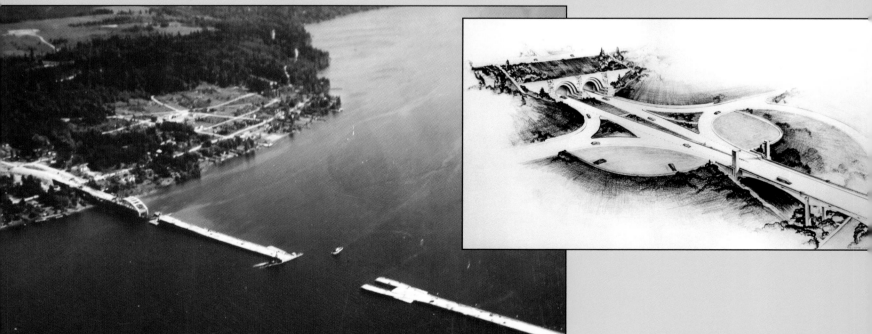

CASCADE
MOUNTAINS

R.M.GILMORE

C-16

Map 444 (*above*).
This rather unusual
bird's-eye map was
included in the final
engineer's report on
the construction of
the Lake Washington
Floating Bridge, issued
in 1940. It does give
a good idea of the
savings in time and
distance that the new
bridge would afford
compared with the
Present Route north
or south of the lake.
Note the *Direction of
Strongest Winds*.

Map 445 (*right*).
An aerial photo made
into a map by the
addition of the new
route across Lake
Washington. Note the
location of the *Toll
Plaza.*

C-4   AERIAL VIEW
           LAKE WASHINGTON BRIDGE PROJECT
     AND RELOCATION OF PRIMARY STATE HIGHWAY No.2.

Galloping Gertie

The second project of the newly created Washington Toll Bridge Authority was a bridge across the Tacoma Narrows. The 2,800-foot-long suspension bridge connected the Kitsap Peninsula to Tacoma and opened to great fanfare on 1 July 1940.

The bridge's tendency to undulate in the frequent strong winds of the Narrows had been noticed even as it was being built, and it quickly acquired the nickname "Galloping Gertie." This worrisome feature was due to a lack of roadbed trusses and the slender proportion of the support towers, narrower than anything designed thus far and narrower than the bridge's contemporary West Coast counterparts in San Francisco and Vancouver, B.C.

In a gale on 7 November 1940, the bridge began its typical oscillation. At about 11 AM, the movement became too much and tore the road deck apart. Several people narrowly escaped; the only fatality was a dog that bit its would-be rescuer, an engineering professor who was trying to take measurements of the bridge's movement.

A new bridge was completed ten years later—one that incorporated the lessons learned from its predecessor's demise. A duplicate span was added in 2007.

MAP 447 (*below*).
The location of the crossing, with new road connections, and the old ferries, are shown on this map of the Tacoma Narrows from a 1938 engineering report.

Above.
The collapsed and torn-apart bridge photographed on 10 November 1940.

Right.
The Tacoma Narrows Bridge under construction toward the end of 1939.

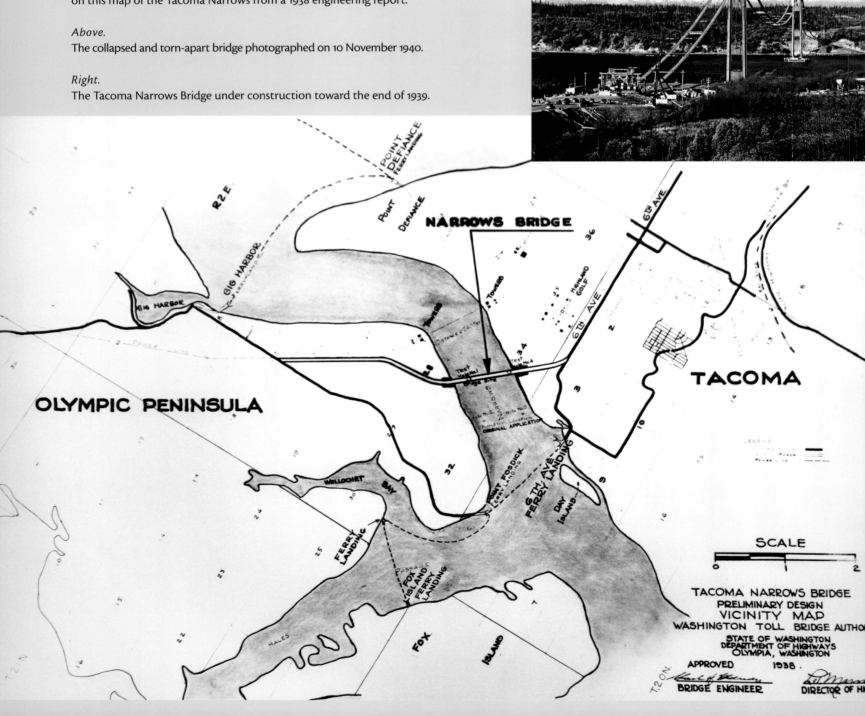

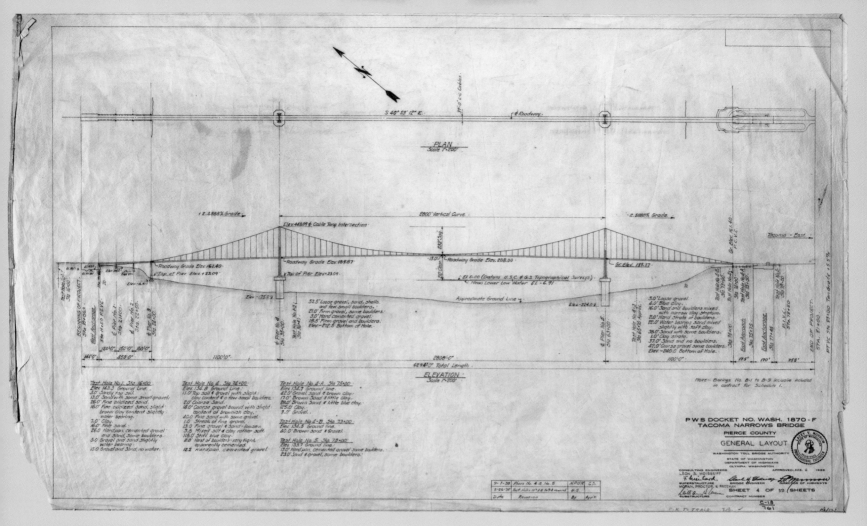

MAP 447 (*above*).
Much studied by bridge engineering students is the set of plans and engineering drawings for the failed bridge. This is the summary sheet, the *General Layout*. At top is the plan and at bottom the side elevation. The drawings have been signed as approved on 6 August 1938 by two consulting engineers, including Leon Salomon Moisseiff; Clark Eldridge, the Department of Highways bridge engineer; and Lacey Murrow, director of the Washington Toll Bridge Authority and the State Department of Highways. One key design fault was that the support towers for the suspension cables were too narrow in relation to their height.

But Eldridge's original design had been further compromised by Moisseiff, a well-known consulting engineer who had been involved in many bridge designs of the 1920s and 1930s and who was a proponent of lighter suspension bridges. He had produced a design that did not include trusses normally required to keep the roadbed rigid. His revised design used less steel and was $4 million cheaper than the original; the federal Public Works Administration, which was loaning the money for the project, had insisted on it. Moisseiff's career was brought to an untimely end by the disaster, and he died only three years later.

Right.
The collapsed bridge is boarded off, and warning signs have been posted—a sad end to an elegant but faulty design. The narrowness of the towers is evident from this view.

A Planned City

Robert A. Long's company, the Long-Bell Lumber Company, was running out of wood. The company had operated for decades in the South, but logs were becoming increasingly hard to find. So in 1918 Long decided to move his operation west, buying up tracts of forest in northern California, southern Oregon, and southwest Washington. He also acquired 70,000 acres of cutting rights from the Weyerhaeuser Timber Company in Lewis and Cowlitz counties in Washington.

By 1921, Long had selected a site where the Cowlitz entered the Columbia for a new lumber-processing plant. Long, being somewhat of a philanthropist, decided to plan and build a model town to house the sizeable workforce his mill would require. The beautiful industrial city he had in mind would, not coincidentally, also allow him to cash in on the expected regional growth his mill would generate.

In 1923 Long applied to the post office for a name, requesting Long View, after a farm he owned in Missouri. The name was rejected because of an existing place with that name—with three families. It was a flag stop on the Spokane, Portland & Seattle Railway. The families agreed to give up the name if Long would build them a covered station to protect mail bags thrown from trains. Longview's name was thus purchased—at a cost of $25.

Long hired a city planner, a real estate developer, and an architect to plan his city and incorporated the Longview Development Company as Long-Bell's real estate subsidiary. Longview was incorporated as a city in 1924. The plan for Longview was distinctly European in style (MAP 449, right), with grand boulevards, esplanades, and a civic center at the city's heart (MAP 448, right, top).

Longview did not achieve the growth Robert Long had hoped for. Planned to be a city of 50,000, by 1930 it had about 10,000 inhabitants. The Long-Bell Lumber Company, badly hit by the Depression, filed for bankruptcy in 1934.

MAP 448 (*above*).

A bird's-eye view of the planned civic center at Longview. Of the buildings shown, only the Hotel Monticello (at center) and the library (at center right) were built, the latter a fine colonial-style building, a gift to the city from Robert Long.

MAP 449 (*below*).

The original plan for Longview in 1923. The grand civic center was to be centrally placed and surrounded by other civic buildings and then housing. A manufacturing area was to be separate, closer to the Columbia River, and Long's lumber mill was on the river itself south of that district. A retail and commercial area was to be located along the Cowlitz riverfront. This plat should be compared with the civic center bird's-eye (MAP 448, *above*) and the advertising bird's-eye sketch opposite (MAP 451, *far right*).

MAP 450 (*above*).
Advertising was critical to Longview's success, and the Longview Company did a lot of it, attracting residents from across the United States. This one was published in 1925. Rail did not meet water and highway until 1928, after Long-Bell built its own railroad—the Longview, Portland & Northern—to service both the mill and the city.

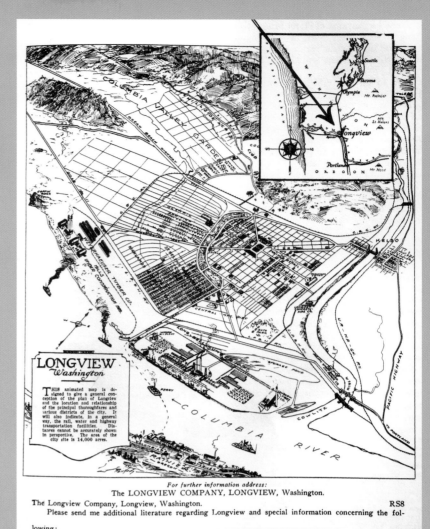

MAP 451 (*above*).
Described as an "animated map" this 1928 advertisement gives a good overall view of the layout and location of Longview. Note the *Weyerhaeuser Timber Co. Mills Under Construction 1928* to the west of the Long-Bell mill. Weyerhaeuser built three sawmills and a shingle mill and in 1933 began production of "Presto Logs," which utilized waste sawdust.

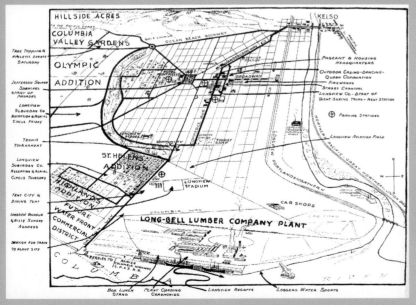

MAP 452 (*above*).
Longview held a four-day Pageant of Progress celebration in July and August 1924 to mark the opening of the *Long-Bell Lumber Company Plant*. This map was in the program. Although some houses had been built by this date, there was still much open space.

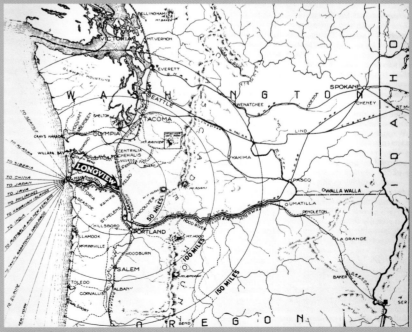

MAP 453 (*above*).
A map from the copious advertising for Longview, this one, from 1924, telling the public just where the city was and promoting its Northwest location.

THE DIRTY THIRTIES

The collapse of the stock market beginning in September 1929 was followed by the Great Depression of the 1930s. By 1932 the stock market had lost 89 percent of its value. As economic activity dropped, workers were laid off in ever-increasing numbers. Some, specially the unskilled, found themselves destitute, and some built shacks in which to live and survive. Many grouped together on waste land. These were the Hoovervilles, disparagingly named after President Herbert Hoover, whom many blamed for the Depression. Seattle had one, which, unusually, was well documented (MAP 454, *below*).

Right.
Seattle's Hooverville, photographed in 1937 from the B.F. Goodrich building shown on MAP 454.

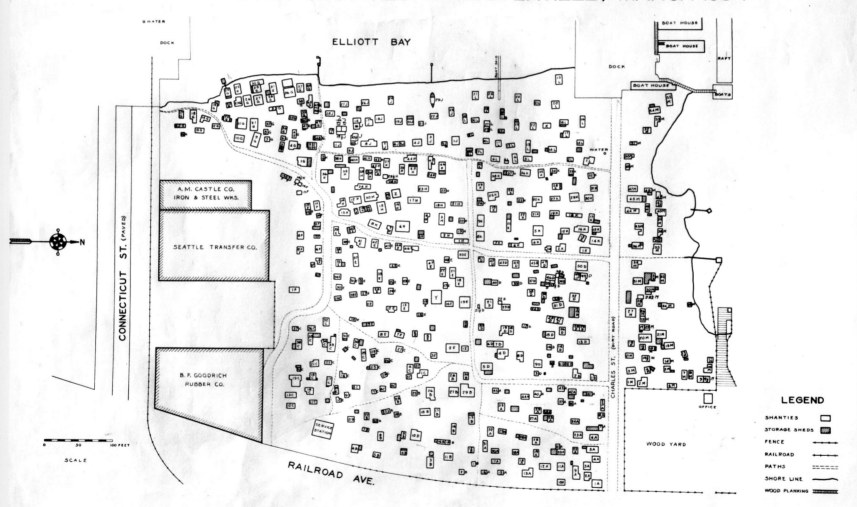

DISTRIBUTION OF SHANTIES IN "HOOVERVILLE", MARCH 1934

MAP 454 (*above*).

Hoovervilles were hardly ever mapped. Not only did they grow up in a haphazard fashion, but they were also expected to be only temporary and were technically illegal in most cases. Seattle's Hooverville was recorded, however, owing to the work of a sociology student, Donald Francis Roy, who not only painstakingly mapped the location of every shack but moved into one as well. Roy wrote that he "wandered for days, pacing off lengths and widths and distances from this and to that and achieved, after a great sacrifice of leather, a fairly accurate map." Roy developed a great rapport with Hooverville's residents and divided the town into twelve lettered parts, identifying each shanty with a whitewashed letter and number, shown on the map. This allowed items like relief payments to be delivered relatively easily. Roy interviewed 650 Hooverville residents and found, not surprisingly, that almost all were unemployed. Most were men; only seven were women. The men were typically single and over forty, and the majority were unskilled laborers. Some 120 were Filipino, 25 were Mexican, and 29 were Afro-American.

Men began building Seattle's Hooverville in October 1931; the shantytown lasted until April 1941, when it burned down. The Seattle Port Commission, on whose land it had been built, quickly ensured that it was not rebuilt.

Map 455 (*right*).
Before the effects of the Depression had come to be sorely felt, this map was published by the Seattle Chamber of Commerce, in 1930, part of a tourist brochure illustrating the attractions of northwest Washington with Seattle at its center. The emphasis is as much on railroads as roads.

Left. With several Seattle landmarks still existing today, this colorful postcard glorifies the city with symbols of progress: the modern train, the automobile, and the plane.

Map 456 (*below*).
This 1937 map of property in Whatcom County, in northwestern Washington, illustrates the effects of the Depression in graphic fashion. The land parcels shown with dots are delinquent with their property taxes. The sea of dots demonstrates the pervasiveness of the inability to pay. For some reason the county tax collector, whom we assume was responsible for this map, has divided delinquent properties into those suitable for agriculture and those not, perhaps with a mind to the land's potential to raise tax monies.

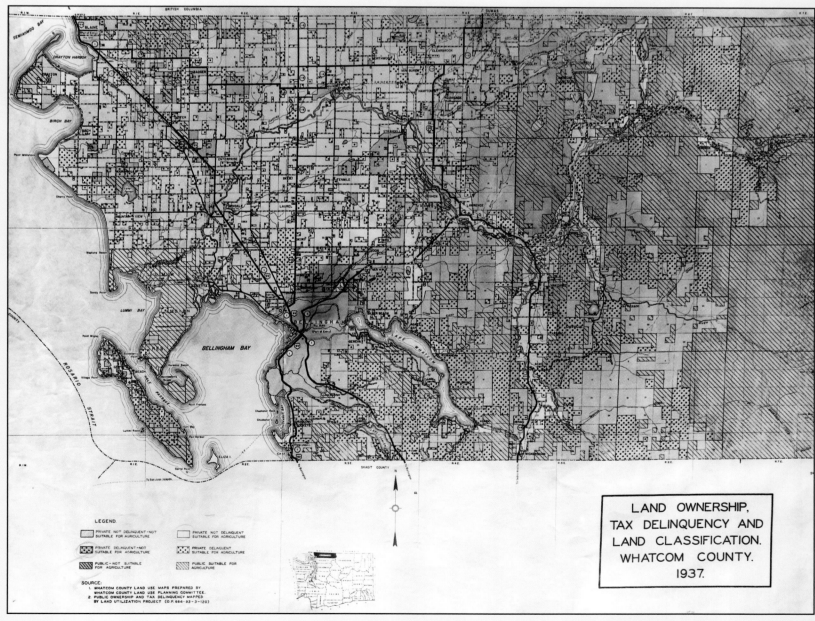

WATER AND POWER

The Northwest has an abundance of rivers, all seeking the sea from the mountainous regions—the Coast Range, the Cascades, and the Rockies. The majority of rivers concentrate into a single basin—that of the Columbia. Couple this geographical fact with the relative paucity of coal or other sources of power (once wood burning was seen as wasteful), the unequal distribution of precipitation, and the fertile yet dry volcanic soils of the interior, and it is not hard to see why hydroelectric generation from dams that could serve double function by holding water for irrigation proved so attractive.

The first long-distance transmission of alternating current electricity in the United States was made in June 1889 from Station A, on the Willamette Falls at Oregon City. The transmission line ran fourteen miles to power the streetlights of Portland. Portlanders could not get enough of this new electrical light; by 1895 a second generating plant, the not-very-imaginatively named Station B (now the Thomas W. Sullivan Plant, named after its designer) was completed, and it still supplies electricity to Portland.

Hydroelectric power was first delivered to Seattle in 1905, from a dam at Cedar Falls, the brainchild of the city's energetic engineer,

MAP 457 (below).
This rather beautiful bird's-eye view was part of a commemorative issue of the *Morning Oregonian* newspaper in December 1896 marking the transmission of hydroelectricity from Station B to Portland. The plant, on an island in the river, is still there today. The transmission line is clearly shown and is also illustrated in the photo, *above*.

Above.
From the same 1937 brochure published by Seattle City Light as the maps at right, this ad extols the virtues of a modern all-electric kitchen.

Reginald Thomson, and a response to a virtual monopoly then held over electrical service and street railways by Puget Sound Traction, Light & Power Company.

The city acquired a new power source in 1917 in the upper Skagit Valley. Here the city built the Gorge Dam, and its first electricity was delivered to Seattle in 1924, initiated by President Calvin Coolidge turning a gold key in the White House, 2,500 miles away. The dam was a wooden structure, an economy measure, and it was not until 1950 that it was replaced by a masonry structure (and a high concrete one ten years later).

Farther up the Skagit the Diablo Dam was begun in 1927 but not completed until 1936 because of an argument between city council and City Light superintendent James D. Ross as to the best site. A further dam, the Ruby Dam, was begun yet higher up the Skagit in 1937, this time using federal funding. It began generating electricity in 1940 and was renamed after Ross, who had died the previous year. The dam was increased in height in several stages and produced its full electrical output in 1954. Because its reservoir, Ross Lake, now backed up and covered five hundred acres in British Columbia, City Light agreed to pay an initial $250,000 and then $5,000 a year to flood Canadian land.

A fourth dam was planned, at Copper Creek, but proved impractical; instead the Gorge Dam was rebuilt higher and in concrete. Proposals to raise the Ross Dam foundered following Canadian objections to the amount of new land that would be flooded. In 1984 the federal government signed the Skagit River Treaty, under the terms of which Seattle dropped plans to raise Ross Dam in return for the right to buy electricity from British Columbia.

SKAGIT RIVER POWER PROJECT
Plan of Ultimate 1,126,000 Horsepower Development.

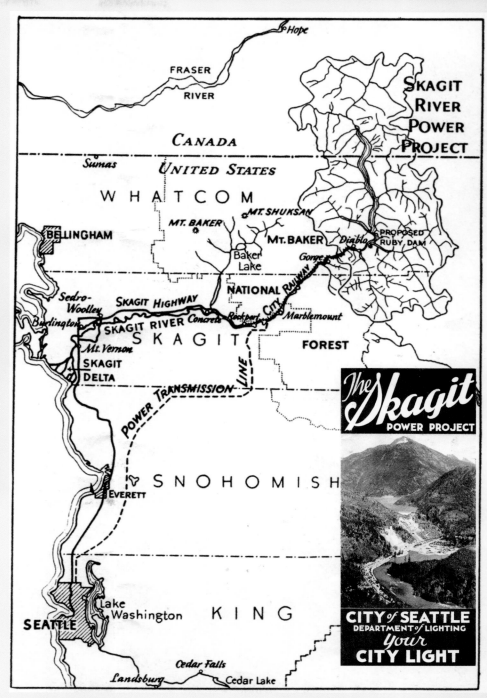

MAP 458 (*above*).
From a 1937 brochure published by Seattle City Light comes this bird's-eye view of the proposed eventual scope of its Skagit hydroelectric project. It shows the *Gorge Dam* and its reservoir, the *Diablo Dam* with its reservoir, and the *Ruby Dam and Power House*, then under construction, with its reservoir. The latter were renamed the Ross Dam and Ross Lake in 1940, following the death of superintendent James Ross.

MAP 459 (*above, right*).
Also from the 1937 brochure, the cover of which is illustrated *inset*, this map shows the *Power Transmission Line* from the *Skagit River Power Project* to *Seattle*. The *Skagit Highway* runs as far as *Rockport*. City Light built a railroad beyond Rockport for construction purposes and also maintained rail passenger cars, which it used to ferry visitors to the dam sites, in two-day outings advertised as part of an extensive public relations effort. Ruby Lake is shown here backed up many miles into Canada, but in the end only 500 acres were flooded beyond the international boundary.

MAP 460 (*right*).
Irrigated areas in 1923 (bright yellow) and the area to be covered by the Columbia River Project at that time (red) are shown on this map published by the Washington Water Power Company that year.

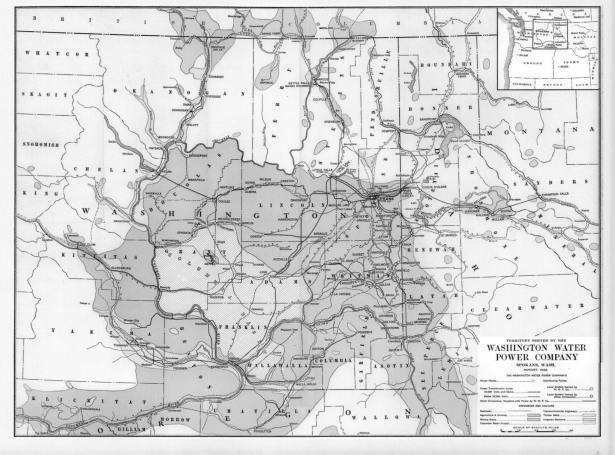

THE BONNEVILLE DAM

ESTIMATED PRINCIPAL QUANTITIES		ESTIMATED PRINCIPAL COST DIVISIONS	
EXCAVATION	4,000,000 Cu. Yds.	INVESTIGATIONS and DESIGNS	$ 300,000
COFFERDAM FILL	300,000 Cu. Yds.	BARGE LOCKS and APPROACH CANAL	2,300,000
CONCRETE	1,000,000 Cu. Yds.	POWER PLANT STRUCTURE	5,400,000
STRUCTURAL STEEL and CASTINGS	15,000 Tons	POWER PLANT MACHINERY and EQUIP'T.	3,100,000
REINFORCING STEEL	12,000 Tons	MAIN DAM	9,900,000
		RAILROAD CHANGES	4,500,000
		HIGHWAY CHANGES	500,000
		FISHWAYS	760,000
		LAND, RIGHTS OF WAY and DAMAGES	3,500,000

BONNEVILLE DAM IS LOCATED IN THE COLUMBIA RIVER 42 MILES EAST OF PORTLAND, OREGON

COST OF DAM, LOCKS AND POWER PLANT WITH TWO POWER UNITS INSTALLED AND SUBSTRUCTURE FOR FOUR ADDITIONAL UNITS: $31,000,000

POWER DEVELOPED WITH TWO UNITS 86,000 KILOWATTS OR 115,000 HORSE POWER

POWER HEAD AT ORDINARY LOW WATER 65 FT. ESTIMATED DATE OF COMPLETION JULY, 1937

MODEL BUILT BY U. S. ENGINEERS

© PHOTO-ART COMMERICAL STUDIOS, PORTLAND

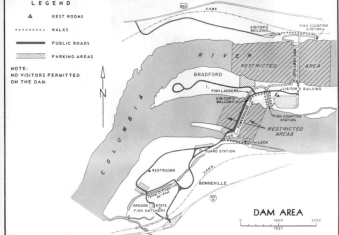

DAM AREA

MAP 461 (*above*).
This model of the completed Bonneville Dam, like the real thing built by the Army Corps of Engineers, seems to date from about 1934. The view is from the Washington side, looking southeast.

MAP 462 (*left*).
A more mundane map shows the Bonneville Dam in 1967. In 1974 another powerhouse was completed by excavating a new channel on the Washington side of the river and creating another island and diverting the railroad and highway on the north bank.

MAP 463 (*below*).
This bird's-eye view of the *Grand Coulee Dam and Power Plant*, the *Columbia Basin Project*, and the *Yakima Irrigation Project* is a southward view that gives an interesting different and downstream perspective. It was reproduced on a postcard in 1942, just after the dam was complete, but some of the features shown (such as Banks Lake Reservoir) were not yet finished.

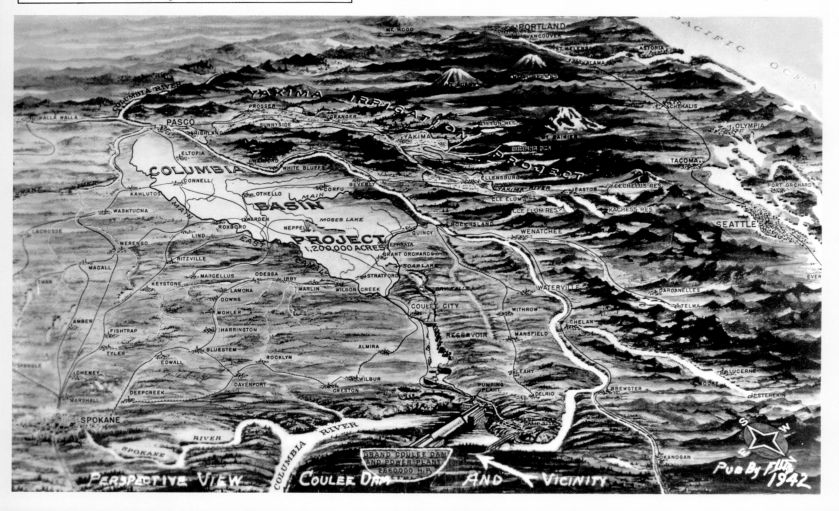

PERSPECTIVE VIEW COULEE DAM AND VICINITY

Pub By Ellis 1942

Schemes for irrigating the vast but dry, otherwise fertile land of the Columbia Basin had been mooted for many years, and small schemes had been built that diverted the water of the Columbia and its tributaries onto fields (see MAP 389, *page 159*). The Grand Coulee of the Columbia, a deep channel cut by an overflow during the Ice Age, was an obvious location for a dam, but the scale of such a project made it very expensive.

By 1920 a scheme had been worked out, but a dam at the Grand Coulee was considered infeasible. Then, in 1926, the federal government initiated a major study of irrigation, flood control, power generation, and navigation on the Columbia above its confluence with the Snake. The U.S. Army Corps of Engineers published a report in 1932. The sale of electricity was to offset the cost of pumping water for irrigation. The Bureau of Reclamation endorsed the report, but, with the onset of the Depression, cost became a major issue. It took a new president, Franklin D. Roosevelt, who included the Grand Coulee project in his New Deal list of public works designed to put America back to work, to allow the project to commence. At the same time, another project far downstream, near Cascade Locks, the Bonneville Dam (MAP 461, *left, top*), was approved, and the Public Works Administration (PWA) funded both projects (MAP 466, *overleaf*).

The Bonneville Dam was completed in 1937, and the first electricity generated was sold a year later. A second powerhouse was added in 1981. Large locks were included, to allow large ships to navigate

MAP 464 (*above*).
This superb bird's-eye map of the entire Columbia Basin upstream of the Snake confluence approaches a conventional map, compared with MAP 463, which is at a more oblique angle. During the Depression the Pacific Northwest was the recipient of two major projects on the Columbia River—the Bonneville Dam and the Grand Coulee Dam. This map was published by the Spokane Chamber of Commerce in 1949 and shows the *Columbia Basin Irrigation Project*, the *Grand Coulee Dam*, and *Lake [Franklin D.] Roosevelt* in relation to the city of *Spokane*. An *Equalizing Reservoir* in the *Grand Coulee* is Banks Lake, at the time actually unfilled. This feature, essentially a massive twenty-seven-mile-long holding reservoir, was created by building earthen dams at either end of the deep canyon of the Grand Coulee, completed in 1951. Water is pumped from Lake Roosevelt into Banks Lake, from where it is distributed to the *Irrigation Area*.

the river. The Grand Coulee Dam, a massive project by any standard, was completed in 1941. It is the largest electricity-generating structure in the United States, and the largest concrete structure, and it is the fifth-largest hydroelectric power generator in the world. Irrigation had been put on hold with the beginning of World War II, and it was not until 1951 that the first irrigation water was applied. After 1948 federal land that was to be irrigated was allocated to applicants. The irrigated area had at first been expected to provide farms for eighty thousand, mainly those relocated from the Dust Bowl of the Midwest, but in the end it was home to six thousand, largely because of improved and larger-scale farming methods.

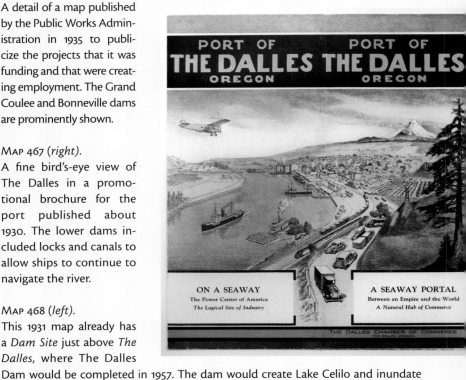

MAP 465 (*left*).
The cover of MAP 464, *previous page*. The Spokane Chamber of Commerce was promoting its city as the headquarters for the Columbia Basin Irrigation Project. The photo is of the Grand Coulee Dam.

MAP 466 (*right, top*).
A detail of a map published by the Public Works Administration in 1935 to publicize the projects that it was funding and that were creating employment. The Grand Coulee and Bonneville dams are prominently shown.

MAP 467 (*right*).
A fine bird's-eye view of The Dalles in a promotional brochure for the port published about 1930. The lower dams included locks and canals to allow ships to continue to navigate the river.

MAP 468 (*left*).
This 1931 map already has a *Dam Site* just above *The Dalles*, where The Dalles

MAP 469 (*below*).
The Chief Joseph Dam was completed in 1955 just above Bridgeport. This map shows the water distribution system to units of irrigable land downstream. Downstream of Wenatchee, the Rock Island Dam of Puget Sound Power & Light is marked. In 1933 this dam became the first to span the Columbia.

Dam would be completed in 1957. The dam would create Lake Celilo and inundate Celilo Falls, site of a longstanding Native settlement and fishery, which would be completed destroyed. Progress for some is not progress to all.

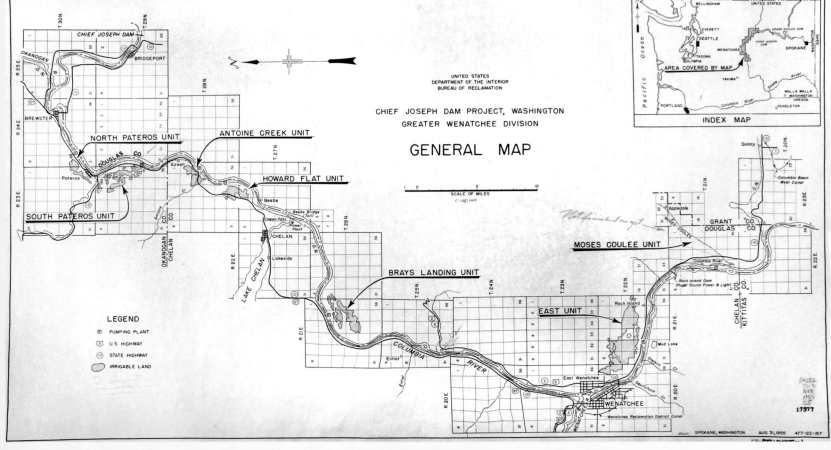

Many other dams have since been constructed on the Columbia, mainly for power generation but also for flood control. One, the Rock Island Dam, just below Wenatchee, actually preceded the Bonneville and Grand Coulee dams, having been completed by Puget Sound Power & Light in 1933. The location of dams and then-projected projects are shown on MAP 472, *right, bottom,* published in 1948. Chief Joseph Dam, at Bridgeport, was completed in 1955 (MAP 469, *left, bottom*). The McNary Dam, at Umatilla, came fully onstream in 1957, as did The Dalles Dam, which created Lake Celilo and flooded Celilo Falls, controversially eliminating the Native fishery. Priest Rapids Dam was completed in 1961; this had been initiated following the Vanport Flood (see page 195).The John Day Dam, east of The Dalles, was completed in 1971. And there were others. The result is that today the Columbia Basin is likely as well controlled as it can be and generates a substantial amount of electricity, reinforcing the Northwest's prosperity.

Many of the early dams were built without regard to the environment, and salmon runs, in particular, were sometimes destroyed. Grand Coulee was built without any provision for the salmon run and cut off the Columbia above it, eliminating the Native fishery and altering a way of life. In the last twenty years or so there has been a move to reverse this process in some areas. On the Olympic Peninsula, for example, the Elwha Dam, on the Elwha River, completed in 1913, is to be dismantled in the hopes of restoring the salmon to this stream.

MAP 471 (*right, top*).
The city of Warden, in the center of the area to be irrigated, promoted itself as the "City with the Golden Future" on this 1947 blueprint. The city has a population of about 2,600 today.

MAP 472 (*right*).
A detailed and colorful summary map of all the existing and proposed power and irrigation projects in the Columbia Basin, published in 1948 by the U.S. Bureau of Reclamation. The numbers refer to the key, *inset.*

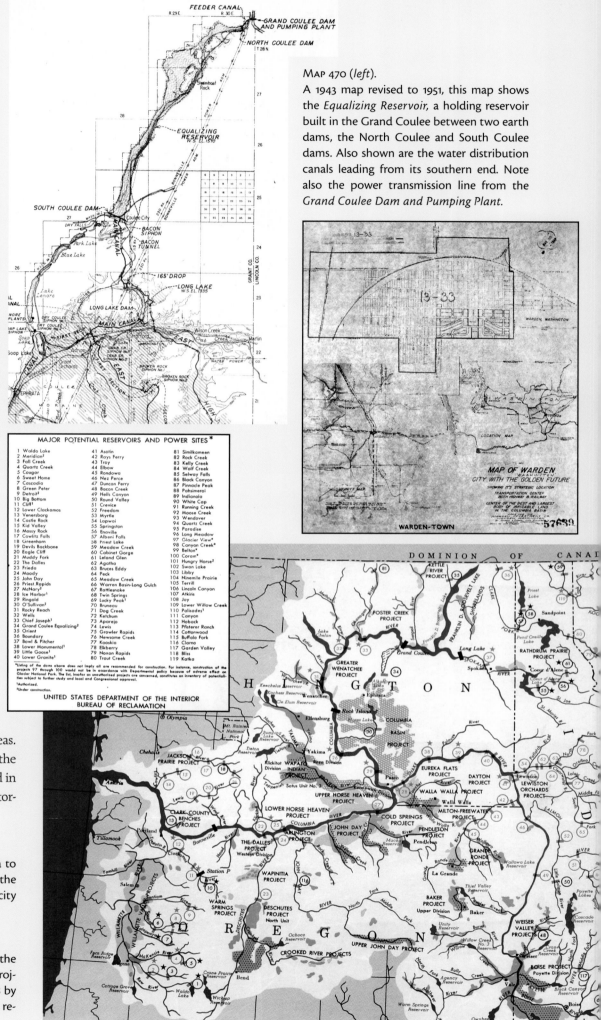

MAP 470 (*left*).
A 1943 map revised to 1951, this map shows the *Equalizing Reservoir,* a holding reservoir built in the Grand Coulee between two earth dams, the North Coulee and South Coulee dams. Also shown are the water distribution canals leading from its southern end. Note also the power transmission line from the *Grand Coulee Dam and Pumping Plant.*

MAJOR POTENTIAL RESERVOIRS AND POWER SITES*

1 Waldo Lake	41 Asotin	81 Similkameen
2 Meridian²	42 Rays Ferry	82 Rock Creek
3 Fall Creek	43 Troy	83 Kelly Creek
4 Quartz Creek	44 Elbow	84 Wolf Creek
5 Cougar	45 Rondowa	85 Selway Falls
6 Sweet Home	46 Nez Perce	86 Black Canyon
7 Cascadia	47 Duncan Ferry	87 Pinnacle Peak
8 Green Peter	48 Bacon Creek	88 Pahsimeroi
9 Detroit²	49 Hells Canyon	89 Indianola
10 Big Bottom	50 Round Valley	90 White Cap
11 Cliff¹	51 Crevice	91 Running Creek
12 Lower Clackamos	52 Freedom	92 Moose Creek
13 Venersborg	53 Myrtle	93 Wendover
14 Castle Rock	54 Lopwai	94 Quartz Creek
15 Kid Valley	55 Springston	95 Paradise
16 Mossy Rock	56 Enaville	96 Long Meadow
17 Cowlitz Falls	57 Albeni Falls	97 Glacier View*
18 Greenhorn	58 Priest Lake	98 Canyon Creek*
19 Devils Backbone	59 Meadow Creek	99 Belton*
20 Eagle Cliff	60 Cabinet Gorge	100 Coram*
21 Muddy Fork	61 Leland Glen	101 Hungry Horse¹
22 The Dalles	62 Agatha	102 Swan Lake
23 Frieda	63 Bruces Eddy	103 Libby
24 Moody	64 Peck	104 Ninemile Prairie
25 John Day	65 Meadow Creek	105 Terrill
26 Priest Rapids	66 Warren Basin-Long Gulch	106 Lincoln Canyon
27 McNary²	67 Rattlesnake	107 Atkins
28 Ice Harbor¹	68 Twin Springs	108 Joy
29 Ringold	69 Lucky Peak¹	109 Lower Willow Creek
30 O'Sullivan²	70 Bruneau	110 Palisades¹
31 Rocky Reach	71 Dog Creek	111 Canyon
32 Wells	72 Ketchum	112 Hoback
33 Chief Joseph¹	73 Aparejo	113 Pfisterer Ranch
34 Grand Coulee Equalizing²	74 Lewis	114 Cottonwood
35 Orient	75 Growler Rapids	115 Buffalo Fork
36 Boundary	76 Newsome Creek	116 Clarno
37 Bowl & Pitcher	77 Kooskia	117 Garden Valley
38 Lower Monumental¹	78 Elkberry	118 Bliss
39 Little Goose¹	79 Naxon Rapids	119 Katka
40 Lower Granite¹	80 Trout Creek	

*Listing of the dams above does not imply all are recommended for construction. For instance, construction of the projects 97 through 100 would not be in accordance with Departmental policy because of adverse effect on Glacier National Park. The list, insofar as unauthorized projects are concerned, constitutes an inventory of potentialities subject to further study and local and Congressional approval.

¹Authorized.
²Under construction.

**UNITED STATES DEPARTMENT OF THE INTERIOR
BUREAU OF RECLAMATION**

WAR IN THE NORTHWEST

World War II revitalized the Northwest, bringing new industries and an influx of population, but disrupted the lives of all. Federal contracts flowed in to Boeing, now well established as a leader in aircraft building (see page 201), and shipyards hummed in many locations.

And soldiers were trained to fight for their country. Enormous camps were created in several locations, but none larger than Camp Adair. In 1942 the U.S. Army took over more than a hundred square miles between Corvallis and Salem (MAP 473, *right*), converting it into a huge military cantonment and training ground. The camp itself (MAP 474, *below*) was briefly the second-largest city in Oregon; today it is the state's largest ghost town. Between 1942 and 1944 an estimated 100,000 troops, most of four Army divisions, trained here. The camp had its own churches, service clubs, a theater, a post office, stores, bus terminal, and telephones at convenient locations throughout the camp. It required all the services of a city, such as electricity and water, and had its own reservoir.

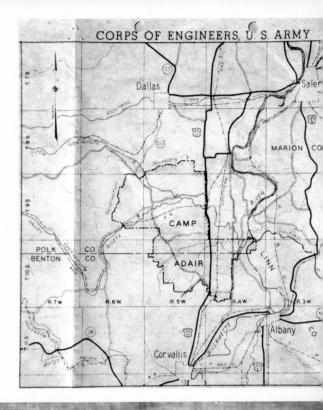

MAP 473 (*above, top*).
The vast area covered by Camp Adair and its training grounds is apparent from this location map, drawn up in 1943. Even part of the Willamette was within the camp, where river crossings could be practiced.

MAP 474 (*above*).
A plan of the building layout of Camp Adair in 1943 shows it to be truly a small city. It had over 1,800 buildings. Today only the streets remain, though in an overgrown state (photo, *left.*) *Right.* A modern sign commemorates the Army divisions that trained at Camp Adair: the 70th, 91st, 96th, and 104th Infantry Divisions.

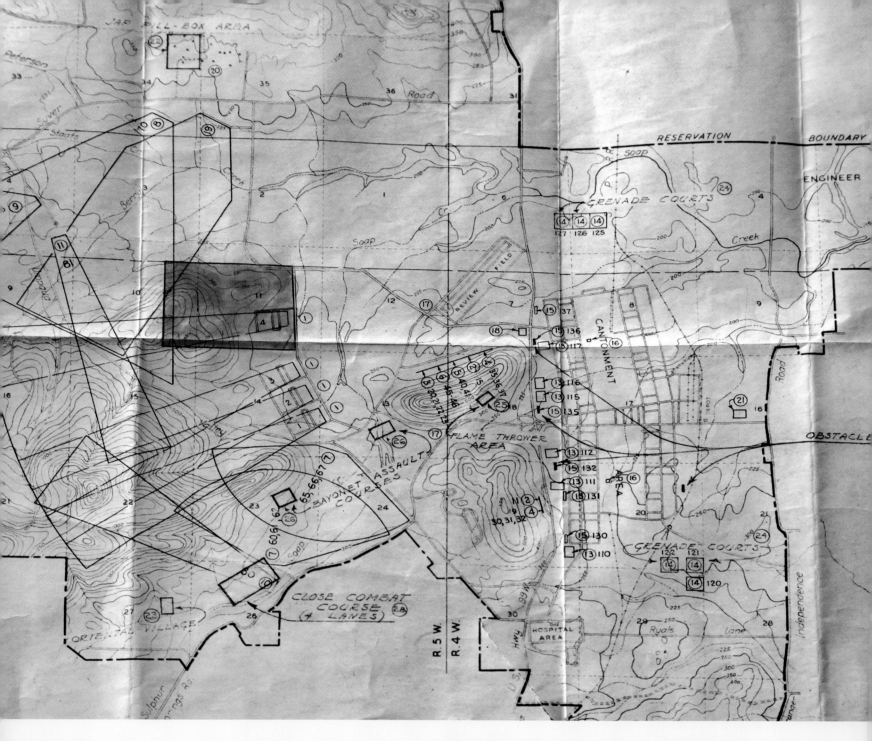

MAP 475 (*above*).

This map of a significant part of Camp Adair locates all the training facilities, many designed to replicate conditions that the troops might find once in action. An entire German village was reproduced, for example, and this map shows an *Oriental Village* (bottom left) and a *Jap Pill-Box Area* (top left). Other training areas shown include a *Flame Thrower Area, Bayonet Assault Courses,* a *Close Combat Course,* and two *Grenade Courts.* The cantonment is at center right, with a *Hospital* south of it, and the inevitable *Review Field* at its north end. There were also pistol and rifle ranges, a machine-gun range, anti-aircraft ranges and other moving-target ranges, and even a gas chamber. Everything necessary to train for potentially horrifying combat to come was prepared for here. The training area was not even confined within the Camp Adair boundaries; from time to time maneuvers were carried out over a six-county region of central Oregon.

Camp Adair was unceremoniously destroyed in March 1947 by a more than a hundred–strong wrecking crew. Today, little remains, and the area has been converted into the State of Oregon's E.E. Wilson Wildlife Area.

The war effort expanded industry. Boeing, which in 1939 had employed four thousand people, by the end of the war employed fifty thousand. Boeing had plants in Seattle, Renton, Everett, Chehalis, Aberdeen, and Bellingham. Governments had to step in to create a great deal of new housing.

Nowhere was the need for housing greater than in Portland, where Henry Kaiser's shipyards geared up to produce ships by the thousands, including the famous Liberty ships, 440-foot-long freighters produced to a standard design that could be mass-produced. Kaiser became known as Sir Launchalot. Kaiser built ships in numerous locations up and down the West Coast, and there were many other shipyards—twenty-nine in Seattle alone, employing more than forty thousand workers. At the Bremerton Navy Yard (MAP 476, *overleaf*) another thirty thousand workers churned out ships.

The aluminum industry got its start in the Northwest during the war because of the availability of low-cost energy from the newly completed Bonneville and Grand Coulee dams. Alcoa built plants at Troutdale, east of Portland, and at Vancouver, across the Columbia, while Reynolds built an aluminum plant at Longview.

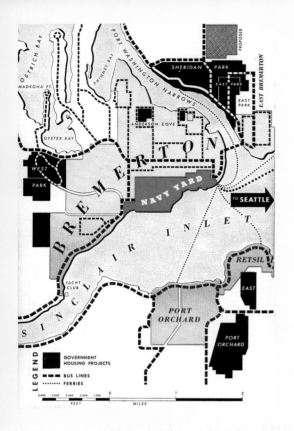

MAP 476 (*left*).

A map from a 1943 brochure designed to attract workers to *Bremerton Navy Yard.* A fleet of ferries was needed to bring workers from Seattle. As the page from the brochure (*right*) suggests, the Navy had buses to bring workers from housing that had been created outside the city.

MAP 477 (*below*).

A map of the new city of Vanport, created across the Columbia from Vancouver on low-lying land protected from flooding only by a railroad dike and with only one way out, via *Denver Avenue*, at right. This was to prove a disastrous combination when the dike gave way on 30 May 1948, wiping out the city in a matter of minutes. The aerial photos (*opposite page*) show Vanport, looking west from above Denver Avenue, in December 1943, soon after it was completed, and during the June 1948 flood; plus a street-level scene (*bottom*). Today the area contains Portland International Raceway, a golf course, industrial buildings—and a lake. Denver Avenue is on the approximate alignment of today's I-5, and the Vanport area is immediately west of the highway.

ONLY A FEW MINUTES FROM HOME TO WORK!

From the time you leave your doorstep until you alight from the bus alongside your work, it need not take any longer than 20 minutes. The homes which the government has built at Bremerton, although located in open country, are all within 5 to 20 minutes of the Navy Yard itself by Navy bus or local transportation systems.

As you can see from the map on the opposite page, Bremerton

is located on one of Puget Sound's many winding inlets. The busy Navy Yard itself is a striking contrast to the thickly wooded slopes of Sinclair inlet, whose sand beaches make it an ideal spot for pleasant summer homes.

Although only an hour from Seattle, Bremerton is close to the Olympic Mountains, with their snow-capped peaks and great national forests.

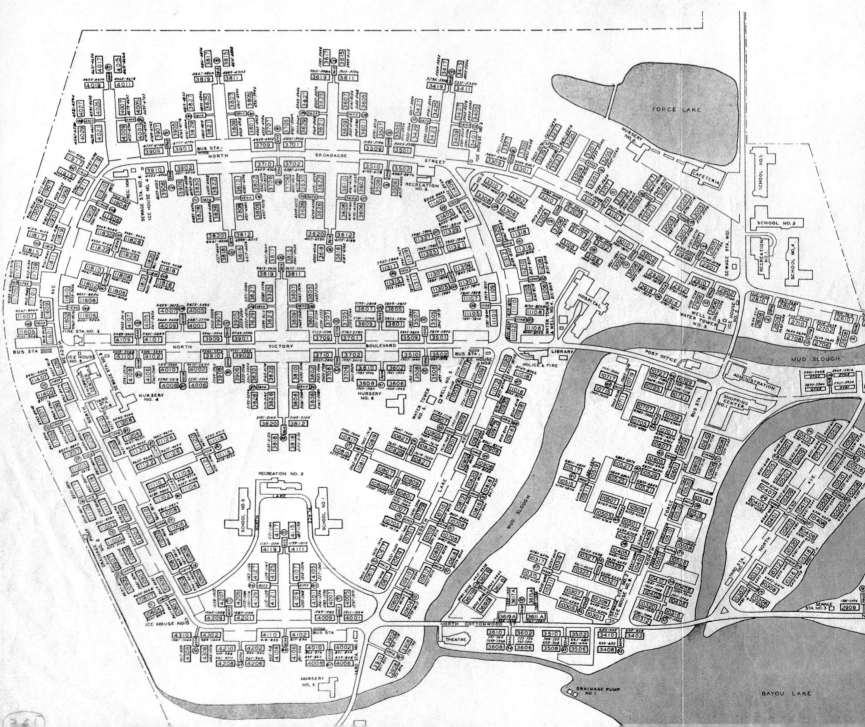

The demand for housing for the workers of wartime industry was met in many ways. Across from the shipyards and aluminum plant of Vancouver, the new city of Vanport was hastily laid out. Initiated by Kaiser, it was soon taken over by the Housing Authority of Portland (HAP), an arm of the Federal Housing Authority. This totally planned community of forty thousand people was significant for two particular reasons: it established the principle of racial integration at a time when it did not exist, and it taught an important lesson—that housing should never be built on land subject to flooding.

About 40 percent of Vanport's population was Afro-American. The housing, as it always was at that time, was racially segregated. But the stores, churches, and other public facilities were not. It was a small wedge that would grow into a national fundamental.

After the war the population of Vanport dropped, and the HAP opened a college to attract veterans. The population was about 18,500, when, in May 1948, the waters of the Columbia started to rise. At 4:17 PM on 30 May 1948, just after the HAP had reassured residents that

the dikes were safe, a dike gave way, sending a ten-foot wall of water into Vanport. Channels and sloughs delayed the water somewhat, giving residents a chance to escape, but the only escape route, Denver Avenue on the west side, soon became choked with vehicles. Fifteen people died in the flood. Vanport was not rebuilt. The college moved to downtown Portland, where today it is Portland State University.

Vanport may have been a beginning lesson in racial tolerance, but another event initiated by World War II was just the opposite. For after the Japanese attacked Pearl Harbor on 7 December 1941 and the United States entered the war, Americans of Japanese origin, even if they were American-born citizens, came under suspicion as enemy agents, and a plan was hatched to ship all Japanese Americans, from California to Washington (and, separately, in British Columbia as well) to camps well away from the coast.

A public paranoia descended on the West Coast following Pearl Harbor, with Japanese aerial attacks being reported—though never substantiated—in Los Angeles, and a Japanese submarine did fire on an oil refinery in California. On the recommendation of General John DeWitt, chief of the Western Defense Command, President Franklin D. Roosevelt signed Executive Order 9066 in February 1942, a week before these incidents. It authorized the designation of military areas from which "any or all persons" could be excluded at the discretion of the military commander. Despite this apparent comprehensiveness, in reality the exclusion was applied only to Japanese Americans. Although some attempts were made to obtain voluntary relocation, this, naturally enough, did not work, and the

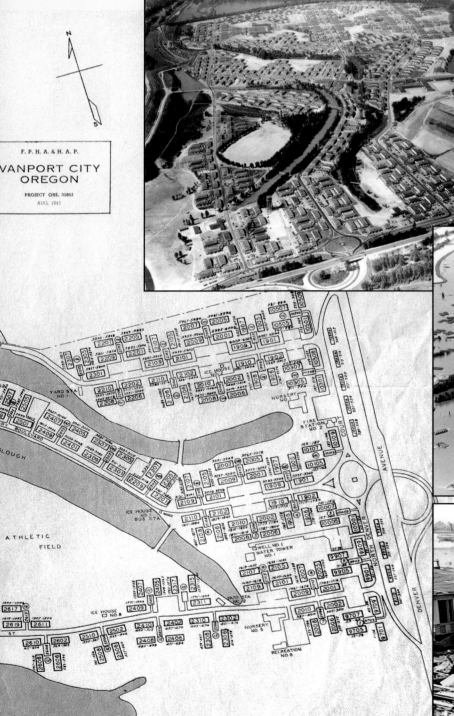

F. P. H. A. & H. A. P.
VANPORT CITY
OREGON
PROJECT ORE. 35053
AUG. 1943

government soon resorted to forcible relocation. All Japanese Americans were required to report to assembly centers, from which they would be allocated to one of ten relocation camps set up throughout the West. In the Northwest, assembly centers were set up at Puyallup and Portland (MAPS 481 and 482, *below, right*), and other Japanese Americans were sent from Northwest locations to Pinedale, near Fresno, in California. From these assembly centers trainloads would be sent, about five hundred people at a time, to the relocation centers at Minidoka, near Twin Falls, Idaho; and Tule Lake, California; with smaller numbers going to Heart Mountain, near Cody, Wyoming; and Manzanar, in the Owens Valley of California. In total, just over fifteen thousand Japanese Americans were sent to relocation camps from Washington and Oregon.

MAP 478 (*left, top*) and MAP 479 (*left, bottom*). From the final report of General John DeWitt in 1943 come these maps detailing the areas from which Japanese Americans were sent to each assembly center, and the relocation centers they were sent to after that. MAP 480 (*below*) shows the location and maximum population at any one time of the Portland and Puyallup assembly centers; MAPS 481 and 482 (*below, bottom*) are location maps of the assembly centers. The photo (*below*) is an aerial view of the Puyallup assembly center. *Above*, Japanese Americans board a train at the Portland assembly center to take them to Tule Lake, Heart Mountain, or Minidoka, the three relocation camps to which people were sent from Portland. According to the report, Portland processed 3,584 persons; Puyallup processed 7,398; and Pinedale 4,011 plus a small undefined number from Bainbridge Island.

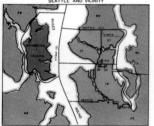
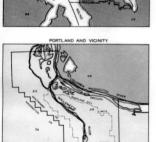

JAPANESE EVACUATION PROGRAM

WESTERN DEFENSE COMMAND AND FOURTH ARMY

WARTIME CIVIL CONTROL ADMINISTRATION

ASSEMBLY CENTER

DESTINATIONS

- COLORADO RIVER
- FRESNO
- GILA RIVER
- MANZANAR
- MARYSVILLE
- MAYER
- MERCED
- PINEDALE
- POMONA
- PORTLAND
- PUYALLUP
- SACRAMENTO
- SALINAS
- SANTA ANITA
- STOCKTON
- TANFORAN
- TULARE
- TULE LAKE
- TURLOCK

LEGEND

---- EXCLUSION AREA BOUNDARY LINES

100 EXCLUSION AREA ORDER NUMBERS

* ASSEMBLY CENTERS

* RELOCATION CENTERS

JAPANESE EVACUATION PROGRAM

WESTERN DEFENSE COMMAND AND FOURTH ARMY

WARTIME CIVIL CONTROL ADMINISTRATION

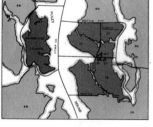
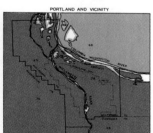

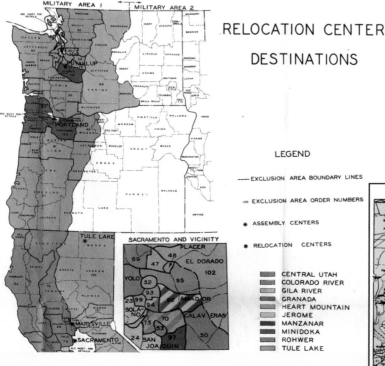

RELOCATION CENTER

DESTINATIONS

LEGEND

---- EXCLUSION AREA BOUNDARY LINES

100 EXCLUSION AREA ORDER NUMBERS

* ASSEMBLY CENTERS

* RELOCATION CENTERS

- CENTRAL UTAH
- COLORADO RIVER
- GILA RIVER
- GRANADA
- HEART MOUNTAIN
- JEROME
- MANZANAR
- MINIDOKA
- ROHWER
- TULE LAKE

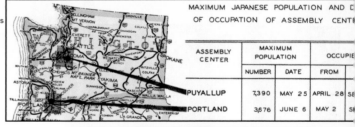

ASSEMBLY CENTER	MAXIMUM POPULATION		OCCUPIE...
	NUMBER	DATE	FROM
PUYALLUP	7,390	MAY 25	APRIL 28 SE...
PORTLAND	3,676	JUNE 6	MAY 2 S...

MAXIMUM JAPANESE POPULATION AND D... OF OCCUPATION OF ASSEMBLY CENT...

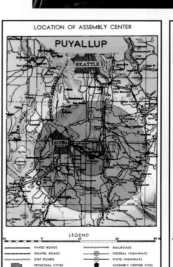

LOCATION OF ASSEMBLY CENTER

PUYALLUP

LOCATION OF ASSEMBLY CENTER

PORTLAND

Above. A military policeman posts Civilian Exclusion Order No 1. on Bainbridge Island about 25 March 1942. The order required all Japanese Americans living in the area to report to a "Civilian Control Station," which was to give "advice" and instructions, deal with disposition of property, and arrange transportation to an assembly center. The photo comes from the 1943 DeWitt report.

MAP 483 (*below*).

This Japanese map, which was reproduced in the *Los Angeles Examiner* on 22 August 1943, reputedly showed the strategy by which the Japanese sought to win the war. It was said to have been obtained by a Korean spy. After capturing islands in the South Pacific and the Aleutians and Hawaii, they would strike the West Coast and the Panama Canal. It seems the Japanese apparently thought they could capture and hold the area of the United States west of the Rocky Mountains while the rest of the country would have to sue for peace. It seems far-fetched now, perhaps, but did not seem so in 1943. After all, unless Japan, a relatively small country, thought such a plan would work, it would hardly have taken on a country with the enormous resources and resolve of the United States.

Above. A Japanese fire balloon is recovered on the beach at Point Roberts, Washington, late in 1944.

Although widely accepted later as being wrong, the relocation was justified in the government report in 1943 as follows: "in emergencies, where the safety of the Nation is involved, consideration of the rights of individuals must be subordinated to the common security." In retrospect, the safety of the nation was not at risk from Japanese Americans, but, as always, the facts are much clearer in hindsight.

Toward the end of the war, the Japanese launched an unusual offensive on the West Coast. A meteorologist had established in the 1920s that the jet stream (high-level wind) blew eastward from Japan, and so the Japanese built and released a series of some nine thousand so-called fire balloons, essentially weather-type balloons carrying a payload of four incendiary bombs each. After the time precalculated as being required for the balloons to reach the Northwest, the incendiary bombs were dropped, one at a time. The idea was to set the forests alight. In fact the balloons were hardly noticed, being directed to remote areas, and most fizzled on wet or snow-covered forests, although six members of a family picnicking on Gearhart Mountain in Southern Oregon in May 1945 were killed when they investigated a balloon that had landed. The public was not informed of the balloon attacks at the time to prevent the Japanese from assessing their success.

The Japanese surrendered on 15 August 1945 after atomic bombs were dropped on Hiroshima and Nagasaki. The Northwest had played its part, for the plutonium for the Nagasaki bomb had been manufactured in Hanford, Washington.

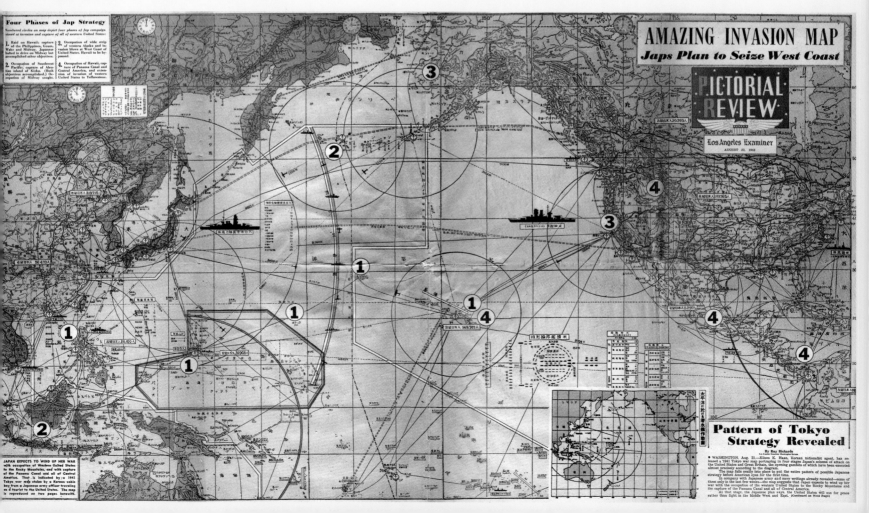

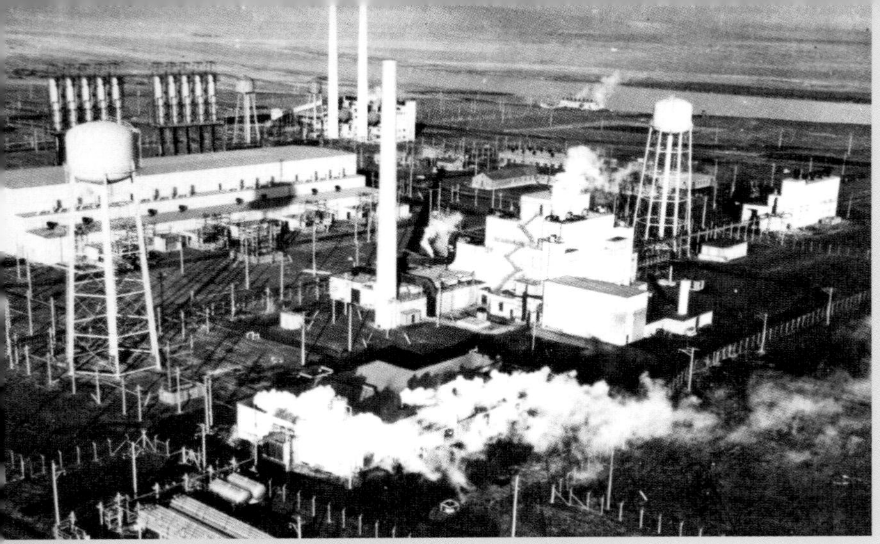

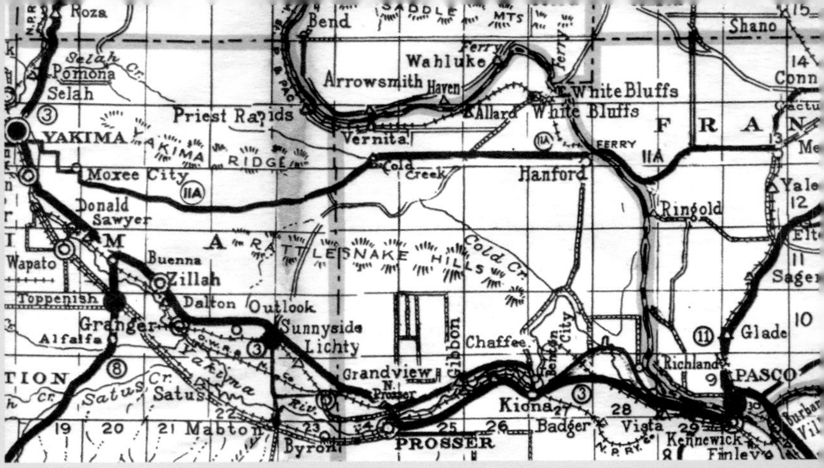

Making Atomic Bombs

The most secretive of all the wartime industries in the United States was located in three places: Oak Ridge, Tennessee; Los Alamos, New Mexico; and Hanford, Washington. And, of course, work continued long after the war had been won.

Hanford was selected to produce enriched plutonium required for the atomic bomb. A site on the banks of the Columbia was where the world's first large-scale nuclear reactor (illustrated *above*) was built.

The government selected the Hanford site in late 1942 and began construction in March 1943. The entire towns of Hanford and White Bluffs were expropriated, along with many square miles of irrigated farms and orchards. The final size of

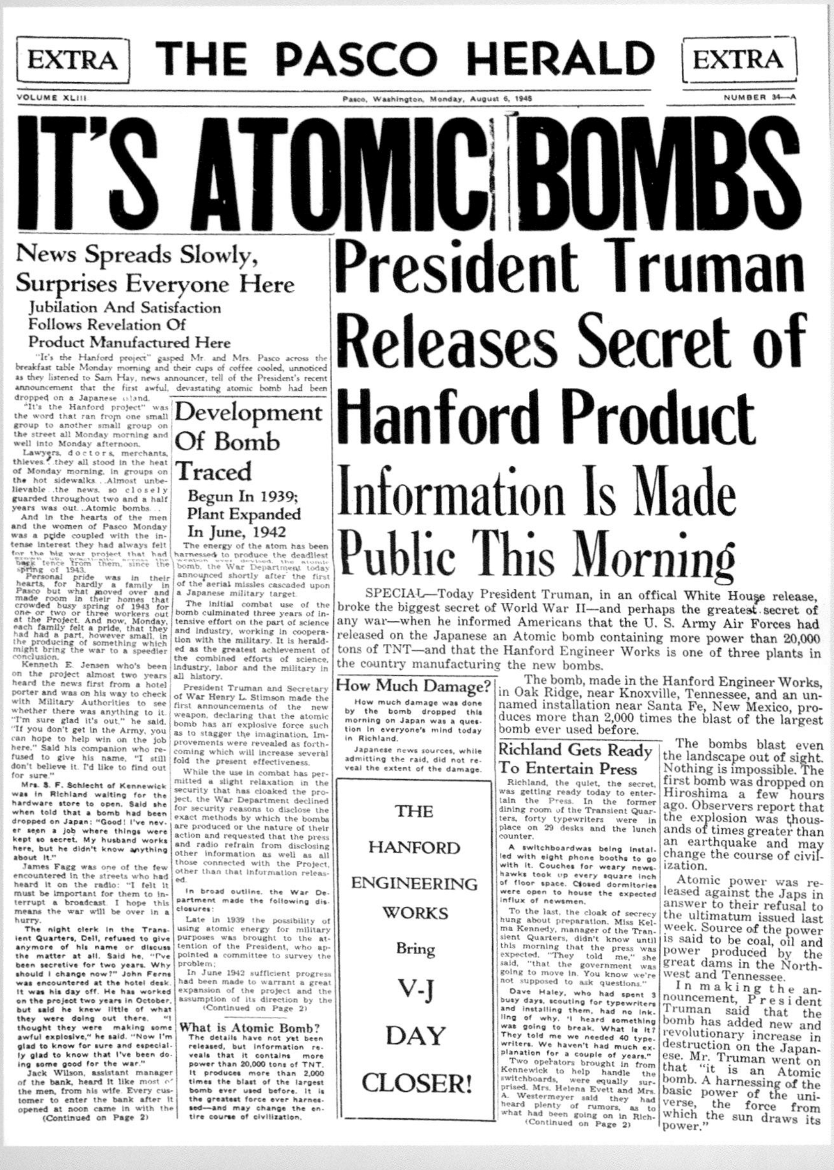
EXTRA **THE PASCO HERALD** EXTRA

VOLUME XLIII — Pasco, Washington, Monday, August 6, 1945 — NUMBER 34—A

IT'S ATOMIC BOMBS

News Spreads Slowly, Surprises Everyone Here

Jubilation And Satisfaction Follows Revelation Of Product Manufactured Here

"It's the Hanford project" gasped Mr. and Mrs. Pasco across the breakfast table Monday morning and their cups of coffee cooled, unnoticed as they listened to Sam Hay, news announcer, tell of the President's recent announcement that the first awful, devastating atomic bomb had been dropped on a Japanese island.

"It's the Hanford project" was the word that ran from one small group to another small group on the street all Monday morning and well into Monday afternoon.

Lawyers, doctors, merchants, thieves... they all stood in the heat of Monday morning, in groups on the hot sidewalks... Almost unbelievable... the news, so closely guarded throughout two and a half years was out... Atomic bombs...

And in the hearts of the men and the women of Pasco Monday was a pride coupled with the intense interest they had always felt for the big war project that had [illegible] back fence from them, since the spring of 1943.

Personal pride was in their hearts, for hardly a family in Pasco but what moved over and made room in their homes that crowded busy spring of 1943 for one or two or three workers out at the Project. And now, Monday, each family felt a pride, that they had had a part, however small, in the producing of something which might bring the war to a speedier conclusion.

Kenneth E. Jensen who's been on the project almost two years heard the news first from a hotel porter and was on his way to check with Military Authorities to see whether there was anything to it. "I'm sure glad it's out," he said. "If you don't get in the Army, you can hope to help win on the job here." Said his companion who refused to give his name, "I still don't believe it. I'd like to find out for sure."

Mrs. S. F. Schlecht of Kennewick was in Richland waiting for the hardware store to open. Said she when told that a bomb had been dropped on Japan: "Good! I've never seen a job where things were kept so secret. My husband works here, but he didn't know anything about it."

James Fagg was one of the few encountered in the streets who had heard it on the radio: "I felt it must be important for them to interrupt a broadcast. I hope this means the war will be over in a hurry.

The night clerk in the Transient Quarters, Dell, refused to give anymore of his name or discuss the matter at all. Said he, "I've been secretive for two years. Why should I change now?" John Ferns was encountered at the hotel desk. It was his day off. He has worked on the project two years in October, but said he knew little of what they were doing out there. "I thought they were making some awful explosive," he said. "Now I'm glad to know for sure and especially glad to know that I've been doing some good for the war."

Jack Wilson, assistant manager of the bank, heard it like most of the men, from his wife. Every customer to enter the bank after it opened at noon came in with the

(Continued on Page 2)

Development Of Bomb Traced

Begun In 1939; Plant Expanded In June, 1942

The energy of the atom has been harnessed to produce the deadliest weapon ever devised, the atomic bomb, the War Department today announced shortly after the first of the aerial missles cascaded upon a Japanese military target.

The initial combat use of the bomb culminated three years of intensive effort on the part of science and industry, working in cooperation with the military. It is heralded as the greatest achievement of the combined efforts of science, industry, labor and the military in all history.

President Truman and Secretary of War Henry L. Stimson made the first announcements of the new weapon, declaring that the atomic bomb has an explosive force such as to stagger the imagination. Improvements were revealed as forthcoming which will increase several fold the present effectiveness.

While the use in combat has permitted a slight relaxation in the security that has cloaked the project, the War Department declined for security reasons to disclose the exact methods by which the bombs are produced or the nature of their action and requested that the press and radio refrain from disclosing other information as well as all those connected with the Project, other than that information released.

In broad outline, the War Department made the following disclosures:

Late in 1939 the possibility of using atomic energy for military purposes was brought to the attention of the President, who appointed a committee to survey the problem;

In June 1942 sufficient progress had been made to warrant a great expansion of the project and the assumption of its direction by the

(Continued on Page 2)

What is Atomic Bomb?

The details have not yet been released, but information reveals that it contains more power than 20,000 tons of TNT. It produces more than 2,000 times the blast of the largest bomb ever used before. It is the greatest force ever harnessed—and may change the entire course of civilization.

President Truman Releases Secret of Hanford Product

Information Is Made Public This Morning

SPECIAL—Today President Truman, in an offical White House release, broke the biggest secret of World War II—and perhaps the greatest secret of any war—when he informed Americans that the U. S. Army Air Forces had released on the Japanese an Atomic bomb containing more power than 20,000 tons of TNT—and that the Hanford Engineer Works is one of three plants in the country manufacturing the new bombs.

The bomb, made in the Hanford Engineer Works, in Oak Ridge, near Knoxville, Tennessee, and an unnamed installation near Santa Fe, New Mexico, produces more than 2,000 times the blast of the largest bomb ever used before.

The bombs blast even the landscape out of sight. Nothing is impossible. The first bomb was dropped on Hiroshima a few hours ago. Observers report that the explosion was thousands of times greater than an earthquake and may change the course of civilization.

Atomic power was released against the Japs in answer to their refusal to the ultimatum issued last week. Source of the power is said to be coal, oil and power produced by the great dams in the Northwest and Tennessee.

In making the announcement, President Truman said that the bomb has added new and revolutionary increase in destruction on the Japanese. Mr. Truman went on that "it is an Atomic bomb. A harnessing of the basic power of the universe, the force from which the sun draws its power."

How Much Damage?

How much damage was done by the bomb dropped this morning on Japan was a question in everyone's mind today in Richland.

Japanese news sources, while admitting the raid, did not reveal the extent of the damage.

THE HANFORD ENGINEERING WORKS Bring V-J DAY CLOSER!

Richland Gets Ready To Entertain Press

Richland, the quiet, the secret, was getting ready today to entertain the Press. In the former dining room of the Transient Quarters, forty typewriters were in place on 29 desks and the lunch counter.

A switchboardwas being installed with eight phone booths to go with it. Couches for weary newshawks took up every square inch of floor space. Closed dormitories were open to house the expected influx of newsmen.

To the last, the cloak of secrecy hung about preparation. Miss Kelma Kennedy, manager of the Transient Quarters, didn't know until this morning that the press was expected. "They told me," she said, "that the government was going to move in. You know we're not supposed to ask questions."

Dave Haley, who had spent 3 busy days, scouting for typewriters and installing them, had no inkling of why. 'I heard something was going to break. What is it? They told me we needed 40 typewriters. We haven't had much explanation for a couple of years."

Two operators brought in from Kennewick to help handle the switchboards, were equally surprised. Mrs. Helena Evett and Mrs. A. Westermeyer said they had heard plenty of rumors, as to what had been going on in Rich-

(Continued on Page 2)

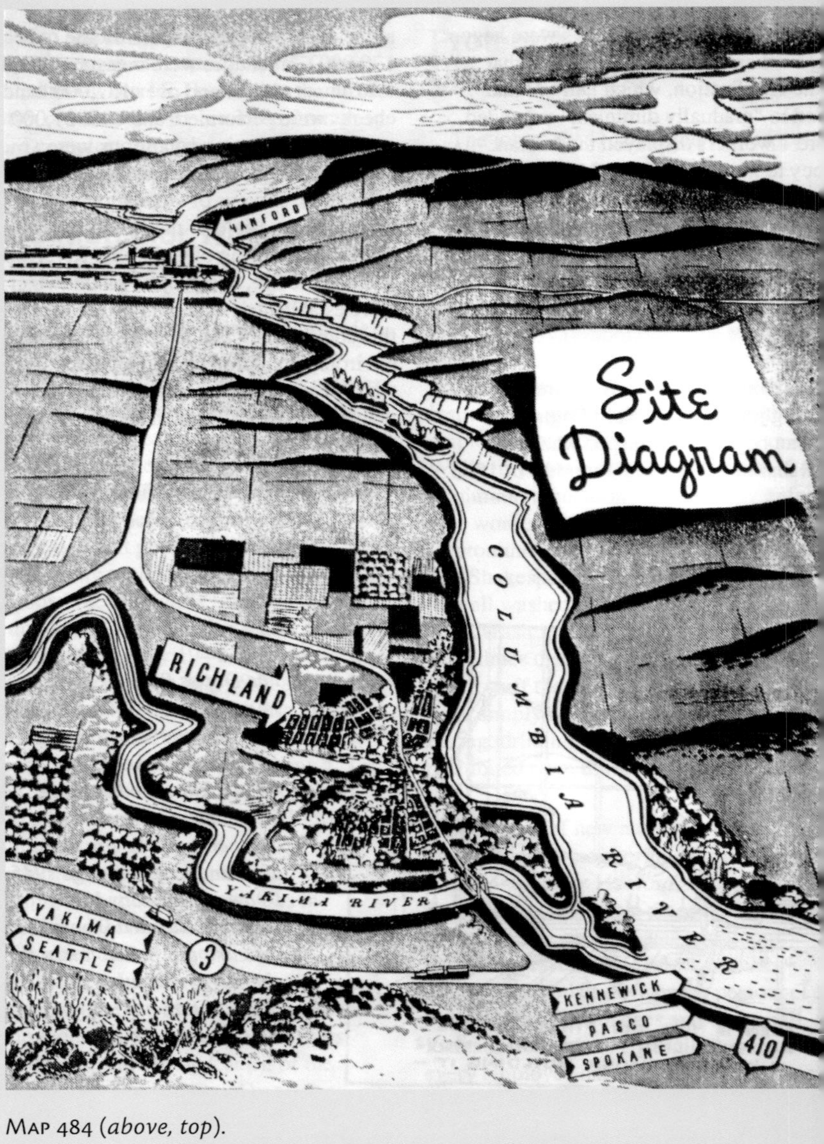

MAP 484 (*above, top*).
This commercially produced map dated 1942 is quite likely the last to show Hanford and White Bluffs as they were before the governmental takeover of the region for the Hanford nuclear plant.

MAP 485 (*above*).
Because of the secrecy and security issues that were involved, published maps of Hanford were few. This bird's-eye sketch, with Richland in the foreground and the Hanford nuclear plant in the distance, was the first official map of any sort to be released to the public. It was offered to journalists after President Harry Truman revealed Hanford's involvement in the Manhattan Project the same day the first atomic bomb was dropped, on Hiroshima on 6 August 1945. *Left,* the local newspaper reports the event.

Left, top. The 100 B Area, site of the world's first large-scale nuclear reactor, the tiered structure at center. It was located on the banks of the Columbia (see MAP 488, *right*), where river water could be used as a coolant.

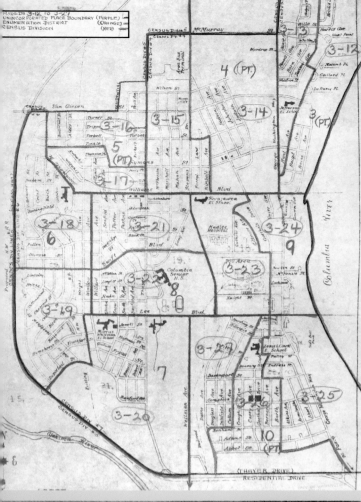

the site was nearly six hundred square miles. The logistics of building such a facility in a hurry were difficult, more so with the need for secrecy. The Hanford site contained 554 buildings, including three nuclear reactors, by war's end.

The influx of workers necessitated the creation of a trailer city at Hanford. The site employed some 51,000 people. After the war, standardized housing designed for one thing—speed of completion—was used to rapidly expand Richland, which became a federally controlled city with restricted access.

Hanford continued to produce plutonium—and electricity—until 1987. The site had nine nuclear reactors by 1964. After shutdown in 1988, a long decommissioning and decontamination period followed. This task, with a $12.2 billion price tag, continues today.

Map 486 (*left*).
Part of the street plan of Richland in 1950. The Army laid out the streets and named them after Army engineers; the streets were then grouped into alphabetical sequences. Standardized house designs—so-called alphabet houses—had designs from A to Z and included duplexes, apartments, and dormitories for single people.

Above is an aerial photo of the trailer city at Hanford, about fifteen miles north of Richland, in 1944. There were 3,600 trailers, and it was said to be the world's largest trailer park.

Left. By February 1948, the building of standardized houses was well under way in Richland.

Map 487 (*left*).
For several decades after World War II the Hanford site was a *Prohibited Area*. This is a 1952 aviation map. Hanford was in the *Seattle Air Defense Identification Zone*. See also Map 501, page 203.

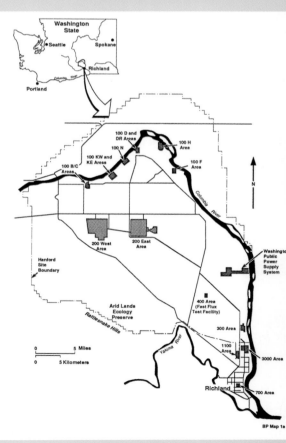

Map 488 (*right*).
Reactor and other sites at Hanford. The *Columbia River* flows through the middle of the site. *100 B/C Areas* is the location of the first nuclear reactor illustrated at *far left, top*. This map is from a 1992 government report.

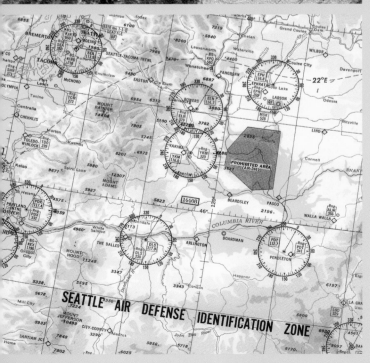

Apples, Apples

Apples grow well on suitably watered land in Central Washington, and this fact was not lost on railroads and other land promoters. Between about 1905 and 1915 what has been called an "apple craze" gripped Washington, and so-called fruit land was sold in large amounts to prospective farmers, attracted by the apparently small amount of work required to produce a fortune. Of course, such tales are never true, and as more land came into apple production, the glut of apples on the market resulted in abandoned orchards.

But Washington apples were good apples, and by the 1920s the state held top place in the nation's production, and cooperatives marketed apples to the East while the railroads carried the produce to the consumer. The planting of large, market-dominating varieties such as Delicious—later Red Delicious—and the development of irrigation in the Yakima Valley, around Wenatchee, and in the Okanogan Valley led to an increase in apple acreage. These locations are today the center of the nation's apple industry.

MAP 491 (*below*).
The Washington State Apple Advertising Commission (now the Washington Apple Commission) was created in 1937 to promote the apple industry. It published this promotional map in 1948 displaying apple varieties then popular.

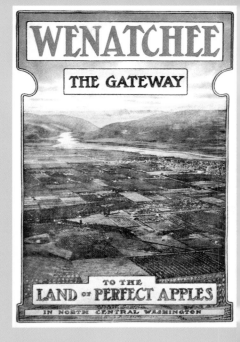

WENATCHEE THE GATEWAY
TO THE LAND OF PERFECT APPLES
IN NORTH CENTRAL WASHINGTON

MAP 489 (*left*).
This apple crate label from the 1940s incorporates a map of Washington.

MAP 490 (*above*).
Promoting apple-growing land near Wenatchee, this 1910 brochure included a bird's-eye view—not an aerial photo—of orchards on its front cover.

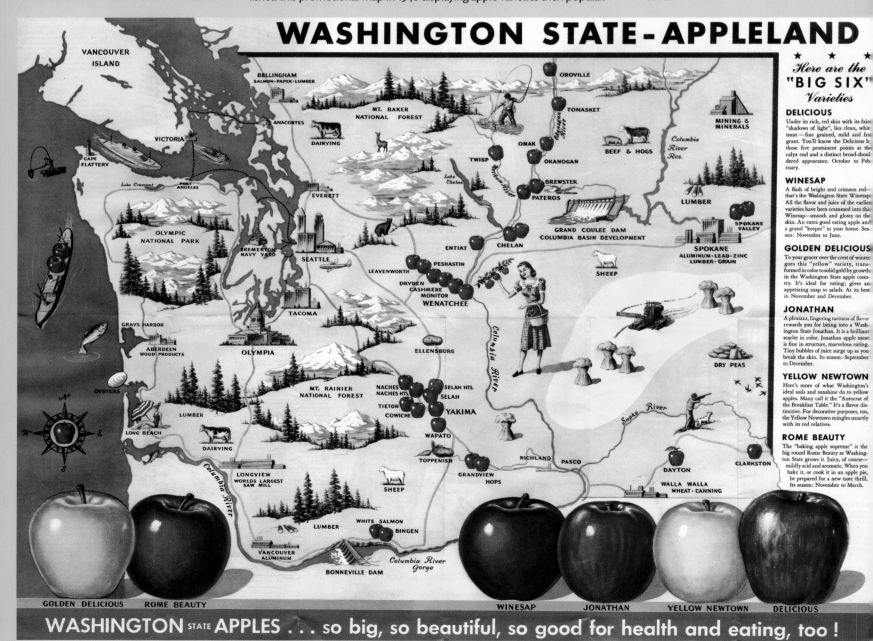

WASHINGTON STATE - APPLELAND

Here are the "BIG SIX" Varieties

DELICIOUS
Under its rich, red skin with its faint "shadows of light," lies clean, white meat—fine grained, mild and fragrant. You'll know the Delicious by those five prominent points at the calyx end and a distinct broad-shouldered appearance. October to February.

WINESAP
A flash of bright and crimson red—that's the Washington State Winesap. All the flavor and juice of the earlier varieties have been crammed into this Winesap—smooth and glossy on the skin. An extra good eating apple and a grand "keeper" in your home. Season: November to June.

GOLDEN DELICIOUS
To your grocer over the crest of winter goes this "yellow" variety, transformed in color to solid gold by growth in the Washington State apple country. It's ideal for eating; gives an appetizing snap to salads. At its best in November and December.

JONATHAN
A pleasant, lingering tartness of flavor rewards you for biting into a Washington State Jonathan. It is a brilliant scarlet in color. Jonathan apple meat is fine in structure, marvelous eating. Tiny bubbles of juice surge up as you break the skin. Its season: September to December.

YELLOW NEWTOWN
Here's more of what Washington's ideal soils and sunshine do to yellow apples. Many call it the "Autocrat of the Breakfast Table." It's a flavor distinctive. For decorative purposes, too, the Yellow Newtown mingles smartly with its red relatives.

ROME BEAUTY
The "baking apple supreme" is the big round Rome Beauty as Washington State grows it. Juicy, of course—mildly acid and aromatic. When you bake it, or cook it in an apple pie, be prepared for a new taste thrill. Its season: November to March.

GOLDEN DELICIOUS   ROME BEAUTY   WINESAP   JONATHAN   YELLOW NEWTOWN   DELICIOUS

WASHINGTON STATE APPLES . . . so big, so beautiful, so good for health and eating, too!

AVIATION NORTHWEST

Early maps for aviation purposes were the responsibility of the Army and Navy until 1925, when they were turned over to the U.S. Coast and Geodetic Survey (USCGS). Many of that agency's first maps were of the Northwest, a special favorite for experimental maps seemingly the Columbia Gorge (MAP 494, *below*). Early planes navigated by matching the terrain beneath them to the map in the cockpit, so an up-to-date map was critical. Some of the developments in early aviation maps of the Northwest are shown on these pages, as aviation went from daytime flying using only roads and rail lines as markers, to night-flying using lights and then radio beams. The modern era of passenger airline service dates from the mid-1940s, when Seattle's Boeing introduced the Boeing 307 Stratoliner, the first airplane to have in-service pressurization, air-conditioning and heating; it was the first practical safe high-altitude passenger airplane.

MAP 492 (*above*).
Part of a 1919 U.S. aviation map, which shows airways only as broad paths many miles wide.

MAP 493 (*right*).
An early strip map for flights between *Portland* and *Spokane*. Published in 1931, it shows the transition from revolving red lights to radio beacons, both of which are shown on this map. *Inset* is an enlargement of the Portland area showing Swan Island Airport—also illustrated in the aerial photo—which the key reveals has a revolving light beacon. This airport was completed in 1930 and required the river's main channel to be switched from the east to the west side of the island. Charles Lindburgh landed here in 1927, before the airport was finished. Also on the inset is *Pearson Vancouver* airport. Pearson was the end point of the first aerial crossing of the Columbia River when in 1905 stunt pilot Lincoln Beachey flew a dirigible to Vancouver Barracks from the grounds of the Lewis and Clark Exposition. Pearson has been the site of a number of milestones in aviation history, notably the end of the first nonstop transpolar flight by the Russian aviator Valery Chkalov in 1937.

MAP 494 (*below*) and MAP 495 (*below, bottom*).
This pair of strip maps of the Columbia Gorge date from 1931 and 1934, respectively, and are some of the earliest created by the USCGS. Early planes flew *in* the gorge. The 1931 map has light beacons, whereas the 1934 map has radio beams. The 1931 light beacons follow the marine navigational practice of green to the left (port) and red to the right (starboard) going upstream—sailors always remember the dictum: Red, Right, Returning. [Continued overleaf.]

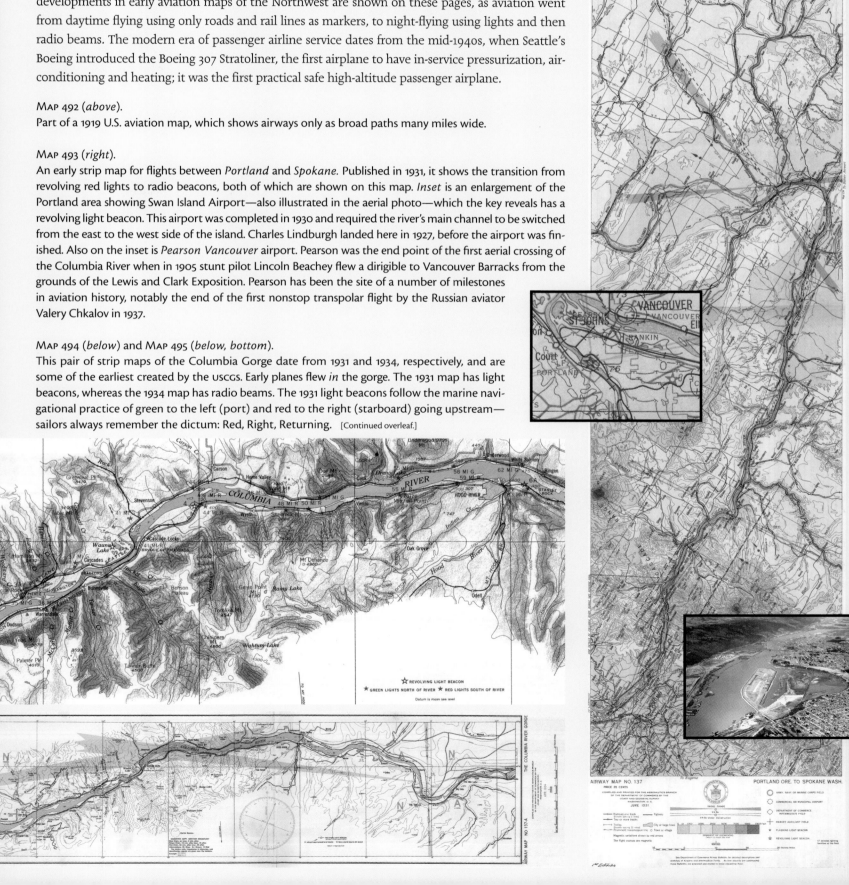

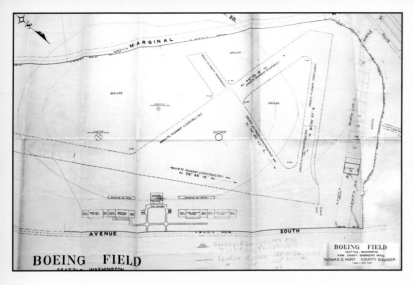

BOEING FIELD
SEATTLE·WASHINGTON

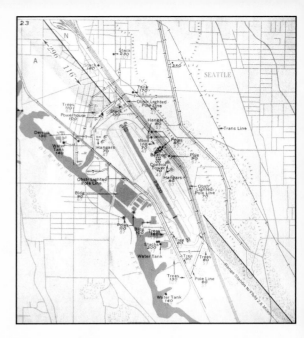

MAP 496 (*left*).
An early map of Boeing Field, in Seattle, showing constructed and proposed runways. The airport was the city's main passenger airport from its construction in 1928 until the military took control during World War II. This map dates from about 1931 and seems to have a lighting plan drawn on it in red pencil. Note the *Warming up Strip*.

Strip maps were discontinued after 1937 because there was too much redundancy and overlap with other maps. The USCGS then developed a comprehensive coverage system based on latitudinal bands, called sectionals. During planning for World War II the system was enlarged and is today still used for the World Aeronautical Charts that cover much of the globe on a 1 to 1 million scale.

MAP 497 (*below*).
The civilian aeronautical chart in 1946, produced by the U.S. Coast and Geodetic Survey. This is part of a sheet covering the western United States. The symbols used are similar to those on the earlier maps, with radio beams used for direction finding prominently displayed in red. Interestingly, although by this time Hanford was in production, there is no restriction noted on planes flying in the area.

MAP 498 (*above*), MAP 499 (*above, right*), and MAP 500 (*above, right, center*).
Maps of Boeing Field, Sea-Tac, and Spokane airports in 1946. All, in the pre-jet age, have considerably shorter runways than they do today. Boeing Field's, here 7,520 feet, is now 10,001 feet. At Sea-Tac, where only north–south runways remain, there are now three, ranging from 8,500 to 11,900 feet in length (the longest having been opened in 2008), compared with the longest at 5,610 feet on this map. Spokane is now a general aviation airport. Spokane International Airport was a field to the west of the city designated Spokane's municipal airport

MAP 501 (*right, center*).
There is a large prohibited zone around the Hanford nuclear plant on this 1956 route map, with the airways avoiding it. The whole is within the *Western Defense Area*, protected by interceptor jets stationed at airports and air bases around the region, such as at *Spokane*. Radio beacons are noted, as on a modern aviation map.

MAP 502 (*far right, top*).
A 1932 Air Corps map of the Seattle–Tacoma area.

MAP 505 (*far right, bottom*).
The nearest thing Washington and Oregon ever had to a regional airline was West Coast Airlines. Here the airline ad shows its route map in 1962. The airline was founded in 1942 and based at Boeing Field. It became Air West through mergers in 1968, and Hughes Air West in 1970. Today it is part of Delta Airlines.

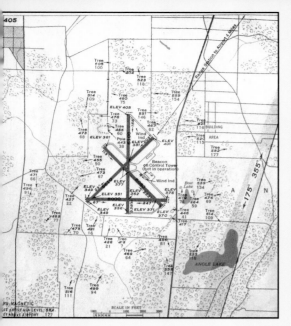

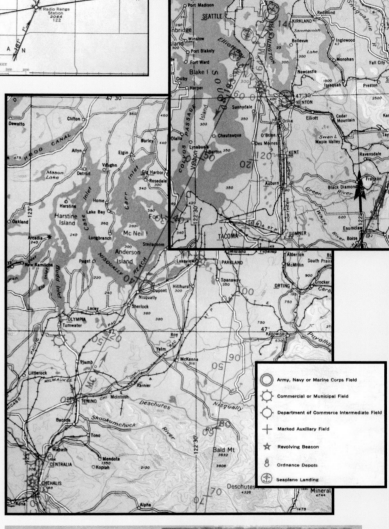

in 1946; it was used during the Cold War to defend Hanford and the Grand Cou-lee Dam. The city airport shown here is now Felts Field. Construction began on Sea-Tac Airport in 1944 after the Boeing Field was taken over for military use. The first scheduled flight from Sea-Tac took place in 1947. The airport was renamed Seattle–Tacoma International Airport in 1949 following the commencement of direct flights to Tokyo by Northwest Airlines that year.

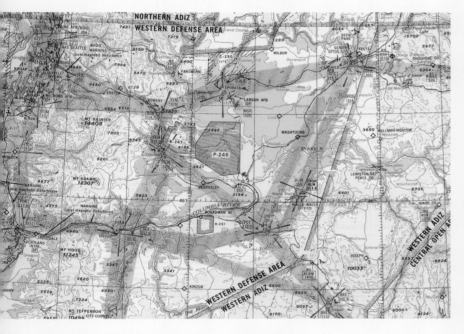

MAP 503 (*below, left*) and MAP 504 (*below, right*).
The USCGS did a lot of experimentation to determine what colors were most useful to pilots flying at night, when they would only have a red light in the cockpit. These were two experiments carried out in 1940, uniquely on the Seattle sheet—violet on buff, and violet on white.

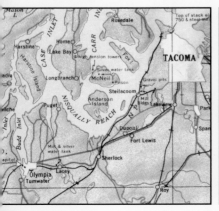

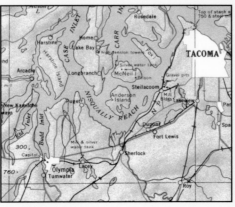

EVACUATE

DON'T SIT UNDER THE MUSHROOM

SEATTLE & KING COUNTY
CIVIL DEFENSE DEPARTMENTS

THE COLD WAR

Signs marking the location of nuclear bomb shelters used to be a ubiquitous feature of all American cities. For a decade or more Americans lived with the threat of nuclear war, a threat that cumulated in October 1962 with the Cuban Missile Crisis.

Civil defense planning was one of the critical components in ensuring that as many people as possible might survive a nuclear attack, up to about 1959 expected to be launched from long-range bombers and, after that, using intercontinental ballistic missiles (ICBMS). Maps and instructions such as those shown here were issued to the populations of most major cities in the United States. In addition, one of the principal rationales for building the interstate freeway system was to permit rapid evacuation of cities under nuclear attack.

The underlying assumption of all these plans was that the city centers would be the primary targets. That may or may not have been the case, and even if targeting city centers were the intention of the attackers, they might not have been able to ensure their missiles would be accurate. Luckily, we never had to find out.

MAP 506 (*right, top*) and MAP 507 (*right*).
The Civil Defense evacuation plan for Seattle and for King County, distributed in a single map and information brochure in 1955. The cover of the plan is shown *above*. The emphasis was on getting people away from the city center, the presumed target, by any means, walking if necessary. The target area was defined as a circle with a thirteen-mile radius and is indicated on MAP 507. If a bomb fell at the city center, more people would survive the farther they were away.

MAP 508 (*below*).
Frank S. Evans, local director of civil defense, stands in the new civil defense control center on Fawcett Avenue in Tacoma on 5 May 1953, built to manage rescue and evacuation efforts in case of a nuclear strike on Tacoma. The center was constructed of reinforced concrete and was considered bombproof. The nameplates would have been placed on the desks of individuals responsible for coordinating each function. The maps were on metal bases so that magnetic markers could be placed on them. Hanging on the wall at right is a transparent set of concentric circles, presumably intended for placing on the map at the location a nuclear bomb fell.

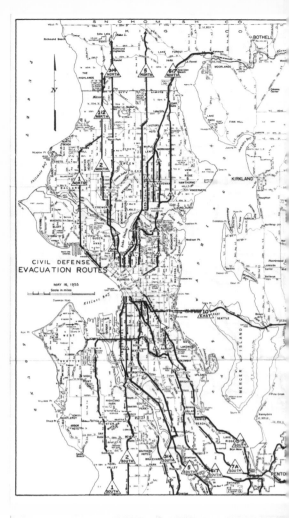

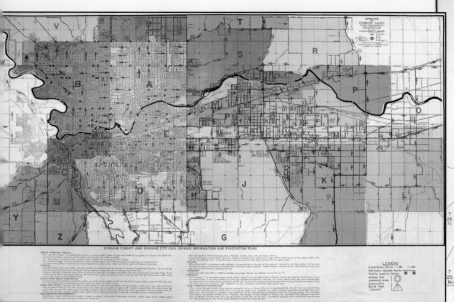

MAP 509 (*above*) and MAP 510 (*right*).
The evacuation plans for Spokane and
Spokane County, distributed in 1955.
Roads would have been converted to
one-way so as to speed the evacuation.

MAP 511 (*below*) and MAP 512 (*below, right*).
Plans for the evacuation of Clark County
and Portland published in early 1956 and
1955, respectively. The Portland brochure
cover is at *right*. Many facets of these
plans seem in retrospect to have been
dubiously feasible. It is likely that one of
their intentions was to prevent panic and
permit an orderly evacuation of as many
people as possible, recognizing that many
would not make it. How precisely the influx of population would have survived
in the surrounding countryside is also
not known; it would have all depended
on the extent of the attack. In all probability millions would have died and the
land laid waste for centuries.

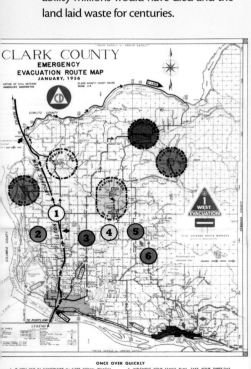

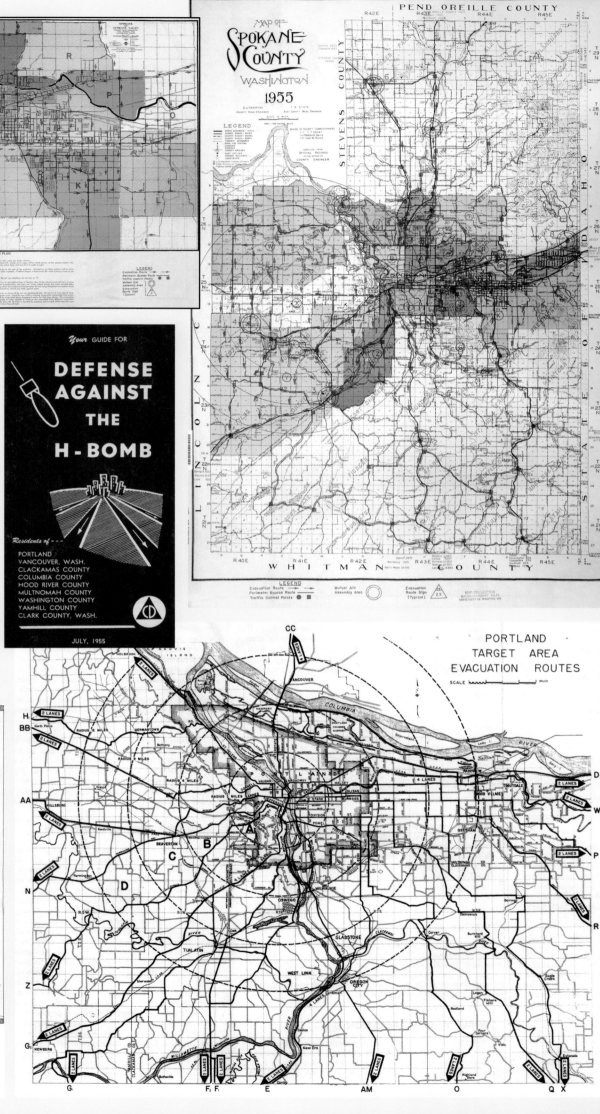

A World's Fair

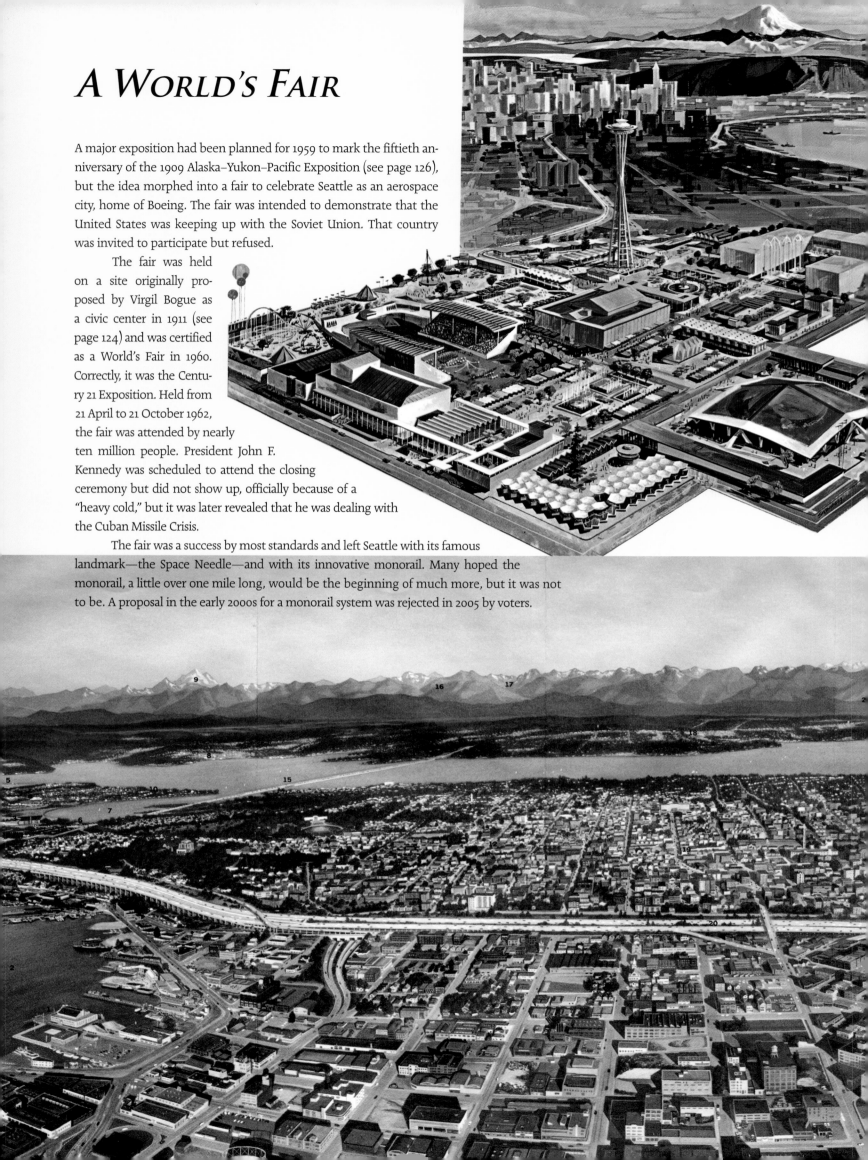

A major exposition had been planned for 1959 to mark the fiftieth anniversary of the 1909 Alaska–Yukon–Pacific Exposition (see page 126), but the idea morphed into a fair to celebrate Seattle as an aerospace city, home of Boeing. The fair was intended to demonstrate that the United States was keeping up with the Soviet Union. That country was invited to participate but refused.

The fair was held on a site originally proposed by Virgil Bogue as a civic center in 1911 (see page 124) and was certified as a World's Fair in 1960. Correctly, it was the Century 21 Exposition. Held from 21 April to 21 October 1962, the fair was attended by nearly ten million people. President John F. Kennedy was scheduled to attend the closing ceremony but did not show up, officially because of a "heavy cold," but it was later revealed that he was dealing with the Cuban Missile Crisis.

The fair was a success by most standards and left Seattle with its famous landmark—the Space Needle—and with its innovative monorail. Many hoped the monorail, a little over one mile long, would be the beginning of much more, but it was not to be. A proposal in the early 2000s for a monorail system was rejected in 2005 by voters.

MAP 513 (*above, left*).
There were seemingly hundreds of brochures, maps, ads, and the like published for the Seattle World's Fair in 1962. This bird's-eye map of the site was one of the best. It was the work of the Washington State Department of Commerce and Economic Development, hoping, no doubt, to leverage the goodwill of the fair into new businesses and new jobs.

MAP 514 (*below, left*), looking east; **MAP 515** (*above*), looking north; **MAP 516** (*below*), looking west; and **MAP 517** (*below, bottom*), looking south.
Four stunning panoramic views in each direction from the top of the Space Needle. Mount Baker can be seen looking north, while Mount Rainier looms large on the horizon looking south. The newly completed Interstate, I-5, is prominent. **MAP 514** looks across Lake Washington, while **MAP 515** shows Lake Union and the Lake Washington Ship Canal prominently across the middle foreground. **MAP 516** largely consists of Elliott Bay. **MAP 517** has downtown Seattle on the left, and the view otherwise looks directly up the Duwamish River, with the reclaimed tidelands, including Harbor Island, very obvious at the river's mouth. The views were all the same size in the original brochure, and together they gave an excellent feel for the city that was Seattle in 1962.

SUPERHIGHWAYS

Drivers have always clamored after better roads. At first it was to get them at all, then to get them "improved"—graded, graveled, or oiled—then paved. And as cars became capable of higher speeds, a demand arose for roads capable of handling them. At first it was for sections of road, junctions, and interchanges where the traffic was particularly heavy or complex, and then for whole roads, a process culminating with what were referred to as superhighways—today's freeways.

The interstate system of freeways dates from 1956, when the federal government passed the Federal-Aid Highway Act, which projected a 41,000-mile network of freeways across the nation. Partly the result of defense planning for possible mass evacuations of cities, the system was renamed in 1993 to recognize this role and its main proponent, becoming the Dwight D. Eisenhower National System of Interstate and Defense Highways. The Northwest got its share of the freeways: I-5, completed about 1972; I-84 by 1975; I-90 by 1984; and I-82, which connects I-84 and I-90 via the Yakima Valley, in 1988. Auxiliary freeway I-405 in Portland was completed in 1973; I-405 in Seattle in 1971; and Portland's I-205 in 1983.

As early as 1938 a superhighway had been proposed for the approximate route now followed

MAP 521 (below, right).
The Northwest section of the proposed *National System of Interstate Highways* from the federal government's "yellow book" of 1955. The roads shown were on a 1947 map reprinted in the 1955 publication. These roads formed the initial interstate freeway network. The routes of Interstates 5, 90, and 84 are shown.

MAP 518 (above).
Drawn in 1943 by Frank Hutchinson of the Oregon Department of Transportation (see page 113) this bird's-eye depicts a proposed interchange at Front Avenue and Harbor Drive onto the Steel Bridge in Portland. In an unusual move, Harbor Drive was replaced by Governor Tom McCall Waterfront Park in 1978.

MAP 519 (below).
This 1938 Department of Transportation planning sketch shows almost exactly the later route of two superhighways, I-5 and I-90.

MAP 520 (bottom left).
Another Hutchinson drawing, this one from 1950, shows main roads leading into Eugene across the Willamette.

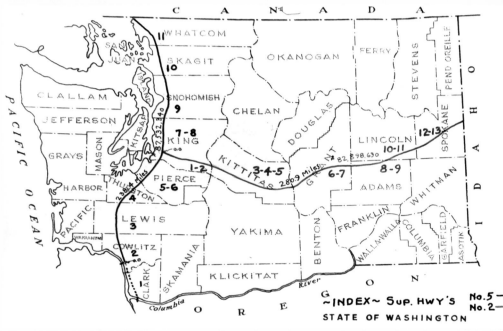

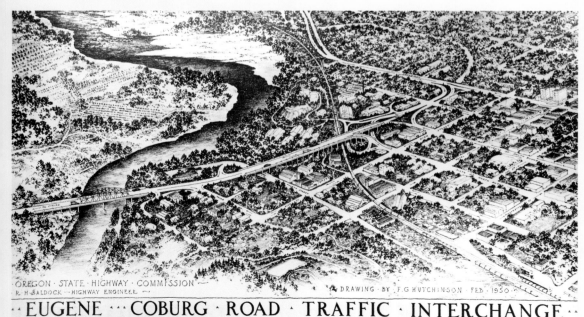

EUGENE ·· COBURG · ROAD · TRAFFIC · INTERCHANGE

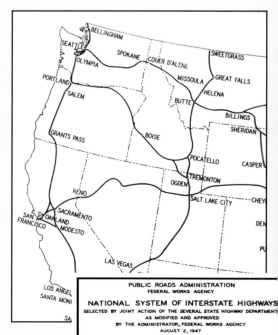

Portland's skyline presents this scene to motorists using the new concrete Baldock approach to the heart of the city.

Aesthetic value of Oregon's farmland is enhanced by concrete ribbon stretching across the McKenzie River North of Eugene.

Final connection of famed coastal highway will be made by bridge across Columbia River at Astoria.

Impressive twin structures at Ontario serve as gateway to Oregon from all points East.

Faster construction is possible through advanced equipment, shown at work on concrete section of Pacific Highway.

Network of interchanges for Medford is typical of interstate links with key Oregon cities.

MAP 522 (*above*).
This promotional pictorial map was created by the Cement Industry of Oregon in 1956, just as work on freeways was beginning.

MAP 523 (*right*).
The location of the *Proposed Tacoma–Seattle–Everett Toll Road*, from a 1955 report.

by I-5 and I-90 (MAP 519, *left*). Both in Washington and Oregon the Pacific Highway (Route 99) already approximated this line, and indeed the interstate to come would in many places follow that alignment, upgrading the road to freeway level.

A superhighway had been proposed as a toll road running through the dense urban agglomeration of the eastern shore of Puget Sound. The Tacoma–Seattle–Everett Toll superhighway was approved by the Washington Legislature in 1953 (MAP 523, *right*). The road was originally intended to be a four-lane divided highway of some sixty-five miles, though from the beginning extra lanes were provided for in heavy-traffic locations, as were reversible lanes.

The use of tolls was declared unconstitutional by the Supreme Court in 1956, but the financial issues were solved by the 1956 Federal-Aid Highway Act, and the first section of the new freeway, between Everett and Seattle, opened in 1965. The rest of the interstate in Washington followed, alternating freeway and Highway 99 segments (MAP 525 to MAP 528, *overleaf*), but the last sections were not fully to freeway standards until 1972. The eastside Seattle freeway, I-405, was likewise an upgrade of largely existing roads. The first freeway section, from Renton to Tukwila, opened in 1965, and in 1971 the entire route was renumbered from SR 405 to I-405 following completion of the upgrade to freeway standards.

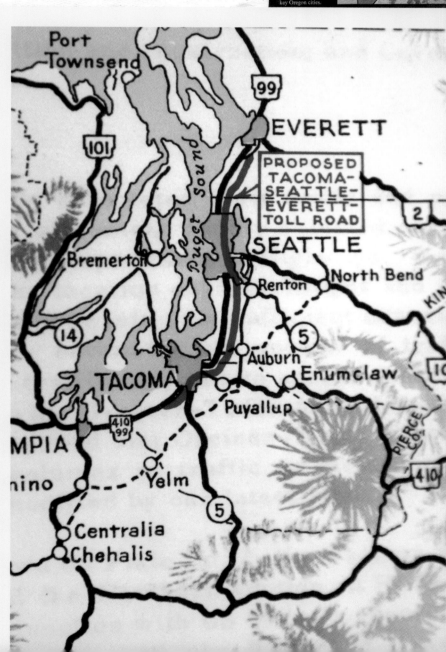

TACOMA–SEATTLE–EVERETT TOLL ROAD
LOCATION MAP–SEATTLE

LEGEND

PROPOSED TOLL ROAD
INTERCHANGES
CITY STREETS

MAP 524 (*above, top*).
The location of the proposed Tacoma–Seattle–Everett Toll Road through Seattle, from a 1955 report.

MAP 525, MAP 526, MAP 527, and MAP 528 (*above, from left to right*). This series of road maps shows how I-5 progressed, though the information on the maps tends to be a little ahead of their dates, the result of the desire to keep them up-to-date as long as possible. The map dates are 1961, 1962, 1962, and 1964.

MAP 529 (*right*).
Although certainly not a freeway, an important addition to the road system of the Northwest was completed in 1972: another crossing of the Cascades. This was the North Cascades Highway (SR 20), which is still typically shut down for some of the winter owing to avalanche danger and the difficulties of keeping the road clear of snow. The road was, like the freeways, an upgrade of an existing route, though this one previously only a trail in many places. A road had been built as far as Diablo in the 1930s to access the Seattle

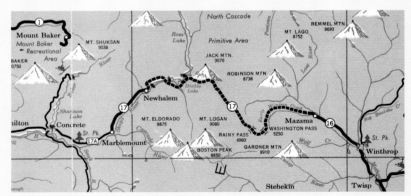

City Light dams constructed during this period (see page 186). State funds were allocated for a road in 1958, and construction of a through road began in 1959. Supporters of the road included logging interests, and it was thought that logging would help pay for the road, but this possibility ended when the North Cascades National Park was created in 1968. Governor Daniel J. Evans opened the road in September 1972—but it was closed again a few weeks later for winter. This map shows the proposed route in 1962.

I-90 in Washington was completed about 1984 and was part of the longest interstate in the United States, totaling some 3,099 miles to Boston, Massachusetts. Most of the Washington route was US-10 from 1926, itself a compilation of earlier roads. Highway 10 was absorbed into I-90 by 1969, but the Seattle section of the freeway was halted by lawsuits in 1970 and was not fully completed for another two decades. The freeway uses twin floating bridges across Lake Washington (see overleaf).

In Oregon, freeways likewise often followed the routes of existing roads. In 1953 the Portland–Salem Expressway had been authorized (MAP 539, *page 214*), and the Federal-Aid Highway Act simply made more federal monies available for it. The expressway, unlike its Washington counterpart, followed Oregon policy and had been planned to be toll free from its inception. An important component of I-5 was the Columbia crossing, and a second span of the Interstate Bridge was completed in 1958 (MAP 538, *page 214*). I-5 in Oregon was completed to four-lane standard from Washington to California by 1966, the same year the Marquam Bridge in Portland was completed to carry I-5 across the Willamette, a bridge so utilitarian that it had sparked a formal protest by the city's arts community.

Also under construction by the time the federal program was announced was the Banfield Expressway, now part of I-84 in Portland, the freeway east along the Columbia Gorge, completed in 1975.

The auxiliary I-405, which loops around the west side of downtown, was effectively finished in 1973 with the completion of the Fremont Bridge across the Willamette, and as a result of the previous protest it was a more aesthetically pleasing bridge to most eyes. The east side auxiliary, the I-205, rejoins I-5 in Washington and required another crossing of the Columbia. The Glenn Jackson Bridge for this link was completed in December 1982 and the final sections of the freeway four months later.

MAP 530 (*above, right*).
The proposed Tacoma–Seattle–Everett Toll Road, from a 1955 report. This was the alignment ultimately followed by I-5.

MAP 531 (*right*).
Planning for a spaghetti of access ramps and junctions for I-90 in Bellevue in 1983.

MAP 532 (*below*).
The complex western end of Interstate 90, which was not completed until 1992, owing to lawsuits outstanding since 1970.

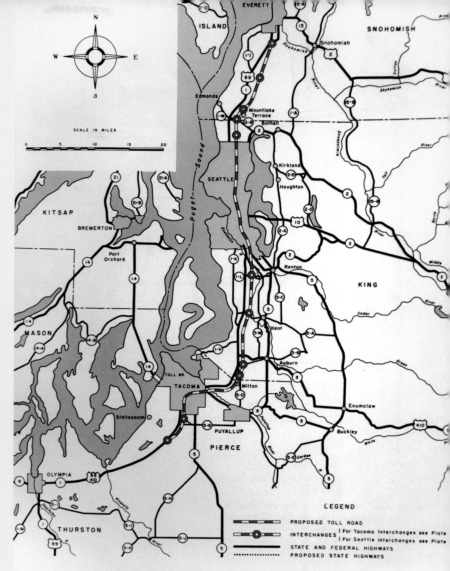

TACOMA - SEATTLE - EVERETT TOLL ROAD
LOCATION MAP

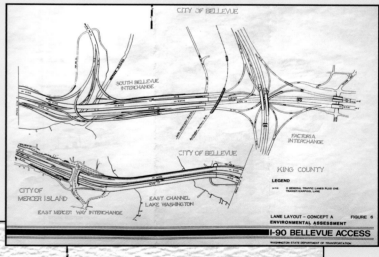

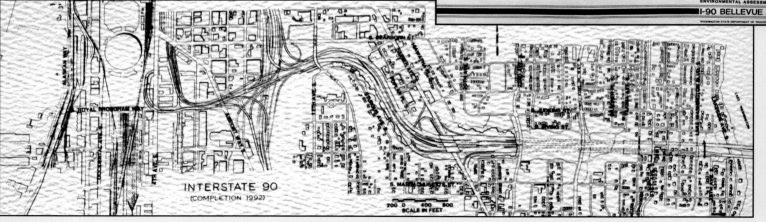

INTERSTATE 90
(COMPLETION 1992)

Floating Bridges to Everywhere

For decades there was tension between Black Ball ferries and the State of Washington, leading to the state takeover of the ferry system in 1951 (see page 161). But for many of the routes, the ferries were intended as just a stopgap measure, for state engineers had discovered a wonderful new device—the floating bridge—that seemed to solve the problem of building bridges where, because of the scouring of glaciation, the water was too deep. One had been used to cross Lake Washington in 1940 (see page 178).

In 1963 a second lake crossing was completed, the Evergreen Point Floating Bridge, and connected with the then-complete central section of I-5 (MAP 533, *right*). The bridge is 1.4 miles long and is one of the longest floating bridges in the world. In 1989 the twin span of the first bridge, the Homer M. Hadley Memorial Bridge, was added to convert the route to the I-90.

Engineers and politicians thought the technology could be used to create a bridge across Puget Sound from the Seattle metro region, an idea that had been around since at least 1948 but for which formal studies were done in 1965 (MAP 535, *right*), when four possible routes were analyzed.

One saltwater floating bridge was built, the one across Hood Canal (and also shown on MAP 535). This 1.47-mile-long bridge was opened in 1961 (MAP 534, *below*). In February 1979 the floating bridge dream began to sink—

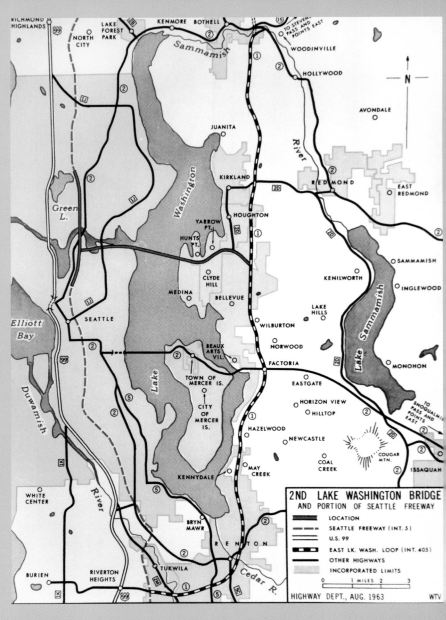

literally. The Hood Canal Bridge sank in a fearsome storm that is thought to have blown open pontoon hatches and allowed water in. It took three years before a replacement could be designed and put in place. And then in November 1990, the original I-90 span sank while undergoing repairs. It took three years to construct a replacement.

MAP 533 (*above*).
The location of the proposed second Lake Washington Bridge is shown on this map in red, from a 1963 study by the Washington State Highway Department.

MAP 534 (*left*).
A fine bird's-eye view of the proposed Hood Canal Bridge produced by the Washington Toll Bridge Authority in 1957.

MAP 535 (*right*).
A map of the cross-sound routes under consideration, from a Washington State Highway Department study published in 1965. Note also *Hood Canal Floating Bridge* at *Port Gamble*.

Above, left, is an image of a floating bridge from the cover of the 1965 cross-sound transportation study.

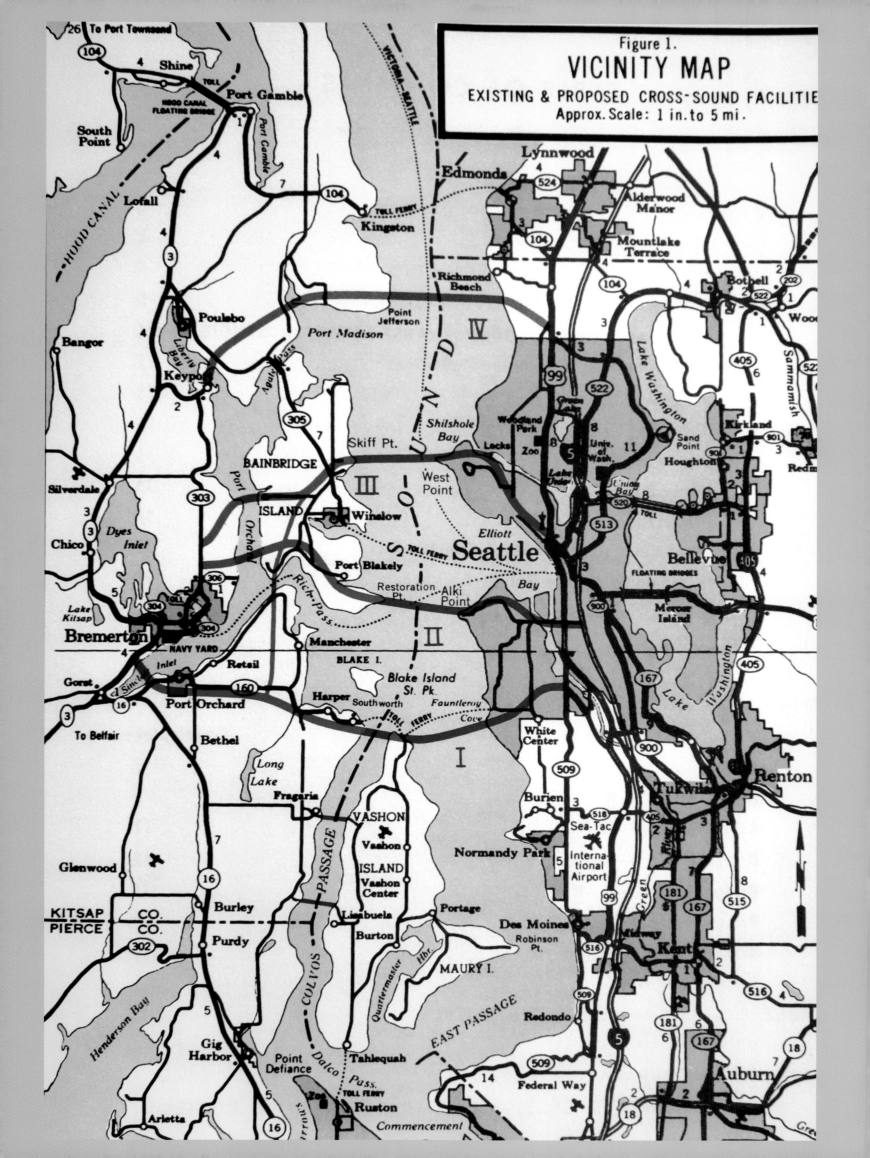

Figure 1.
VICINITY MAP
EXISTING & PROPOSED CROSS-SOUND FACILITIES
Approx. Scale: 1 in. to 5 mi.

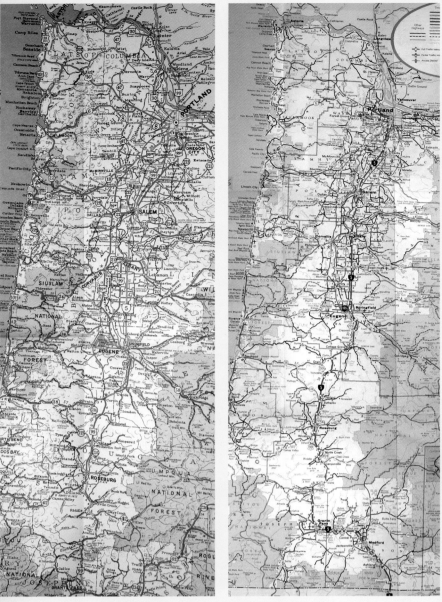

MAP 536 (*above, left*) and MAP 537 (*above, right*).
Road maps from 1961 (*left*) and 1965 (*right*) show the progress of the construction of the I-5 in Oregon.

COMPREHENSIVE DEVELOPMENT PLAN
PORTLAND, OREGON
PORTLAND CITY PLANNING COMMISSION
JULY, 1966   SCALE   NORTH ▲

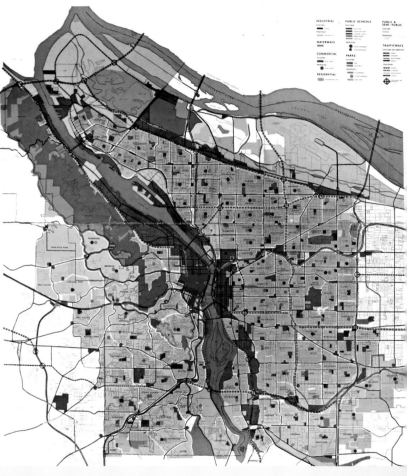

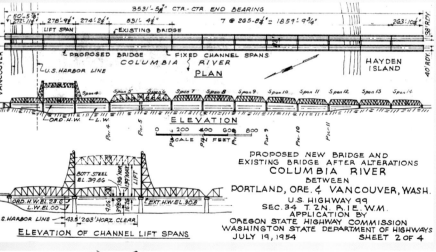

MAP 538 (*left*).
The proposed doubling-up of the Interstate Bridge across the Columbia between *Vancouver* (at left) and Portland (at right) is shown in 1954. The bridge, originally built in 1917, was doubled in 1958. Today it carries I-5, but engineers now consider the bridge design obsolete.

MAP 539 (*left*).
A sketch map of the Portland–Salem Expressway in 1953, showing the diversion from Highway 99E. The map is from a newspaper.

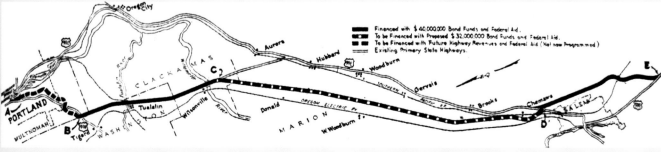

A considerable system of freeways was planned for Portland in the 1960s, but a citizens' revolt against them in the 1970s put an end to most. One catalyst was the proposed Mount Hood Freeway, which would have run parallel to the alignment of Highway 26 (SE Powell Boulevard). It has been calculated that 1 percent of the Portland housing stock would have been sacrificed to build this freeway. The plans for this freeway were finally withdrawn in May 1976. A short section of the freeway was built in Gresham and is today part of the Mount Hood Highway.

MAP 540 (*left, top*).
A model, made in 1962, of the proposed freeway connections at the east end of the Fremont Bridge. The Eastbank Freeway, in the foreground, is I-5, and the connection from the Fremont Bridge became the connection to I-405. The proposed Fremont Freeway, which runs to the right on this model, would have run along the alignment of NE Prescott Street, but the project was cancelled.

MAP 541 (*left, center*).
This comprehensive development plan for Portland summarizes all the freeways proposed to cut through the city in 1966. Four bridges span the Columbia. The existing Interstate Bridge is at Vancouver, Washington, and the easternmost bridge was constructed in 1983, on a similar alignment to the one shown here, as the Glenn Jackson Bridge, carrying the I-205 across the river; the other two proposals were never built.

MAP 542 (*right, top*) and MAP 543 (*right, center*).
A plan, and a dramatic bird's-eye view, of the proposed Mount Hood Freeway in Portland in 1973, slated at that time to be I-80N. Both maps give a very good idea of the degree of disruption of the existing residential neighborhoods that would have occurred. Some mitigation of noise would have been achieved by running the freeway lower than the surrounding residential neighborhoods. This is one of four alignments that were considered.

MAP 544 (*right, bottom*).
The Mount Hood Freeway was to have been carried across the Willamette, where it would have connected with the freeways on the west bank. This fine photograph-like bird's-eye view of the proposed bridge looks east; the gas-holders visible at top can also be seen in MAP 543 in the foreground. The bridge is the Marquam Bridge, which was completed in 1966 to carry I-5 across the river. It is a 1,043-foot, three-span, double-deck continuous truss section bridge, and there were no provisions for pedestrians whatsoever.

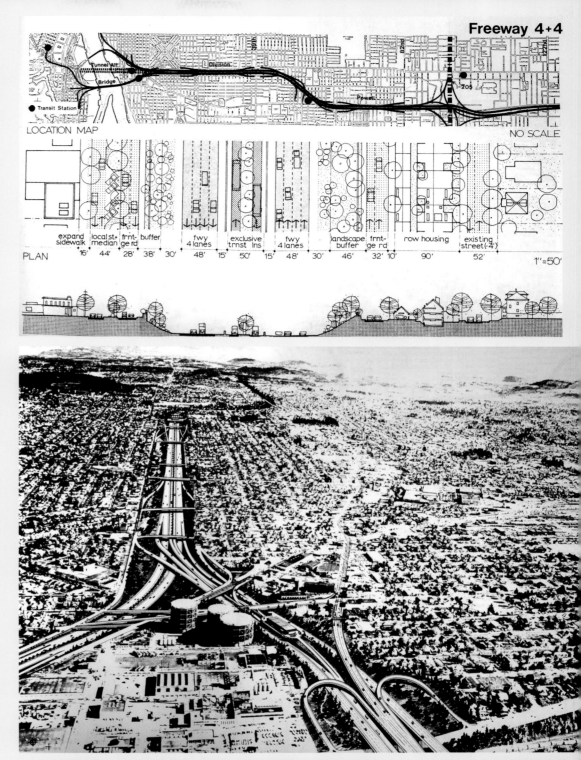

The Northwest Environment

At 8:32 AM on Sunday, 18 May 1980, Mount St. Helens, one of the several volcanoes along the Cascades of the Northwest, erupted in what was the most catastrophic volcanic eruption in U.S. history. Fifty-seven people died; 250 houses, 47 bridges, 15 miles of railroad, and 185 miles of road were destroyed (MAP 549, *right, bottom*).

It was a reminder that we inhabit the Northwest by virtue of nature's acquiescence. We may have molded the environment to suit us, but it is but a superficial molding. The Nisqually Earthquake of 2001, which damaged some of the old buildings in Seattle so carefully rebuilt with brick after the 1889 fire, was another reminder of our fragile hold.

The Northwest remains dependent on harnessing water resources for power and irrigation (MAP 546, *below*) and is now looking more critically at harnessing other natural resources such as wind power or the geothermal energy concomitant with the volcanoes of the Pacific "Rim of Fire" (MAP 548, *right*).

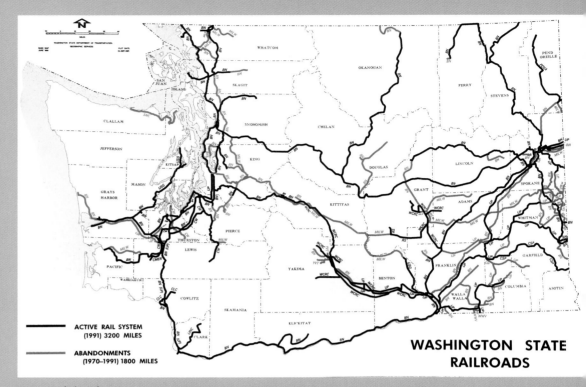

ACTIVE RAIL SYSTEM
(1991) 3200 MILES

ABANDONMENTS
(1970–1991) 1800 MILES

WASHINGTON STATE RAILROADS

MAP 545 (*above*).
The Northwest has continued to lose rail lines as the railroads consolidate and concentrate their businesses on bulk and container transport, having given passenger service to the government's Amtrak in 1971, although some smaller operations have been taken over by so-called short line railroads that serve local needs and feed the larger operations. At the same time, commuter railroads have become more popular, with Seattle and Tacoma, and especially Portland, developing long rail commuter lines such as Portland's MAX system.

MAP 546 (*below*).
A summary of the use of Oregon's water resources published in 1976.

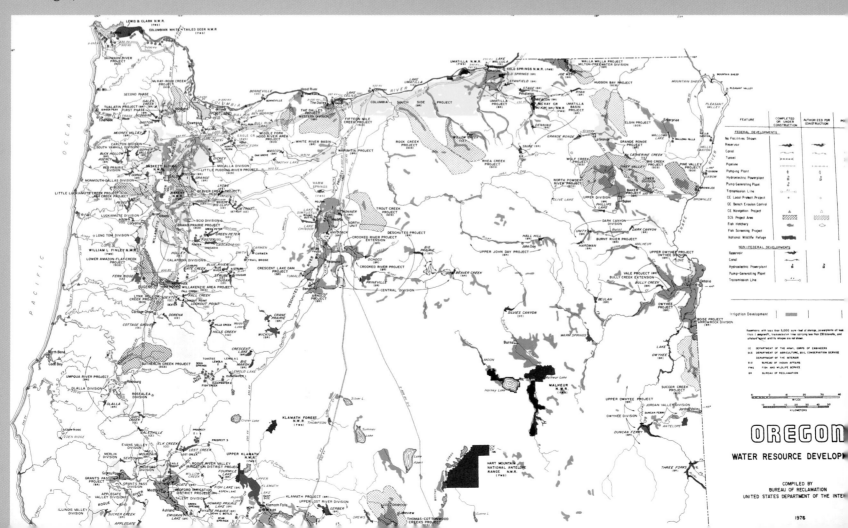

OREGON
WATER RESOURCE DEVELOPMENT

COMPILED BY
BUREAU OF RECLAMATION
UNITED STATES DEPARTMENT OF THE INTERIOR

1976

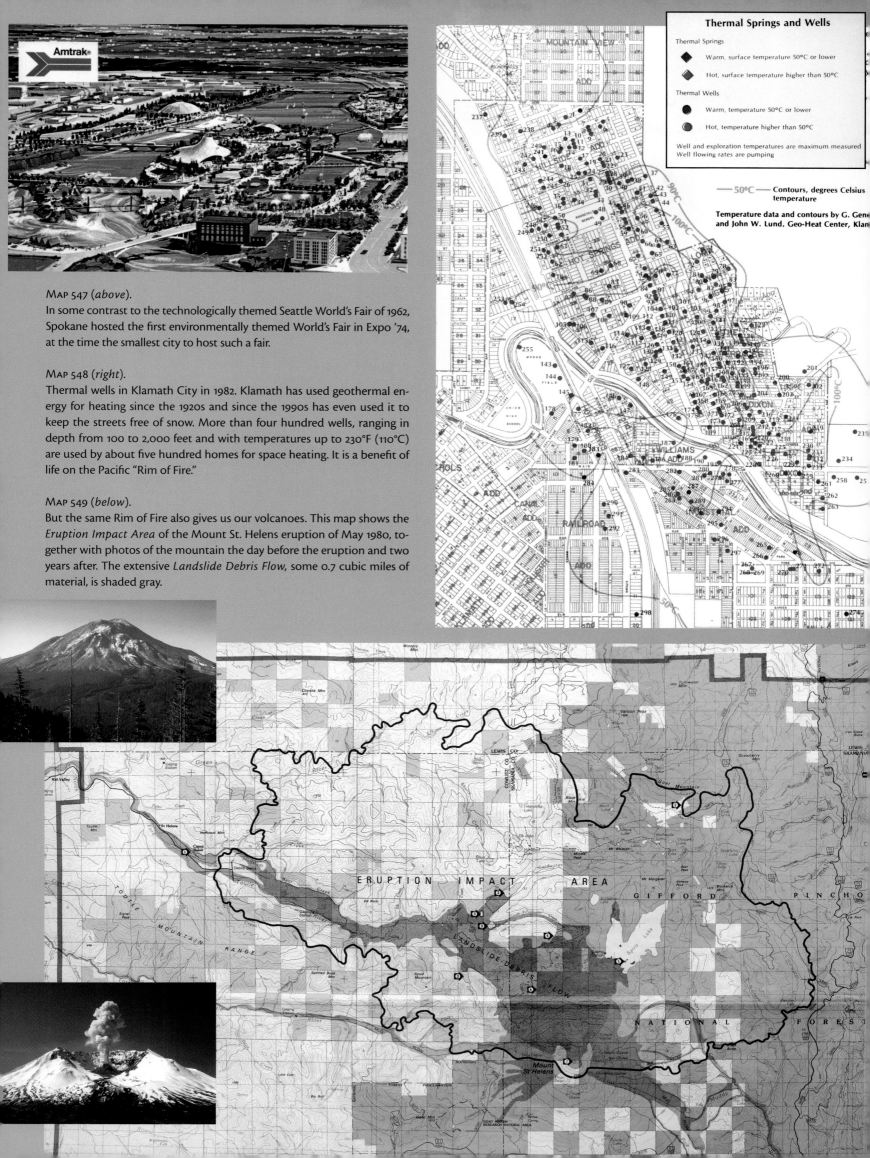

Map 547 (*above*).
In some contrast to the technologically themed Seattle World's Fair of 1962, Spokane hosted the first environmentally themed World's Fair in Expo '74, at the time the smallest city to host such a fair.

Map 548 (*right*).
Thermal wells in Klamath City in 1982. Klamath has used geothermal energy for heating since the 1920s and since the 1990s has even used it to keep the streets free of snow. More than four hundred wells, ranging in depth from 100 to 2,000 feet and with temperatures up to 230°F (110°C) are used by about five hundred homes for space heating. It is a benefit of life on the Pacific "Rim of Fire."

Map 549 (*below*).
But the same Rim of Fire also gives us our volcanoes. This map shows the *Eruption Impact Area* of the Mount St. Helens eruption of May 1980, together with photos of the mountain the day before the eruption and two years after. The extensive *Landslide Debris Flow,* some 0.7 cubic miles of material, is shaded gray.

Thermal Springs and Wells

Thermal Springs

◆ Warm, surface temperature 50°C or lower

◈ Hot, surface temperature higher than 50°C

Thermal Wells

● Warm, temperature 50°C or lower

◐ Hot, temperature higher than 50°C

Well and exploration temperatures are maximum measured Well flowing rates are pumping

— 50°C — Contours, degrees Celsius temperature

Temperature data and contours by G. Gene and John W. Lund, Geo-Heat Center, Klam

MAP CATALOG & SOURCES

Many maps do not have sources quoted because they are from the collection of the author; some are from other private collections.

MAP 1 (*half-title page*).
De Vereenigde Staten van Noord Amerika
Charles van der Linde (?), c. 1838
James Walker Collection

MAP 2 (*title page*).
A Map Exhibiting all the New Discoveries in the Interior Parts of North America (with extensive manuscript additions)
Aaron Arrowsmith, 1824, with additions by HBC employees
Hudson's Bay Company Archives G4/31

MAP 3 (*copyright page*).
Sketch map of Oregon, with letter
James O. Raynor, 1859
James Walker Collection

MAP 4 (*page 6*).
Map of the State of Oregon
Title Guarantee and Trust Co./Huber and Maxwell Civil Engineers, 1904

MAP 5 (*page 7*).
Washington, The Evergreen State
State of Washington, 1931

MAP 6 (*page 7*).
1836 Oregon Territory 1936
U.S. postage stamp, 1936

MAP 7 (*page 9*).
Map of Oregon Showing the location of Indian Tribes 1852
Washington State University wsu588

MAP 8 (*page 10*).
Map of the Indian Nations and Tribes of the Territory of Washington and of the Territory of Nebraska West of the Mouth of the Yellowstone. Made under the direction of Isaac I. Stevens, Gov. of Wash. Terr. & Sup't of Ind. Affairs, March 1857.
William H. Carlton, 1857
University of Washington Map Libraries uwm9

MAP 9 (*page 11*).
Distribution of Tribes of the Upper Columbia Region in Washington, Oregon and Idaho: including all those of the Smohalla and Shaker religions
James Mooney, 1894
University of Washington Map Libraries uwm112

MAP 10 (*page 11*).
Map Showing the Distribution of the Indian Tribes of Washington Territory
Department of the Interior, 1876
National Oceanic and Atmospheric Administration Central Library

MAP 11 (*page 12*).
Univeri Orbis Sev Terreni Globi Plano Effigies
From: Gerard de Jode, *Speculum Orbis Terrarum*, 1578
Library of Congress

MAP 12 (*page 12*).
Universalis Cosmographia Secundum Ptholomaei Traditionem et Americi Vespucci Aliou[m]que Lustrationes
Part of main map and inset map
Martin Waldseemüller and Matthias Ringmann, 1507
Library of Congress G3200 1507 .W3 Vault

MAP 13 (*page 12*).
Limes Occidentis Quivira et Anian
From: Cornelius Wytfliet, *Descriptionis Ptolemaicae Augmentum sive Occidentis*, 1597

MAP 14 (*page 13*).
Quiviræ Regnu cum alijs versus Borea
From: Cornelis de Jode, *Speculum Orbis Terrarum*, 1593

MAP 15 (*page 14*).
Typus Orbis Terrarum
Abraham Ortelius, 1570
From: *Theatrum Orbis Terrarum*
Library of Congress
G1006 .T5 1570b Vault

MAP 16 (*page 14*).
Vera Totius Expeditionis Nauticae descriptio D. Franc. Draci
Joducus Hondius, 1595
Library of Congress
G3201.S12 1595 .H6 Vault

MAP 17 (*page 15*).
Hydrographic chart of the Washington Coast
U.S. Coast and Geodetic Survey, 1932

MAP 18 (*page 15*).
Untitled map illustrating the voyage of Juan Francisco de la Bodega y Quadra, 1775
From: Daines Barrington, *Miscellanies*, 1781

MAP 19 (*page 15*).
Carta particolare della stretto di Iezofra l'America e l'Isola Iezo D'America
From: Robert Dudley, *Dell'Arcano del Mare*, 1647

MAP 20 (*page 15*).
New Map of Part of North America
Joseph La France, 1744
From: Arthur Dobbs, *An Account of the Countries Adjoining to Hudson's Bay*, 1744

MAP 21 (*page 16*).
North America with Hudson's Bay and Straights Anno 1748
Richard Seale, 1748
Hudson's Bay Company Archives G4/20b

MAP 22 (*page 16*).
Amerique Septentrionale
Gilles and Didier Robert de Vaugondy, 1762

MAP 23 (*page 16*).
Amerique Septentrionale
Jean Janvier, 1762
Library of Congress G3300 1762 J31 vault

MAP 24 (*page 17*).
Carte Générale des Découvertes de l'Amiral de Fonte
Gilles and Didier Robert de Vaugondy, 1755
From: Denis Diderot, *Encyclopédie*, 1755

MAP 25 (*page 17*).
Mappe Monde ou Globe Terrestre en deux Plans Hemispheres
Jean Covens and Corneille Mortier, c. 1780

MAP 26 (*page 17*).
Amerique Septentrionale
From: Jean B. Nolin, 1783, *Atlas Général a L'Usage des Colleges et Maisons d'Education*
Library of Congress G1015 .N68 1783

MAP 27 (*page 18*).
Chart of North and South America
From: Thomas Jefferys, *A General Topography of North America and the West Indies*, 1768
Library of Congress G1105.J4 1768

MAP 28 (*page 18*).
Carte Marine entre Californie/Carte Marine de L'Amerique Septentrional
Isaac Brouckner, 1749
Library and Archives Canada NMC 14044/14043

MAP 29 (*page 18*).
A New Map of North America from the Latest Discoveries
From: Jonathan Carver, *Travels through the Interior Parts of North America in 1766, 1767, and 1768*, 1778

MAP 30 (*page 18*).
A Plan of Captain Carver's Travels in the Interior Parts of North America in 1766 and 1767
From: Jonathan Carver, *Travels through the Interior Parts of North America in 1766, 1767, and 1768*, 1778

MAP 31 (*page 19*).
Carta reducida de las Costas, y Mares Septentrionales de California
Juan Francisco de la Bodega y Quadra, 1775
Archivo General de Indias MP Mexico 581

MAP 32 (*page 19*).
Carta Reducida del Oceano Asiatico ò Mar del Sur que contiene la Costa de la California comprehendida desde el Puerto de Monterrey
José de Cañizares, 1774
U.S. National Archives RG 77 "Spanish maps of unknown origin" No. 67

MAP 33 (*page 20*).
Plano de la Rada de Bucareli
Bruno de Hezeta y Dudagoitia, 1775
Archivo General de Indias

MAP 34 (*page 20*).
Carta contiene parte de la costa de la California
Bernabe Muñoz, 1787
Library of Congress G4362.C6 1787.MS TIL vault

MAP 35 (*page 20*).
Plano de la Bahía de la Asunción
Bruno de Hezeta y Dudagoitia, 1775
Archivo General de Indias

MAP 36 (*page 21*).
Sketch of the Entrance of the Strait of Juan de Fuca by Charles Duncan August 15, 1788
Charles Duncan, copy of chart by Charles Barkley
Alexander Dalrymple, 1790

MAP 37 (*page 21*).
Track from first making the Continent, March 7th, to Anchoring in King George's Sound
Journal of James Burney, 1778
U.K. National Archives ADM 51/4528

MAP 38 (*page 21*).
Chart of the NW Coast of America and the NE Coast of Asia explored in the years 1778 and 1779
From: James Cook, *Voyage to the Pacific Ocean*, 1784

MAP 39 (*page 21*).
*A Chart of the Interior Part of North America
Demonstrating the very great probability of an Inland
Navigation from Hudsons Bay to the West Coast*
From: John Meares, *Voyages Made in the Years 1788 and
1789 from China to the North West Coast of America*, 1790

MAP 40 (*page 22*).
Chart of the World on Mercator's Projection
Aaron Arrowsmith, 1790
British Library

MAP 41 (*page 22*).
*A Map Exhibiting All the New Discoveries in the Interior
Parts of North America*
Aaron Arrowsmith, 1795
Hudson's Bay Company Archives G.4/26

MAP 42 (*page 22*).
*Plano del Estrcho de Fuca reconicido y le bantado en
ano 1790*
Gonzalo López de Haro (attrib.), 1790

MAP 43 (*page 22*).
Carta que comprehende
José María Nárvaez, 1791
Library of Congress G3351.P5 1799.C vault, Map 12

MAP 44 (*page 22*).
Plano de la Bahia de Nuñez Gaona
Manuel Quimper, 1790
From: Henry Wagner, *Spanish Explorations in the Strait
of Juan de Fuca*, 1933

MAP 45 (*page 22*).
Puerto de la Bodega y Quadra
Manuel Quimper, 1790
From: Henry Wagner, *Spanish Explorations in the Strait
of Juan de Fuca*, 1933

MAP 46 (*page 23*).
*Carta Esferica de la parte de la Costa No. de America
Comprehende entre la Entrada de Juan de Fuca y la
Salidas de las Goletas con algunos Canales interiores*
Dionisio Alcalá Galiano, 1792
U.K. National Archives FO 925 1650 (13)

MAP 47 (*page 23*).
*Chart of the Coast of N W America and islands
adjacent north Westward of the Gulf of Georgia as
explored by His Majesty's ships* Discovery & Chatham
in the months of July & August 1792
George Vancouver/Joseph Baker, 1792
U.K. National Archives MPG 557 (3)

MAP 48 (*page 23*).
Chart Shewing Part of the Coast of N.W. America
From: George Vancouver, *A Voyage of Discovery to the
North Pacific Ocean and Round the World 1791–1795*,
Atlas.

MAP 49 (*page 24*).
*Chart of the Coast of N W America and islands
adjacent north Westward of the Gulf of Georgia as explored by His Majesty's ships* Discovery & Chatham *in
the months of July & August 1792* (Puget Sound detail)
George Vancouver/Joseph Baker, 1792
U.K. National Archives MPG 557 (3)

MAP 50 (*page 24*).
*A Sketch of the Columbia River Explored in His
Majesty's Arm'd Brig Chatham Lieut Broughton
Commander in October 1792*
William Broughton, 1792
U.K. Hydrographic Office 229 on Rv

MAP 51 (*page 24*).
Gray's Harbour Situated on the Coast of New Albion
Joseph Whidbey, 1792
U.K. Hydrographic Office

MAP 52 (*page 25*).
A Chart of the Western Coast of N. America
Joseph Baker, 1792
U.K. Hydrographic Office 228 on 82

MAP 53 (*page 25*).
*The Entrance of Columbia River from the North Pacific
Ocean Discovered [sic] By Capt. Vancouver*
James Winter Lake, 1805
Hudson's Bay Company Archives A1/220, fo.36d

MAP 54 (*page 25*).
Columbia's River
Robert Gray, 1792 (Copy given to George Vancouver)
U.K. National Archives MPG 557 (1)

MAP 55 (*page 26*).
*Copy of a Map presented to the Congress by Peter Pond,
a native of Milford in the State of Connecticut*
Anon. (Peter Pond, 1784), 1785
Service historique de la Marine Recueil 67 No. 30

MAP 56 (*page 26*).
*A Topogra[phical] Sketch of the Missouri and Upper
Missisippi Exhibiting the various Nations and tribes of
Indians who inhabit the Country*
Antoine Soulard, 1795 (English copy)
Beinecke Library, Yale University

MAP 57 (*page 26*).
*A Map Exhibiting all the New Discoveries
in the Interior Parts of North America*
Aaron Arrowsmith, 1802
LC: G3300 1802 A7 Vault Casetop

MAP 58 (*page 27*).
Carte du Mississipi et ses embranchements
James Pitot, 1802
Service historique de la Marine

MAP 59 (*page 27*).
A Map of Part of the Continent of North America
(composite of several William Clark maps)
Nicholas King, 1805
LC: G3300 1805 .C5 Vault Oversize

MAP 60 (*page 27*).
*A Map of Lewis & Clark's Track, Across the
Western Portion of North America from
the Mississippi to the Pacific Ocean*
William Clark, engraved by Samuel Lewis, 1814
From: Meriwether Lewis, *A History of the Expedition
under the Commands of Captains Lewis and Clark*, 1814

MAP 61 (*page 28*).
*A Map of the Discoveries of Capt. Lewis & Clark from
the Rockey Mountain and the River Lewis to the Cap
of Disappointment or the Columbia River at the North
Pacific Ocean by observation of Robert Frazer*
Robert Frazer, c. 1807
Library of Congress: G4126.S12 1807 .F5 Vault Oversize

MAP 62 (*page 28*).
Sketch map of the mouth of the Columbia River
William Clark, 1806
Beinecke Library, Yale University

MAP 63 (*page 29*).
Great Falls of Columbia River
William Clark, 1805
Fort Clatsop National Monument

MAP 64 (*page 29*).
Cape Disappointment and the north shore of the
Columbia River
Page 152, William Clark journal, November 1805
American Philosophical Society, Codex I, p. 152, Neg 913

MAP 65 (*page 29*).
A Map of part of the Continent of North America
William Clark, 1810
Beinecke Library, Yale University

MAP 66 (*page 29*).
Map of the Columbia and Multnomah *rivers furnished
Capt. C. by an old and inteligent Indian*
Meriwether Lewis, 3 April 1806
American Philosophical Society, Lewis and Clark
journals, Codex K, pp. 28–29

MAP 67 (*page 30*).
Routes of Hunt and Stuart
From: Washington Irving, *Astoria*, 1836

MAP 68 (*page 30*).
*Map of the North-West Territory of the Province of
Canada from actual survey during the years 1782 to 1812*
David Thompson, 1814
Archives of Ontario: AO 1541

MAP 69 (*page 31*).
*Map of the United States of America with the
Contiguous British and Spanish Possessions*
John Melish, 1816
LC: G3700 1816 M4f Mel

MAP 70 (*page 31*).
Map of the Oregon Territory
David Thompson, 1843 or before
U.K. National Archives FO 925/4622, sheets 15, 16, part
of 17, and 19

MAP 71 (*page 32*).
*A Map Exhibiting all the New Discoveries in the
Interior Parts of North America*
Aaron Arrowsmith, 1824, with manuscript additions in
red ink presumed to be by Hudson's Bay Company
employees
University of British Columbia G3300 1795 .A7 1824

MAP 72 (*page 32*).
*A Map Exhibiting all the New Discoveries in the
Interior Parts of North America* (with extensive manu-
script additions on paper patches)
Aaron Arrowsmith, 1824, with additions by HBC
employees
Hudson's Bay Company Archives G4/31

MAP 73 (*page 33*).
*Map of the United States of North America
with parts of the adjacent countries*
David H. Burr, 1839
From: J. Arrowsmith, *The American Atlas*, 1829

MAP 74 (*page 33*).
P. S. Ogdens Camp Track 1829
Peter Skene Ogden, 1829
Hudson's Bay Company Archives B202/a/8, fo. 1

MAP 75 (*page 34*).
Map of the Columbia River Basin
Alexander Ross, 1821 and 1849
British Library Add. MS. 31,358 B

MAP 76 (*page 35*).
Map of the Oregon Territory
David Thompson, before 1843
U.K. National Archives FO 925/4622 sheet 15

MAP 77 (*page 35*).
Plan of the Entrance of the Columbia River
Mervin Vavasour, 1845
U.K. National Archives MPK 1/59

MAP 118 (*page 53*).
Untitled map of the Pacific Northwest
From: *Maps of the Land Boundary Between the British Possessions in North America and the United States, 1846* (published 1869)
U.K. National Archives FO 925/1623B

MAP 119 (*page 53*).
Topographical Map of the Road from Missouri to Oregon, commencing at the Mouth of the Kansas and ending at the Mouth of the Wallah-Wallah in the Columbia
Charles Preuss, 1846
Library of Congress G4127.O7 1846 .F7 TIL Vault

MAP 120 (*page 53*).
Map of an Exploring Expedition to the Rocky Mountains in the Year 1842 and to Oregon and North California in the Years 1843–44
John Charles Frémont, 1845
Library of Congress G4051.S12 1844 .F72 Vault

MAP 121 (*page 55*).
Sketch showing the site of the Oregon City on the Willamette River
Henry Warre and Mervin Vavasour, 1845
U.K. National Archives MPK 1/59

MAP 122 (*page 55*).
Plan of Oregon City
John McLoughlin and others, 1850, with manuscript additions to 1945
Museum of the Oregon Territory

MAP 123 (*page 56*).
Photographs of "full-scale map" of Champoeg
2008

MAP 124 (*page 56*).
Champoeg and area
General Land Office survey map, 1852
Bureau of Land Management

MAP 125 (*page 56*).
Map of Donation Land Claims
General Land Office, 1860
Bureau of Land Management

MAP 126 (*page 57*).
Oregon City and area
General Land Office, 1852 (1915 copy)
Bureau of Land Management

MAP 127 (*page 57*).
Map of Peoria Linn Cᵒ· Oregon
Anon., c. 1851
Barry Lawrence Rudermann

MAP 128 (*page 57*).
Salem
General Land Office, 1852
Bureau of Land Management

MAP 129 (*page 58*).
Sketch of the Wallamette Valley, showing the purchases and reservations made by the Board of Commissioners, appointed to treat with the Indians of Oregon, April and May 1851
George Gibbs and Edmund A. Starling, 1851
U.S. National Archives, RG 75, Bureau of Indian Affairs, Map 195

MAP 130 (*page 59*).
Diagram of a Portion of Oregon Territory
John Preston, 1852
U.K. National Archives FO 925/1650

MAP 131 (*page 60*).
Township No. 1 S.R. No 1 East (Portland)
Anon., 1850
Beineke Rare Book Library, Yale University

MAP 132 (*page 60*).
Portland, Oregon
E.S. Glover, 1879
Library of Congress G4294.P6A3 1879 .G6

MAP 133 (*page 61*).
Portland and area to north
General Land Office, 1852 (1914 copy)
Bureau of Land Management

MAP 134 (*page 61*).
Portland and area to south
General Land Office, 1852
Bureau of Land Management

MAP 135 (*page 62*).
Map of the United States of America
George Woolworth Colton, J.H. Colton Co., 1850
Library of Congress G3700 1850 .C6 TIL

MAP 136 (*page 62*).
Map of the United States West of the Mississippi
D. McGowan and George H. Hildt
Library of Congress G4050 1859 .M2 RR 176

MAP 137 (*page 63*).
Map of Oregon, Washington, and Part of British Columbia
Samuel Augustus Mitchell, 1860

MAP 138 (*page 63*).
Map of Oregon, Washington, and Part of Idaho
Samuel Augustus Mitchell, 1864

MAP 139 (*page 63*).
Postcard map of Oregon
Arbuckle Bros. Coffee Company, 1889

MAP 140 (*page 63*).
Postcard map of Washington
Arbuckle Bros. Coffee Company, 1889

MAP 141 (*page 64*).
Plan of Seattle 1855–6
Thomas S. Phelps, 1856

MAP 142 (*page 64*).
Indian War Battlefields
From: Clinton A. Snowden, *History of Washington: The Rise and Progress of an American State*, 1909–11

MAP 143 (*page 65*).
Map of Military Reconnaissance from Fort Dalles, Oregon, via Fort Wallah-Wallah to Fort Taylor, Washington Territory
John Mullan, 1858

MAP 144 (*page 65*).
Plan of Col. Steptoe's Battlefield on the Ingossomen Creek
Theodore Kolecki, 1858
From: *Report of the Secretary of War*, 15 February 1859
Washington State Library

MAP 145 (*page 66*).
Plan of the Battle of Four Lakes Sept. 1st 1858 and the Battle of the Spokane Plains Sept. 5th 1858 fought by the U.S. Troops under Col. Geo. Wright 9th Infy with the Northern Indians, Palouses, Spokanes, Coeur d'Alenes Etc.
Theodore Kolecki, 1858
From: *Report of the Secretary of War*, 15 February 1859
Washington State Library

MAP 146 (*page 67*).
Indian Land Cessions in the United States Washington and Oregon (composite)
Charles Royce, 1896–97
From: Smithsonian Institution, Bureau of American Ethnology, *Eighteenth Annual Report, 1896–97*
Library of Congress E51 .U55 18th

MAP 147 (*page 67*).
Map . . . showing . . . Constitutional Population . . . To Which is Added a Sketch of the Principal Indian Reservations and Ranges
Francis A. Walker, 1871
From: U.S. Census Office, *Statistical Atlas of the United States*, 1874

MAP 148 (*page 68*).
Lower Oregon and Upper California
Thomas Tennant, 1853
David Rumsey Collection

MAP 149 (*page 68*).
Diagram of a Portion of Oregon Territory
John Preston, 1852
U.K. National Archives FO 925/1650

MAP 150 (*page 69*).
Jacksonville and One-Horse Town
General Land Office survey map, 1854
Bureau of Land Management

MAP 151 (*page 69*).
Foster & Gunnell's Mining Map of Southern Oregon
Foster & Gunnell, 1904

MAP 152 (*page 70*).
Diagram of a Portion of Oregon Territory
John Preston, 1852
U.K. National Archives FO 925/1650

MAP 153 (*page 71*).
Preliminary Survey of Duwamish Bay W.T.
William A. McMurtrie and George Farquhar
United States Coast Survey, 1854
National Oceanic and Atmospheric Administration
Central Library

MAP 154 (*page 72*).
Steilacoom and vicinity
General Land Office survey map, 1855
Bureau of Land Management

MAP 155 (*page 72*).
Olympia and vicinity
General Land Office survey map, 1854
Bureau of Land Management

MAP 156 (*page 73*).
Map of Public Surveys in the Territory of Washington to Accompany Report of Surveyor General. 1865.
Surveyor General's Office. 1865.
Washington State University Library.

MAP 157 (*page 73*).
Plan of Fort Bellingham
U.S. National Archives

MAP 158 (*page 73*).
Plan of the Town of Sehome. Bellingham Bay W.T.
Plat as filed, 1858
Western Washington University

MAP 159 (*page 74*).
Mouth of Columbia River from a Preliminary Survey
[William McMurtrie], U.S. Coast Survey, 1851
American Geographical Society

MAP 274 (page 114).
Seattle Harbor, Puget Sound, Washington Territory
U.S. Coast and Geodetic Survey, 1879
University of Washington Map Library

MAP 275 (page 114).
Bird's-eye View of the City of Seattle, Puget Sound, Washington Territory, 1878
E.S. Glover, 1878
Library of Congress G4284.S4 A3 1878 .G6

MAP 276 (page 115).
Bird's Eye View of the City of Seattle, WT. Puget Sound. County Seat of King County. 1884
J.J. Stoner, 1884
Library of Congress G4284 .S4A3 1884 .W4

MAP 277 (page 115).
This bird's eye view of Seattle has been prepared especially by Llewellyn, Dodge & Co.
Llewellyn, Dodge & Co., 1889 (with later addition of fire boundary)
Museum of History and Industry

MAP 278 (page 116).
Map Showing Territorial Growth of the City of Seattle
Annual Report of the Seattle City Engineer, 1891 (1892)
University of Washington Special Collections
G4284.S4 G46 1891.S4

MAP 279 (page 116).
Map of West Seattle, King County, Washington
L.T. Merry, 1899
University of Washington Special Collections
G4284 S4:2W46 1899 M4

MAP 280 (page 117).
Birds-eye View of Seattle and Environs, King County, Wash., 1891
Augustus Koch, 1891
Library of Congress
G4284.S4A3 1891 .K6 Oversize

MAP 281 (page 118).
Track System of Puget Sound Traction, Light and Power Co. Seattle Division
Puget Sound Traction, Light & Power Co., 1917
Washington State Archives, Puget Sound Office

MAP 282 (page 118).
Puget Sound Electric Railway
Puget Sound Electric Railway, 1910
From: *Official Guide of the Railways*, 1910

MAP 283 (page 118).
Lake Washington, Washington
U.S. Coast and Geodetic Survey, 1905
University of Washington Map Library

MAP 284 (page 119).
Seattle Harbor
Harbor Line Commission survey, October 1890
Washington State Archives, Puget Sound Branch

MAP 285 (page 119).
Seattle Harbor
Harbor Line Commission survey, October 1890
Washington State Archives, Puget Sound Branch

MAP 286 (page 119).
Anderson's New Guide Map of the City of Seattle and Environs, Washington
O.P. Anderson & Co., 1890
Library of Congress G4284.S4 1890 .O6 TIL

MAP 287 (page 120).
Guide Map of Seattle Showing Tide Lands to be Filled and Canal to be Constructed by the Seattle and Lake Washington Waterway Company
Seattle & Lake Washington Waterway Co., c. 1895, on a base map by O.P. Anderson & Co.
University of Washington Special Collections
G4284.S4 P53 1895.S4

MAP 288 (page 120).
Seattle the Gateway—All Great Rail Lines Lead to Seattle
Unknown, 1897
From: Lisa Mighetto and Marcia Babcock Montgomery, *Hard Drive to the Klondike: Promoting Seattle during the Gold Rush*, 1998.

MAP 289 (page 121).
New Guide Map of Seattle, Washington, 1907
Jackson Realty & Loan Co., 1907
University of Washington Special Collections
G4284.S4 1907.H2

MAP 290 (page 121).
Anderson Map Co's 1909 Official Map of Greater Seattle
Anderson Map Co., 1909
Washington State Archives, Puget Sound Branch

MAP 291 (page 122).
Main Business District Periscopic Seattle
Ross W. Tulloch, Periscopic Map Company, 1903
Library of Congress G4284.S4A3 1903 .T8

MAP 292 (page 122).
Seattle's Coming Retail and Apartment-House District
B. Dudley Stuart, 1917
University of Washington Special Collections
G4284.S4:2D46 A3 1917.S7

MAP 293 (page 123).
Relief Map of Civic Improvement Showing Denny Regrade Area
Relief model, George Nelson Contractor, c. 1930
Museum of History and Industry

MAP 294 (page 123).
Cheap Factory Sites in Seattle's Manufacturing Center
Newspaper advertisement, 1908

MAP 295 (page 124).
Map of the City of Seattle and Adjacent Territory accompanying report of the Municipal Plans Commission showing Existing and Proposed Parks and Park Boulevards
Virgil G. Bogue, 1911
From: *Plan of Seattle: Report of the Municipal Plans Commission submitting Report of Virgil G. Bogue Engineer*, 1911

MAP 296 (page 124).
King County Government Building & Land Development Commercial Waterway Development No. 1.
King County, 1911
Washington State Archives, Puget Sound Branch

MAP 297 (page 124).
Map of the City of Seattle and Adjacent Territory accompanying report of the Municipal Plans Commission showing Existing and Proposed Street Railways
Virgil G. Bogue, 1911
From: *Plan of Seattle: Report of the Municipal Plans Commission submitting Report of Virgil G. Bogue Engineer*, 1911

MAP 298 (page 124).
Civic Center Group, looking south on Central Avenue
David J. Myers, 1911
From: *Plan of Seattle: Report of the Municipal Plans Commission submitting Report of Virgil G. Bogue Engineer*, 1911

MAP 299 (page 124).
Map of the City of Seattle and Adjacent Territory accompanying report of the Municipal Plans Commission showing Existing and Proposed Arterial Highways
Virgil G. Bogue, 1911
From: *Plan of Seattle: Report of the Municipal Plans Commission submitting Report of Virgil G. Bogue Engineer*, 1911

MAP 300 (page 125).
Birdseye View of Great Seattle and Vicinity
Kennedy Company, c. 1922
Washington State Archives, Puget Sound Branch

MAP 301 (page 125).
Baist's Real Estate Atlas of Surveys of Seattle
G.W. Baist, 1912
Washington State Archives, Puget Sound Branch

MAP 302 (page 126).
Alaska–Yukon–Pacific Exposition
Postcard, 1907
University of Washington Special Collections
G4284.S4:2D42 A3 1909.C3

MAP 303 (page 126).
Anderson Map Co's 1909 Official Map of Greater Seattle
Anderson Map Co., 1909
Washington State Archives, Puget Sound Branch

MAP 304 (page 126).
Alaska–Yukon–Pacific Exposition Seattle, Washington, 1909, Plan of Grounds and Buildings
Olmsted Brothers Landscape Architects, 1909
University of Washington Special Collections
G4284.S4:2A42 1906.C3

MAP 305 (page 127).
Authorized Birds Eye View of the Alaska–Yukon–Pacific Exposition, Seattle, U.S.A. 1909
Alaska–Yukon–Pacific Exposition, 1909
University of Washington Special Collections
G4284.S4:2D42 A3 1909.O4

MAP 306 (page 127).
New Guide Map of Seattle, Washington, 1907
Jackson Realty & Loan Co., 1907
University of Washington Special Collections
G4284.S4 1907.H2

MAP 307 (pages 128–29).
Proposed Route of Canal to Connect Lakes Union and Washington with Puget Sound
U.S. Army Corps of Engineers, 1891
University of Washington Special Collections
G4284.S4:2L3 1891.U5

MAP 308 (page 128).
Map Showing Territorial Growth of the City of Seattle
Annual Report of the Seattle City Engineer, 1891 (1892)
University of Washington Special Collections
G4284.S4 G46 1891.S4

MAP 309 (page 128).
Lake Washington
U.S. Coast and Geodetic Survey, 1905
University of Washington Map Library

MAP 441 (*page 178*).
*Relocation of Primary State Highway No. 2 and
the Lake Washington Bridge Project*
Washington State Highways Department, c. 1938
Washington State Archives

MAP 442 (*page 178*).
Projected route of Highway 2 across the Lake Washington
Floating Bridge
Washington State Highways Department, c. 1938
Washington State Archives

MAP 443 (*page 178*).
*Lake Washington Bridge Project E. Tunnel Plaza at 35
Ave*
From: *Final Progress Construction Report, the Lake
Washington Floating Bridge Project*
Washington Toll Bridge Authority, 1940
Washington State Archives

MAP 444 (*page 179*).
Bird's-eye map of the new route across the Lake
Washington Floating Bridge
From: *Final Progress Construction Report, the Lake
Washington Floating Bridge Project*
Washington Toll Bridge Authority, 1940
Washington State Archives

MAP 445 (*page 179*).
*Aerial View Lake Washington Bridge Project and
Relocation of Primary State Highway No 2*
From: *Final Progress Construction Report, the Lake
Washington Floating Bridge Project*
Washington Toll Bridge Authority, 1940
Washington State Archives

MAP 446 (*page 180*).
*Tacoma Narrows Bridge Preliminary Design Vicinity
Map*
Washington Toll Bridge Authority, 1938
Washington State Archives

MAP 447 (*page 181*).
PWS Docket No. Wash. 1870-F Tacoma Narrows Bridge
(Sheet 4 of 39 sheets)
Washington Toll Bridge Authority, 1938
Washington State Archives

MAP 448 (*page 182*).
Bird's-eye view of proposed Longview civic center
Hoit, Prince & Barnes, architects, c. 1923

MAP 449 (*page 182*).
*Plat of Longview, Washington, Showing Business District
and Portion of Residential and Manufacturing Districts*
Hare & Hare, city planners; George E. Kessler, city plan-
ning consultant; and B.L. Lambuth, realtor, 1923
University of Washington Special Collections
G4284.L7645 1923.H3

MAP 450 (*page 183*).
*Where Rail, Water and Highway Meet—Longview,
Washington*
Longview Company ad, 1925

MAP 451 (*page 183*).
Longview, Washington
Longview Company ad, 1928

MAP 452 (*page 183*).
Map of Longview for Pageant of Progress
Longview Company, 1924

MAP 453 (*page 183*).
Map showing location of Longview
Longview Company ad, 1924

MAP 454 (*page 184*).
Distribution of Shanties in "Hooverville" March 1934
Donald Francis Roy, 1934
University of Washington Library

MAP 455 (*page 185*).
Seattle and the Northwest
Seattle Chamber of Commerce, 1930

MAP 456 (*page 185*).
*Land Ownership, Tax Delinquency and Land
Classification. Whatcom County. 1937*
Anon., 1937(?)
University of Washington Map Library

MAP 457 (*page 186*).
*Portland and the Manufacturing District at Willa-
mette Falls, Oregon City*
Morning Oregonian, 16 December 1896

MAP 458 (*page 187*).
Skagit River Power Project
City of Seattle Department of Lighting, 1937

MAP 459 (*page 187*).
Map of City Light Transmission Lines
City of Seattle Department of Lighting, 1937

MAP 460 (*page 187*).
*Territory Served by the Washington Water Power
Company*
Washington Water Power Company, 1923
Washington State University wsu259

MAP 461 (*page 188*).
The Bonneville Dam
U.S. Army Corps of Engineers' model, c. 1934
Photo of display at the Bonneville Dam Visitors Center,
2008

MAP 462 (*page 188*).
Dam Area
Bonneville Dam visitor's brochure
U.S. Army Corps of Engineers, 1967

MAP 463 (*page 188*).
Perspective View Coulee Dam and Vicinity
Undecipherable, 1942

MAP 464 (*page 189*).
*Panoramic Perspective of the Spokane Region
Including the Geological and
Scenic Wonderland Embracing the Columbia Basin
Irrigation Project and Grand Coulee Dam*
Spokane Chamber of Commerce, 1949

MAP 465 (*page 190*).
Columbia Basin Irrigation Project
Brochure and map cover,
Spokane Chamber of Commerce, 1949

MAP 466 (*page 190*).
PWA Rebuilds the Nation
Public Works Administration, 1935
David Rumsey Collection

MAP 467 (*page 190*).
Port of The Dalles, Oregon
Port of The Dalles, c. 1930

MAP 468 (*page 190*).
Airway Map No. 137-A The Columbia River Gorge
U.S. Coast and Geodetic Survey, 1931
National Oceanic and Atmospheric Administration
Central Library

MAP 469 (*page 190*).
*Chief Joseph Dam Project, Washington, Greater
Wenatchee Division General Map*
U.S. Department of the Interior, Bureau of Reclamation,
1955
University of Washington Map Library

MAP 470 (*page 191*).
Columbia Basin Project, Washington, Irrigation System
U.S. Department of the Interior, Bureau of Reclamation,
1943
University of Washington Map Library

MAP 471 (*page 191*).
*Map of Warden, City with the Golden Future, Showing
its Strategic Location*
Dodson and Eason, 1947
University of Washington Map Library

MAP 472 (*page 191*).
Comprehensive Plan for the Columbia Basin
U.S. Department of the Interior, Bureau of Reclamation,
1948
Portland State University Map Library

MAP 473 (*page 192*).
Vicinity Map
Inset map in: *Camp Adair, Corvallis, Oregon.
Training Aids General Layout*
U.S. Engineer Office, Portland, Oregon, 1943
with revisions to 1945
Oregon State Archives Map Drawer 13, 27

MAP 474 (*page 192*).
Camp Adair, Oregon. Building Layout
Office of the Post Engineer, 1943
Oregon State Archives Map Drawer 13, 26

MAP 475 (*page 193*).
*Camp Adair, Corvallis, Oregon. Training Aids
General Layout*
U.S. Engineer Office, Portland, Oregon, 1943 with
revisions to 1945
Oregon State Archives Map Drawer 13, 27

MAP 476 (*page 194*).
Bremerton, showing location of Navy Yard
From: U.S. Navy brochure *Keep the Ships Fighting,* 1943
University of Washington Special Collections
VA70.B74 P848 1943

MAP 477 (*pages 194–95*).
Vanport City, Oregon
U.S. Federal Public Housing Authority, 1943
University of Oregon Library G4294. V34 1943 .U5

MAP 478 (*page 196*).
*Japanese Evacuation Program Assembly Center
Destinations*
From: *Final Report Japanese Evacuation from the
West Coast 1942*
Final report of Lieutenant General John L. DeWitt,
Western Defense Command
U.S. Government Printing Office, 1943

MAP 479 (*page 196*).
*Japanese Evacuation Program Relocation Center
Destinations*
From: *Final Report Japanese Evacuation from the
West Coast 1942*
Final report of Lieutenant General John L. DeWitt,
Western Defense Command
U.S. Government Printing Office, 1943

MAP 480 (*page 196*).
Maximum Japanese Population and Dates of Occupa-tion of Assembly Centers
From: *Final Report Japanese Evacuation from the West Coast 1942*
Final report of Lieutenant General John L. DeWitt, Western Defense Command
U.S. Government Printing Office, 1943

MAP 481 (*page 196*).
Location of Assembly Center Puyallup
From: *Final Report Japanese Evacuation from the West Coast 1942*
Final report of Lieutenant General John L. DeWitt, Western Defense Command
U.S. Government Printing Office, 1943

MAP 482 (*page 196*).
Location of Assembly Center Portland
From: *Final Report Japanese Evacuation from the West Coast 1942*
Final report of Lieutenant General John L. DeWitt, Western Defense Command
U.S. Government Printing Office, 1943

MAP 483 (*page 197*).
Amazing Invasion Map: Japs Plan to Seize West Coast
Reproduced in the *Los Angeles Examiner*, 22 August 1943

MAP 484 (*page 198*).
Map of the State of Washington
D.H. White, 1942

MAP 485 (*page 198*).
Site Diagram
DuPont Corporation sketch for press release, 1945
From: U.S. Department of Energy, Office of Environmental Restoration and Waste Management, *Legend and Legacy: Fifty Years of Defense Production at the Hanford Site*, 1992

MAP 486 (*page 199*).
Richland–Uninc.
Anon., c. 1950
University of Washington Map Library

MAP 487 (*page 199*).
Route Chart 2202 Northwest United States
U.S. Coast and Geodetic Survey, June 1949, revised to May 1952
National Oceanic and Atmospheric Administration
Central Library

MAP 488 (*page 199*).
Location and Regional Map of the Hanford Site
From: U.S. Department of Energy, Office of Environmental Restoration and Waste Management, *Legend and Legacy: Fifty Years of Defense Production at the Hanford Site*, 1992

MAP 489 (*page 200*).
Welch Apples
Apple crate label, c. 1940s

MAP 490 (*page 200*).
Wenatchee, The Gateway to the Land of Perfect Apples in North Central Washington
Wenatchee Commercial Club, 1910

MAP 491 (*page 200*).
Washington State—Appleland
Washington State Apple Advertising Commission, 1948

MAP 492 (*page 201*).
Official Map of American and Canadian Airways and Aerial Mail Routes
Aeronautic Maps Association, 1919
National Oceanic and Atmospheric Administration
Central Library

MAP 493 (*page 201*).
Airway Map No. 137, Portland to Spokane, Wash.
U.S. Coast and Geodetic Survey, June 1931
National Oceanic and Atmospheric Administration
Central Library

MAP 494 (*page 201*).
Airway Map No. 137-A, The Columbia River Gorge
U.S. Coast and Geodetic Survey, June 1931
National Oceanic and Atmospheric Administration
Central Library

MAP 495 (*page 201*).
Airway Map No. 137-A, The Columbia River Gorge
U.S. Coast and Geodetic Survey, April 1934
National Oceanic and Atmospheric Administration
Central Library

MAP 496 (*page 202*).
Boeing Field
King County Engineer's Office, 1931
Washington State Archives, Puget Sound Branch

MAP 497 (*page 202*).
No. 3060d (Civilian aeronautical chart of Western U.S.)
U.S. Coast and Geodetic Survey, 1946
National Oceanic and Atmospheric Administration
Central Library

MAP 498 (*page 202*).
King County Airport—(Boeing Field) Seattle, Wash. Airport Chart
U.S. Coast and Geodetic Survey, 1946
University of Washington Library

MAP 499 (*page 203*).
Seattle–Tacoma Airport, Seattle, Wash. Airport Chart
U.S. Coast and Geodetic Survey, 1946
University of Washington Library

MAP 500 (*page 203*).
Felts Field, Spokane, Wash. Airport Chart
U.S. Coast and Geodetic Survey, 1946
University of Washington Library

MAP 501 (*page 203*).
(RC 2212) Chicago to Seattle Route Chart
U.S. Coast and Geodetic Survey, 1956
National Oceanic and Atmospheric Administration
Central Library

MAP 502 (*page 203*).
Air Corps Map. Vancouver to Seattle, Washington. Air Navigation Map No.43 (Experimental)
Air Corps, 1932
University of Washington Map Library

MAP 503 (*page 203*).
Seattle Sectional Aeronautical Chart
Experimental sheet, violet on buff (with added comment in pencil)
U.S. Coast and Geodetic Survey, 1940
National Oceanic and Atmospheric Administration
Central Library

MAP 504 (*page 203*).
Seattle Sectional Aeronautical Chart
Experimental sheet, violet on white (added comment in pencil)
U.S. Coast and Geodetic Survey, 1940
National Oceanic and Atmospheric Administration
Central Library

MAP 505 (*page 203*).
WCA is going your way
Ad for West Coast Airlines in *Official Guide Book, Seattle World's Fair 1962*, 1962
University of Washington Library

MAP 506 (*page 204*).
Civil Defense Evacuation Routes (Seattle)
From: *Evacuate: Don't Sit Under the Mushroom*,
Seattle & King County Civil Defense Departments, 1955

MAP 507 (*page 204*).
Civil Defense Evacuation Routes (King County)
From: *Evacuate: Don't Sit Under the Mushroom*,
Seattle & King County Civil Defense Departments, 1955

MAP 508 (*page 204*).
Photo of Frank S. Evans, director of civil defense, in new civil defense control center, 5 May 1953
Richards Studio photo, Tacoma Public Library

MAP 509 (*page 205*).
Spokane County and Spokane City Civil Defense Information and Evacuation Plan (Spokane City)
Spokane County and Spokane City Civil Defense, 1955

MAP 510 (*page 205*).
Map of Spokane County, Washington, 1955
Spokane County and Spokane City Civil Defense Information and Evacuation Plan (Spokane County)
Spokane County and Spokane City Civil Defense, 1955

MAP 511 (*page 205*).
Clark County Emergency Evacuation Route Map
Clark County Civil Defense, 1956

MAP 512 (*page 205*).
Portland Target Area Evacuation Routes
Clark County Emergency Evacuation Route Map
From : *Your Guide for Defense against the H-Bomb*
Portland and Clark County Civil Defense, 1955

MAP 513 (*page 206*).
Seattle World's Fair 1962
Washington State Department of Commerce and Economic Development, 1962

MAP 514 (*pages 206–07*).
Panoramic view from the top of the Space Needle, looking east
World's Fair brochure, 1962

MAP 515 (*page 207*).
Panoramic view from the top of the Space Needle, looking north
World's Fair brochure, 1962

MAP 516 (*page 207*).
Panoramic view from the top of the Space Needle, looking west
World's Fair brochure, 1962

MAP 517 (*page 207*).
Panoramic view from the top of the Space Needle, looking south
World's Fair brochure, 1962

III

OTHER ILLUSTRATIONS

All modern photographs and some modern photographs of older items are by the author, unless credited
below.

(t = top; c = center; b = bottom.)

IIIIIIIIIIIIIIIIIIIIIIIIIIIIIIIIIIIIIIIIIIIIIIIIIIIIIIIIIIIIIIIIIIIIIIIIIIIIIIIII

FURTHER READING

Abbott, Carl. *Portland: Gateway to the Northwest.* Northridge, California: Windsor Publications, 1985.

Armbruster, Kurt E. *Orphan Road: The Railroad Comes to Seattle, 1853–1911.* Pullman: Washington State University Press, 1999.

Atwood, Kay. *Chaining Oregon: Surveying the Public Lands of the Pacific Northwest, 1851–1855.* Blacksburg, Virginia: McDonald & Woodward, 2008.

Baker, John H. *Camp Adair: The Story of a World War II Cantonment.* Newport, Oregon: self-published, 2003.

Barlee, N.L. *Gold Creeks and Ghost Towns of Northeastern Washington.* Blaine: Hancock House, 1999.

Berner, Richard C. *Seattle 1900–1920: From Boomtown, Urban Turbulence, to Restoration.* Seattle: Charles Press, 1991.

Bogue, Virgil G. *Plan of Seattle: Report of the Municipal Plans Commission Submitting Report of Virgil G. Bogue Engineer 1911.* Seattle: Lowman & Hanford Co., 1911.

Bottenberg, Ray. *Bridges of the Oregon Coast.* Charleston, South Carolina: Arcadia Publishing, 2006.

———. *Bridges of Portland.* Charleston, South Carolina: Arcadia Publishing, 2007.

Bowen, William A. *The Willamette Valley: Migration and Settlement on the Oregon Frontier.* Seattle: University of Washington Press, 1978.

Burkhardt, D.C. Jesse. *Railroads of the Columbia River Gorge.* Charleston, South Carolina: Arcadia Publishing, 2004.

Culp, Edwin D. *Early Oregon Days.* Caldwell, Idaho: The Caxton Press, 1987.

Dodds, Gordon B. *The American Northwest: A History of Oregon and Washington.* Arlington Heights, Illinois: The Forum Press, 1986.

Dryden, Cecil. *Dryden's History of Washington.* Portland: Binford & Mort, 1968.

Fahey, John. *Shaping Spokane: Jay P. Graves and His Times.* Seattle: University of Washington Press, 1994.

Ficken, Robert E. *Washington Territory.* Pullman: Washington State University Press, 2002.

Ficken, Robert E., and Charles P. LeWarne. *Washington: A Centennial History.* Seattle: University of Washington Press, 1988.

Glassley, Ray Hoard. *Pacific Northwest Indian Wars.* Portland: Binford & Mort, 1953.

Goetzmann, William H. *Exploration and Empire: The Explorer and the Scientist in the Winning of the American West.* Austin, Texas: Texas State Historical Association, 2000.

Hayes, Derek. *Historical Atlas of the Pacific Northwest.* Seattle: Sasquatch Books, 1999.

———. *Historical Atlas of the North Pacific Ocean.* Seattle: Sasquatch Books, 2001.

———. *Historical Atlas of the United States.* Berkeley: University of California Press, 2006.

———. *Historical Atlas of the American West.* Berkeley: University of California Press, 2009.

———. *Historical Atlas of the North American Railroad.* Berkeley: University of California Press, 2010.

Hedges, James Blaine. *Henry Villard and the Railways of the Northwest.* New York: Russell & Russell, 1930 (republished 1967).

Hidy, Ralph W., et al. *The Great Northern Railway: A History.* Boston: Harvard Business School Press, 1988.

Highsmith, Richard M. (ed.). *Atlas of the Pacific Northwest: Resources and Development.* Corvallis: Oregon State University Press, 1962.

HistoryLink.org. The Free Online Encyclopedia of Washington State History. http://www.historylink.org/

Jeffcott, P.R. *Chechaco and Sourdough, or the Mount Baker Gold Rush.* Bellingham: self-published, 1963.

Kirk, Ruth. *Sunrise to Paradise: The Story of Mount Rainier National Park.* Seattle: University of Washington Press, 1999.

Klein, Maury. *Union Pacific: The Rebirth, 1894–1969.* New York: Doubleday, 1990.

Lewty, Peter J. *To the Columbia Gateway: The Oregon Railway and the Northern Pacific, 1879–1884.* Pullman: Washington State University Press, 1987.

———. *Across the Columbia Plain: Railroad Expansion in the Pacific Northwest, 1885–1893.* Pullman: Washington State University Press, 1995.

Loy, William G. (ed.). *Atlas of Oregon.* Eugene: University of Oregon Press, 2001 (second edition).

McArthur, Lewis A. *Oregon Geographic Names.* Portland: Oregon Historical Society Press, 1982.

McClelland, John M. *R.A. Long's Planned City: The Story of Longview.* Longview: WestMedia Corp., 1998.

Marschner, Janice. *Oregon 1859: A Snapshot in Time.* Portland: Timber Press, 2008.

Martin, Albro. *James J. Hill and the Opening of the Northwest.* New York: Oxford University Press, 1976.

Mighetto, Lisa, and Marcia Babcock Montgomery. *Hard Drive to the Klondike: Promoting Seattle during the Gold Rush.* A Historical Resource Study for the Seattle Unit of the Klondike Gold Rush National Historic Park. Seattle: Historical Research Associates, 1998.

Morgan, Murray. *Skid Road, Seattle: Her First 125 Years.* Sausalito, California: Comstock Editions, 1978 (revised edition).

———. *Puget's Sound: A Narrative of Early Tacoma and the Southern Sound.* Seattle: University of Washington Press, 1979.

Oregon Encyclopedia, The. http://www.oregonencyclopedia.org/

Powell, Fred Wilbur (ed.). *Hall J. Kelley on Oregon.* Princeton: Princeton University Press, 1932.

Puget Sound Maritime Historical Society. *Maritime Seattle.* Charleston, South Carolina: Arcadia Publishing, 2002.

Robbins, William G. *Landscapes of Promise: The Oregon Story 1800–1940.* Seattle: University of Washington Press, 1997.

Robertson, Donald B. *Encyclopedia of Western Railroad History: Volume III. Oregon–Washington.* Caldwell, Idaho: The Caxton Printers, 1995.

Roe, Joann. *North Cascades Highway: Washington's Popular and Scenic Pass.* Seattle: The Mountaineers, 1997.

———. *Stevens Pass: Gateway to Seattle.* Caldwell, Idaho: Caxton Press, 2002.

Ruby, Robert H., and John A. Brown. *Indians of the Pacific Northwest: A History.* Norman, Oklahoma: University of Oklahoma Press, 1981.

———. *A Guide to the Indian Tribes of the Pacific Northwest.* Norman, Oklahoma: University of Oklahoma Press, 1986.

Schwantes, Carlos A. *Railroad Signatures across the Pacific Northwest.* Seattle: University of Washington Press, 1999.

Schwantes, Carlos A., et al. *Washington: Images of a State's Heritage.* Spokane: Melior Publications, 1988.

Scott, George W. *The Politics of Transportation.* In: *Columbia, The Magazine of Northwest History,* Vol. 9, No. 1, Spring 1995. Washington State Historical Society, Tacoma.

Seattle Historical Society. *Seattle Century.* Seattle: Superior Publishing, 1952.

Smalley, Eugene V. *History of the Northern Pacific Railroad.* New York: G.P. Putnam's Sons, 1883.

Snyder, Eugene. *Early Portland: Stump-Town Triumphant.* Portland: Binford & Mort, 1970.

Speroff, Leon. *The Deschutes River Railroad War.* Portland: Arnica, 2007.

Thompson, Richard. *Portland's Streetcars.* Charleston, South Carolina: Arcadia Publishing, 2006.

———. *Willamette Valley Railways.* Charleston, South Carolina: Arcadia Publishing, 2008.

United States Government Printing Office. *Final Report Japanese Evacuation from the West Coast 1942.* Washington: U.S. Government Printing Office, 1943.

Utley, Robert M., and Wilcomb E. Washburn. *Indian Wars.* New York: American Heritage, 1985.

Wagner, Henry R. *Spanish Explorations in the Strait of Juan de Fuca.* New York: AMS, 1971 (reprint of 1933 edition).

Warren, James R. *King County and its Emerald City: Seattle: An Illustrated History.* [Sun Valley, California]: American Historical Press, 1997.

Map 550 (*left*).
Frank McCaffrey's pictorial bird's-eye map of Seattle was published in 1931, at a time when it would likely have found few purchasers. The map sings the praises of Seattle at every opportunity: there is *Boeing Airport* and the note *Boeing is One of the Largest Airplane Factories in the U.S.*; the *Government Locks* are *Second Largest Locks in the World!* At *Smith's Cove*—above a sea serpent—is the *Largest Commercial Dock in the World.* And it seems that an unfortunate motorist on the way to Tacoma and Portland on the *Pacific Highway* is getting a speeding ticket. Note also the map at bottom *To simplify Seattle's Street Arrangement.*

INDEX

Italicized page numbers indicate captions to maps or other illustrations.

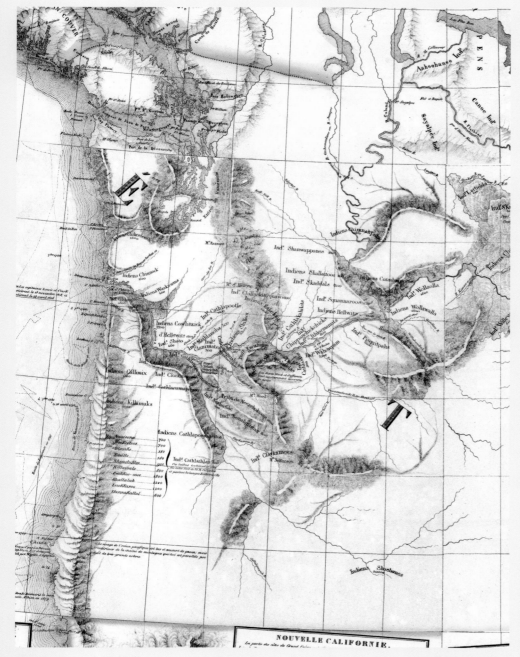

MAP 551.
In 1827 the Flemish geographer Philippe Vandermaelen published his *Atlas Universel* in six volumes, the first atlas to depict every region of the world at the same scale. It was also the first lithographed world atlas. This was his rendering of the portion of North America that is now Washington and Oregon.

plans for Puget Sound, *132, 141*; portage railroad (1863), *76*; in Tacoma, *133, 134*; Villard's plans for Northwest coast, *90–91, 91, 92*

United States, federal government: boundary disputes in Pacific Northwest, 42–48, *49, 50–51*; Highway Act (1921), 172; joint occupancy agreement (1818), 38, 56; land reserve in Port Angeles, 141, *141*; Oregon Question, 42–48; overland route to coast needed, 26. *See also entries starting with "U.S."*

United States Coast Survey, 74–75

University of Washington, 126

urban transportation, 107, 110, 117, *118*

U.S. Army Corps of Engineers, 77, 128, *188, 189*

U.S. Coast and Geodetic Survey, 74–75, 108, 131, 202, *203*

U.S. Exploring Expedition, 43, *44–45*

U.S. Forest Service, 152, *153*

U.S. National Park Service, 153

U.S. Navy, at Bremerton, 134, *135*, 161, 193, *194*

Utah & Northern Railway, 98

Valdés, Cayetano, *23*

Valerianos, Apostolos (Juan de Fuca), 14–15, *16*, 21

Vancouver, BC, *101, 162*, 174

Vancouver, George: explored Point Roberts, 49; and hypothesized Northwest Passage, *13*, 23; maps (1792), *23, 24*; missed and later explored Columbia River, 23, *25, 25*

Vancouver, WA, *106*, 107, *201*

Vancouver Island, *23*, 48, *50–51*

Vandermaelen, Philippe, *239*

Vanport, OR, *194, 195, 195*

Vavasour, Mervin, *36, 37*, 48, *54, 55*

Villard, Henry, 84, *84*, 90, 95

Vizcaíno, Sebastián, *15, 18*

wagon roads: Barlow Road, 53, 57, *59*; in eastern Oregon (1885), *103*; military, 78–79, 156. *See also* Oregon Trail

Waiilatpu, 40

Waldseemüller, Martin, *12*

Walla Walla, WA, 7, *73*, 84, 114, 144, *144*

Walla Walla & Columbia River Railroad, 81, 84, *84*, 144

Walla Walla people, 65, 67

Walla Walla Valley, *34*, 84

Wallula, WA, 84, *84, 85*

Warbass, Edward, *70*

Warden, WA, *191*

War of 1812, 31

Warre, Lieutenant Henry, *35, 36*, 48, *54, 55*

Washington (before Territory status), 19, *20, 21*, 40

Washington, Treaty of (1846), 48

Washington Apple Commission, *200*

Washington State (1889 to present): achieved statehood (1889), 62; Native peoples of (1894), *11*; postcard map (1889), *63*; railroad development, *96, 101, 169*; roads and highways, 172–77, *173, 177, 211*; seal of, *62*; tourist brochure (1931), *7*

Washington State Ferries, 161, *212*

Washington Territory: achieved Territory status (1853), 62; map (1857), *10*; map (1859), *62*; map (1860), *63*; map (1865), *73*; Native peoples, *11*, 64–67; Northern Pacific, proposed routes, *84*; Olympia as capital, 70; population of, 72; portage roads and railroads, *76*; and San Juan Islands, 51; Yakima's dispute with Northern Pacific, 144. *See also* North Oregon

Washington Toll Bridge Authority, 178–79

Washington Water Power Company, *187*

water-driven mills, *55, 56, 61*

water resources, in Oregon, *216*. *See also* dams; hydroelectricity; irrigation; water-driven mills

Wenatchee, WA, *96*, 200, *200*

West Coast Airlines, *202*

Western Defense Area, *202*

West Okanogan Valley Irrigation Project, 158–59, *159*

Weyerhaeuser, Frederick, 100

Weyerhaeuser Timber Company, 100, 182, *183*

Whatcom, WA, 104, 138–39, *138, 139*

Whatcom County, WA, 104, *105, 185*

Whatcom Creek, 72

Wheatdale, WA, *95*

Whidbey, Joseph, 23, *24*

Whidbey Island, *24*

Whitcomb, Lot, *61*

White, Harry, *135*

White Bear Lake, *18*

White Bluffs, WA, 198, *198*

Whitman, Marcus: and Samuel Parker, 36; traveled Oregon Trail (1836), 7, *52–53*; Whitman Mission, and Native attack, 10, 39–40, *39, 52*, 64

Whitney, Asa, 80, *81*

Wilder, Walter, *135*

Wilkes, Charles, 43, *44–45*

Wilkeson, Samuel, 88

Wilkeson, WA, 82, *82*, 88, *95*

Willamette Falls, 11, 77, *186*

Willamette National Forest, *152*

Willamette River and Valley: in British/American negotiations (1846), *46*; Broughton's map (1792), *24*; Burr's map (1839), *33*; Camp Adair, *192*; Champoeg, 56; Corvallis, 59, *147*; development of (1840s and 1850s), *146*; Eugene and Springfield, *145*; French Prairie settlement, 39, 40, *59*; Gibbs and Starling's map (1851), *59*; Hudson's Bay map, *32*; Kelley's map (1830), *38*; Lewis and Clark maps, *27, 29*; Melish's map (1816), *31*; Native peoples of (1876), *11*; Oregon Central map (1869), *93*; Oregon City, *54, 59*; Portland bridges, 107; Preston's map (1852), *59*; reached by Oregon Trail, 53; Salem, 57; settling of (1840s and 1850s), 54–59; steamboats on, 77

Willamette Valley and Cascade Mountain Military Wagon Road, 78, 79, *79, 157*

Willapa Bay, 163, *163*

Williamson, Henry, *106*

Wilson River Highway, *164*

Wind River Mountains, 30, *30*

Winship, Nathaniel, Jonathan & Abiel, 35

Wishram, WA, *169*

Wolf Creek Highway, *164*

Works Progress Administration, *164*

World Aeronautical Chart system, *202*

World War II, 192–99, *202*

Wright, Colonel George, 66, 67

Wyeth, Nathaniel J., 39, 52

Wyld, James, *46*

Wytfliet, Cornelius, *12*

Yakima, WA (formerly North Yakima), 144, *144*

Yakima Irrigation Project, *188*

Yakima people, 65, *67, 73*

Yakima Reservation, 87

Yakima Valley, *83*, 88, 200, *208*

Yakima War, 64, 67

Yaquina Bay, *75*

Yesler, Henry, 72

Young, Ewing, 38, 56

Back endpaper map.
This is the second edition of Philadelphia mapmaker S. Augustus Mitchell's *New Map of Texas, Oregon, and California,* published in 1846. Mitchell had to produce a second edition in 1846 because negotiations between the American and British governments had finally produced an international boundary west of the Rocky Mountains. The boundary followed the forty-ninth parallel except for Vancouver Island, which became British. An earlier proposed boundary cutting through Vancouver Island remains on the map. Other earlier boundary proposals can be seen on the map, following the Columbia River and cutting off the Olympic Peninsula to create an "island" of United States territory that would permit occupation of the west shore of Puget Sound. Charles Wilkes's report of 1844, based on his 1841 visit, convinced the American government of the value of Puget Sound as a harbor, and this was a British attempt to satisfy this requirement without giving too much away. Some American settlement in the Willamette Valley is noted at *Oregon City.* The Hudson's Bay Company trading establishments *Fort Vancouver, Ft. Nisqually, Ft. Walla Walla, Ft. Okonagan,* and *Ft. Colville* are shown.